D1233804

Genomic Politics

Genomic Politics

How the Revolution in Genomic Science Is Shaping American Society

JENNIFER HOCHSCHILD

OXFORD
UNIVERSITY PRESS

Oxford University Press is a department of the University of Oxford. It furthers the University's objective of excellence in research, scholarship, and education by publishing worldwide. Oxford is a registered trade mark of Oxford University Press in the UK and certain other countries.

Published in the United States of America by Oxford University Press
198 Madison Avenue, New York, NY 10016, United States of America.

Library of Congress Control Number: 2021936956
ISBN 978–0–19–755073–1

DOI: 10.1093/oso/9780197550731.001.0001

1 3 5 7 9 8 6 4 2

Printed by LSC Communications, United States of America

Dedicated to Melissa, who I miss all the time,
and to Barbara Hochschild,
who lived through the century in which
genetics moved into society

There is in biology at the moment a sense of barely contained expectations reminiscent of the physical sciences at the beginning of the 20th century. It is a feeling of advancing into the unknown and [a recognition] that where this advance will lead is both exciting and mysterious. . . . The analogy between 20th-century physics and 21st-century biology will continue, for both good and ill.

—*The Economist*, 2007

We possess the ability to . . . direct the evolution of our own species. This is unprecedented in the history of life on earth. It is beyond our comprehension. And it forces us to confront an impossible but essential question: What will we, a fractious species whose members can't agree on much, choose to do with this awesome power?

—Jennifer Doudna and Samuel Sternberg, 2018

Genomic science is too big of a train to derail, the work will happen, so social scientists need to be involved in order to contribute to its appropriate interpretation.

—Social science expert on genomics, 2018

Table of Contents

Preface and Acknowledgments

A 2020 advertisement by Memorial Sloan Kettering Cancer Center promotes innovative, genomics-based treatment. Its video version asserts that "there will never be one cure for cancer. There will be millions," and promises "more precise, personal, innovative care." The slogan is "More science. Less fear." The aim of *Genomic Politics* is to probe that slogan. I do so not through science, examination of medical cases, analyses of the economics of cancer care, or the many other frames employed by many excellent authors, but as a political scientist: I examine the politics and ideology surrounding the societal use of genomic science in the United States. The book analyzes disputes in which people who seek the same outcome come close to accusing one another of murder, and probes judicial decisions bringing together Supreme Court justices who agree in fewer than 5 percent of their other cases. It explores racial differences in views of "biogeographic ancestry" testing, and the politics around research on whether Covid-19 disproportionately attacks African Americans or men because of their genetic makeup. Genomics is, as Memorial Sloan Kettering emphasizes, science—but as the hospital well knows, it also encompasses fear and scorn, hope and excitement. Like any hugely important enterprise, its societal uses are intrinsically political and ideological—and like Memorial Sloan Kettering Cancer Center (though without its financial incentive), I want to bring what I know about genomics into public visibility.

First things first: what is genomics? The National Human Genome Research Institute (NHGRI) defines it as "the study of all of a person's genes (the genome), including interactions of those genes with each other and with the person's environment." The genome, in turn, is "an organism's complete set of DNA, . . . [which] contains the information needed to build the entire human body." A metaphor helps. A single strand of DNA, residing in almost every human cell, is made up of about 3 billion base pairs, each a two-unit combination of the chemicals adenosine, thymine, cytosine, and guanine; we can analogize the base pairs to letters of the alphabet. The 3 billion letters combine into words, about 20,000 of which are "readable" genes. Genetics is the study of individual words, small combinations of words in phrases and sentences, or words embedded in not-yet-interpreted "gibberish." The words, in turn, are strung together in such a way that they comprise an intelligible, meaningful book that has uninterpretable sections but is nonetheless much more than the sum of its parts. Genomics is the study of that book. Genomic scientists may examine individual letters, words, the supposed gibberish between words, paragraphs and chapters, concepts and insights that run through the whole volume, and the ways in which that book slots into the whole "library" of the environment and other organisms. When we understand the book and its context, from each letter through all the ideas and the interactions among ideas in many books, we will understand human heredity.

None of us, of course, comes close to understanding human heredity; perhaps we never will. But each analyst is seeking a partial answer to the question posed by sociologist Jenny Reardon: "Now that we have 'the human genome sequence,' what does it mean?" Her search focuses on the sociology of science and the moral values associated with intervention in local communities.[1] Memorial Sloan Kettering's search is for the "precise, personal" patterns of base pairs, genes, and genomic interactions that will cure particular patients of their distinctive cancers. As a political scientist, I search for answers to Reardon's question in the exercise of power, struggles around governance, choices among policies, the distribution of outcomes, and ideological worldviews—that is, genomic politics.

I would never have made it far into this search had it not been for Larry Jacobs's casual suggestion that I apply for a Robert Wood Johnson Foundation Investigator Award in Health Policy Research, something he has surely forgotten. It was a surprising suggestion: I am a scholar of American racial, class, and immigration politics. But after attending conferences on the medication BiDil and serving as a discussant for Mary-Claire King's 2006 Tanner Lectures at Harvard University, I was hooked. Maya Sen, then a Ph.D. student in Harvard's Government and Social Policy Program, joined me as co-principal investigator, and with the advice of many experts, we received an RWJF Award in 2010. So in addition to Larry Jacobs, my first and deepest thanks go to the Robert Wood Johnson Foundation, which took a chance on an eager but fairly ignorant applicant. I hope *Genomic Politics*, though it may not bear much resemblance to the proposal's promises, turns out to be worth the grantor's confidence.

Of course, the book rests on the shoulders of very many others as well. Primary is Maya Sen, who was a partner in developing the focus, designing the GKAP survey, identifying interview subjects, analyzing data, getting control over the scholarly literature, specifying the theory, and everything else involved in pushing a large project forward. She is co-author on all five of the articles published so far out of this research. Maya's work has moved in a different direction since I began to focus intently on this book, and she cannot be held responsible for anything in it—but I could not have written it without our cherished collaboration.

Several others have contributed invaluable ideas, words, data analyses, theoretical framings, and other forms of intellectual companionship. They are co-authors on one or more articles, with more collaboration to come, I hope: Alex Crabill, Meredith Dost, Mayya Komisarchik, and Elizabeth Suhay.

A chief virtue of teaching at Harvard is the amazing talent and energy of its students. If I occasionally get frustrated in trying to keep up with them, it is nonetheless always rewarding to make the attempt. A not-so-small army of undergraduate and graduate research assistants made the later chapters of *Genomic Politics* possible, along with contributing the essential and ephemeral element of belief that there really was a project here worth doing. First among equals are Chris Chaky, Mara Roth, and Ryan Zhang, to whom I am very grateful. They are followed in close succession by Andrew Benitez, Raphael Broh, Angelo Dagonel, Kaneesha Johnson, Scott Kall, Gabriel Karger, Layla Kousari, Cara Kupferman, Shom Mazumder, Natalie Padilla, Ben Polk, Anna Remus, Nicholas Short, Claire Sukumar, Mikael Tessema, Ifeoma Thorpe-White, Bruno Villegas, Max Weiss, and Claire Wheeler.

In addition to those already named, Angie Bautista-Chavez, Kristina Brandt, Graciela Carrasco, Medha Gargeya, Michael Anthony George, Ben Gruenbaum, Seth Henderson, Mayya Komisarchik, Brendan McElroy, Gabriela Malina, Angela Primbas, and Shanna Weitz all deserve gratitude and perhaps an apology for their heroic work coding articles by social science experts in genomic science. Chris Chaky and Ryan Zhang were primarily responsible for pulling thousands of coded items on dozens of spreadsheets into a coherent and usable database.

Along with learning about genomics, I learned a lot about writing surveys over the past decade. For that fascinating and useful education, I thank Patrick Moynihan, Chase Harrison, Poom Nikulkij, Wendy Mansfield, and Tom Smith.

A wonderful group of genomics experts served as the Advisory Committee during the years of the RWJF award. I expect that RWJF would not have had half as much trust in Maya's and my ability to pull this project off if those committee members had not agreed to help it develop; I certainly would not have done so. I thank David Altshuler, Michele Caggana, Hank Greely, Evelynn Hammonds, Steven Pinker, and Patrick Sturgis for their wise comments. I am still using notes from our conversations to help shape my arguments.

Many colleagues have listened to, read parts of, or argued about some or much of this book. In addition to co-authors and student assistants, they include C. Anthony Broh, Henry Louis Gates Jr., Evelynn Hammonds, Macartan Humphreys, and Wendy Roth.

Cynthia Morton brought me (as "the ELSI component") into a medical genomics research project that has taught me a great deal about how clinicians incorporate genomic medicine into practice. We should all be so lucky as to have medical professionals with the skill, commitment, and creativity of this group. The project is Sequencing a Baby for an Optimal Outcome (SEQaBOO), Cynthia Morton, Ph.D., Principal Investigator, Brigham and Women's Hospital. It is funded by the National Institutes of Health (http://seqaboo.bwh.harvard.edu/).

Many institutions and their individual embodiments have made it possible for me to complete this project, albeit too slowly. Harvard University, especially the Departments of Government and of African and African American Studies, the Harvard Kennedy School, and the Center for American Political Studies, provide—as well as way too many distractions—arguably the best environment in the world to teach, do research, learn, and engage in all of the other activities that make an academic life so privileged. Harvard's research librarians are amazing; I took full advantage of the expertise of Diane Sredl, Valerie Weis, and Michelle Pearse. The Eli and Edythe L. Broad Institute of MIT and Harvard took the unusual step of admitting me, a social scientist, as an Affiliated Member; much of what I know about genomic science I have learned from its extraordinary seminars, publications, and community. The John W. Kluge Chair in American Law and Governance at the Library of Congress (2011) and a Fellowship at the Straus Institute for the Advanced Study of Law and Justice, New York University School of Law (2013–2014) provided the settings and collegiality so essential for success in the strange enterprise of staring at a blank computer screen and willing it to fill with words and numbers.

The National Science Foundation, through its support of TESS (Time-Sharing Experiments for the Social Sciences) and of the General Social

Survey, provided the structures for survey experiments and questions. My thanks to the American taxpaying public.

Participants in many seminars and panels encouraged, criticized, doubted, speculated, and pushed me into what I hope is a better book than it would have been without their help. Some events were held at annual meetings of RWJF Investigator Award recipients and annual meetings of the Midwest Political Science Association and the American Political Science Association. Others were held at Columbia University, the Harvard Kennedy School, Harvard's Weatherhead Center for International Affairs, New York University Abu Dhabi, the University of British Columbia, Vanderbilt University, and an annual meeting of the American Association for the Advancement of Science. A year as a Phi Beta Kappa Visiting Scholar in 2016–17 enabled me to meet, and try out ideas on, professors and students at Clark University, Furman University, Macalester College, University of Missouri, Randolph Macon College, Virginia Polytechnic Institute and State University, Sewanee: The University of the South, and Skidmore College. Delivering a keynote speech at the WZB Berlin Social Science Center enabled a seminar at which members gently strove to update my knowledge of popular culture beyond *GATTACA*.

I am in debt to the social scientists, genomic scientists, and other experts who gave their time and expertise so generously in interviews and survey responses. Directly asking experts for their views deepens one's sense of genomics as both a fascinating research subject and a Rorschach test of hopes, fears, and mental maps. The observations of these especially knowledgeable and committed actors, who care enough about the politics and ideology of genomic science to yield to my importuning for some of their valuable time, are fascinating and revealing. Since many spoke openly after being promised confidentiality, I cannot thank them by name. I hope I have done justice to their ideas.

The various surveys, and questionnaires to experts, received the following approvals or exemptions from Harvard University's Institutional Review Board: F-19007-101; F-19928-101; F-19209-101; F-19209-102; F19209-103; F22804-101; F19928-103; IRB 17-0938; IRB19-0704.

I am deeply grateful to my editor at Oxford University Press, David McBride. I was thrilled by his immediate and continuing enthusiasm for *Genomic Politics*, and he sat on my metaphorical shoulder throughout revisions. I have worked with David in various guises; he is a friend as well as a trusted guide to the mysteries of publishing. Holly Mitchell, the assistant editor, is friendly, helpful, and just a little stern about deadlines, exactly as she should be. Cheryl Merritt expertly and efficiently guided me and the manuscript through the many stages of production. Dalton Conley and three anonymous reviewers raised the difficult questions and made the sweeping suggestions that every author dreads and cherishes. That they all concentrated so well in the middle of a frightening pandemic makes their help that much more valued.

Finally, my family. Tony Broh continues to be at the center of my life. While writing this book, I have watched Eleanor Broh, Raphael Broh and Sarah Hutcherson, and Henry Hochschild grow into adulthood, overcome terrible obstacles, find meanings for their adult lives, and love and laugh. I am so privileged to have them as my family. This book is dedicated to my sister Melissa Trier Hochschild, whose life was too short, and to my mother, Barbara Elisabeth Hochschild, who at age 99½ still watched CNN all day and still asked me when that genome book was ever going to be finished. She aimed to live until it was done, and she almost made it.

1
Introduction
"There Be Monsters"?

> The use of science and technology in an effort to enhance human beings is taking us beyond the outer edges of the moral, ethical, and religious maps bequeathed to us by previous generations. We are in terra incognita, where the ancient maps sometimes noted, "There Be Monsters."
>
> —Al Gore, 2013

At the turn of the twenty-first century, few Americans had heard of genomics. The term and its cognates had never appeared in a public opinion poll. The phrase "Human Genome Project"—an enterprise to which taxpayers in several countries were contributing almost $3 billion—appeared only 125 times (less than a dozen times a year) in American newspapers between May 8, 1989, and January 1, 2000.[1] Twenty years later, genomic science is everywhere. DNA tests determine whose dog soiled the sidewalk, and whether a dog of questionable ancestry is acceptable to a co-op board in New York. DNA tests determine if the fish in your sushi is really salmon, who poached deer or elephants, and who stole sheep. Genomicists are working to reverse the extinction of wooly mammoths and to restore genetic diversity to the cancer-ridden Tasmanian devil. The consulting firm Battelle reports that oil and gas developers are tracking polar bear activity through DNA left on their paw prints. The dating service GenePartner sells the possibility of finding one's mate through genomic tests ("Love is no coincidence: matching people by analyzing your DNA"). At home during a Covid-19 lockdown, you can play the board game *Metanon: The Biocode Adventure* with your children while glancing at the poster of your own or your loved one's DNA that is the focal point of your living room art collection (Figure 1.1).[2]

Figure 1.1 "From life comes art"
Source: DNA11

In a less frivolous vein, DNA tests determined if the person assassinated by American troops in 2011 really was Osama bin Laden. DNA tests were used to identify 7,000 victims of the 1995 Srebrenica massacre and to reassemble scattered body parts. They can identify bone fragments of soldiers lost in World War II. DNA tests using stored rape kits can point to perpetrators, or exonerate people convicted of a felony, years after a sexual assault occurred. Genomic research can find the cause of rare diseases, and gene therapies contribute to fighting the "monster inside" of cancer, brain disease, retinal dystrophy, or junctional epidermolysis bullosa. Genetically modified foods deliver vitamins to malnourished babies; genetically modified mosquitoes might eliminate dengue fever or malaria; genetically engineered bacteria might clean up oil spills. Prenatal genetic therapy might prevent devastating diseases in newborns; germline gene editing might someday do the same for generations to come. Genetic ancestry tests permit a New York resident named Mika Stump, who grew up in foster homes, to declare, "[Now] I have a place where I can go back and say, 'This is who I am; this is my home.' That's something I never, ever expected to say." An economist shows that DNA databases can help to convict the guilty, exonerate the innocent (sometimes after decades of imprisonment), "deter crime by profiled offenders, reduce crime rates, and [be] more cost-effective than traditional law enforcement tools."[3] A smartphone can use gene editing techniques to test for Covid-19 infection and measure its severity.

Most vividly, in 2020 genomic sequencing of the SARS-CoV-2 virus enabled unprecedentedly rapid development of highly efficacious vaccines against Covid-19. Using mRNA (messenger RNA) as the foundation of a new strategy for developing a vaccine that primes the immune system against the targeted virus, scientists in half a dozen countries created, tested, and distributed vaccines in less than a year. That compares with the average over the past half century of a decade to develop a new vaccine.

Genomic science is not all entertaining or lifesaving; its societal uses can be unsettling, even dangerous. Perhaps there really are monsters on the perimeters of the map of genomics technology; as *The Economist* notes in an observation quoted as the first epigraph for this book, biology's advance into the unknown is "for both good and ill." Forensic biobanks may create a system of what law professor Jeffrey Rosen calls "permanent genetic surveillance" of all Americans. Disseminating genetically modified food could introduce "food totalitarianism . . . monocultures, deadness," according to environmental activist Vandana Shiva. If genetically modified mosquitoes are released into dengue-ridden communities, warn biologists James Bull and Harmit Malik, "escape of a GDS [gene drive system—a genetic modification designed to spread rapidly through a targeted population] into a beneficial species could spell its doom. . . . Stopping a GDS once it is released is not easy—maybe not possible." Prenatal genetic interventions and gene editing raise the threat of everything from designer babies to the transformation of gorillas into weapons of mass destruction to the reinvention of humankind (as in, respectively, the movies *GATTACA*, *Rampage*, and *Elysium*). Neuroscientist Diane DiEuliis and her co-author "posit the strong likelihood that development of genetically modified or created neurotropic substances . . . represents a novel—and realizable—path to creating potential neuroweapons." Identifying distinctive genetic traits to partly explain the disproportionate vulnerability of people of color to Covid-19, writes bioarcheologist Sonia Zakrzewski, is "problematic and shocking. . . . Ascribing the disparity of Covid-19 impacts on BAME [Black, Asian, and minority ethnic background] groups as *genetic* masks the more likely, and more pernicious, causes: systemic inequalities." Most simply, wrote a reader to the editor of the *New York Times*, "we don't want our children or their grandchildren to suffer the horrors of a poorly run experiment gone outrageously amok."[4]

Behind these lists, which could be doubled or tripled, lies the most unsettling point: unraveling "the mysteries of the DNA molecule," wrote the pediatrician Jean-François Mattei, "raises the question of the place of mankind in the universe." If geneticist George Church is right that "synthetic biology will reinvent nature and ourselves," is that a reason to celebrate, shut it down—or create commissions, write regulations, and encourage citizen scientists? Exciting and mysterious indeed, as *The Economist* says.[5]

People will respond to this Janus-faced advance into the unknown in different, perhaps antagonistic, ways. Deep, wide, and long-lasting societal contention around important issues is inevitably political and probably ideological. To make sense of such contention, I begin by outlining a basic framework of four stances toward genomic science that, I will argue, collectively capture Americans' main positions about genomics as a whole and its particular uses. Unlike most other political disputes in the United States today, these stances do not line up with partisan affiliation (although later in *Genomic Politics* I explore ways in which partisanship might be creeping in). Nor are any of them obviously "left-wing" or "right-wing"; they diverge along different dimensions. Religious conviction about genetics may permit an alliance between African Americans and Republicans; the drive to find solutions to policy problems may bring social constructionists and geneticists much closer than either set of actors can envision.

After I outline the basic framework that structures the book, the Introduction offers brief discussions of morally and empirically fraught disputes in order to make more concrete the breadth, depth, and unusual nature of contention over genomics. These illustrative disputes range widely, from a race-based medication called BiDil to DNA ancestry testing, databases used in the criminal justice system, and prenatal gene therapy or editing. I then begin the process of mapping the disputes onto the basic framework, in order to signal to readers what lies ahead in the next eight chapters. As I proceed, we will explore these disputes in more detail, examine opinions held by the public and by genomics experts, and consider strategies for handling these political hot potatoes. Many pages from now, I conclude by using all of this material to show how the United States can advance into the exciting and mysterious genomic unknown in a way that is both democratic and responsible, despite the fact that we are a fractious species whose members can't agree on much.

The Basic Framework

The basic framework that shapes the narrative of *Genomic Politics* is a well-tested classic: a 2 × 2 grid with four quadrants, each of which sits in the intersections of two independent dimensions. The logic is easily grasped if we consider two dimensions of color: light to dark, and red to blue (Figure 1.2).

The four intersections of those two dimensions—the quadrants—are light red (which we call "pink"), dark red (which we can call "crimson"), light blue ("baby blue"), and dark blue ("navy blue"). Like my basic framework, the 2 × 2 color framework is imperfect, but good enough to work with: pink and baby blue have one dimension in common; pink and crimson have a different dimension in common; pink is more different from navy blue than from either crimson or baby blue.

Following that simple logic, Figure 1.3 shows the skeleton of the basic framework of *Genomic Politics*.

The vertical dimension, corresponding with hue, focuses on one's knowledge of or belief about the impact of genetics on behaviors, traits, or physical conditions. It ranges from the assertion that "genetics is really important" to the assertion that "genetics is relatively unimportant or not relevant at all." The horizontal dimension, corresponding with color, focuses on a person's judgment or preferences about risk-taking (the basic framework applies to other entities as well as people, but we save that for later). It ranges from the sense that "overall, new technologies are beneficial, although we must beware possible harms" to "new technologies risk serious harms, even if they also carry some benefits."

The four stances toward genomic science that I explore throughout this book correspond to the colors. The top left quadrant "Enthusiasm" occupies the position of pink; its proponents perceive genetics to explain a lot

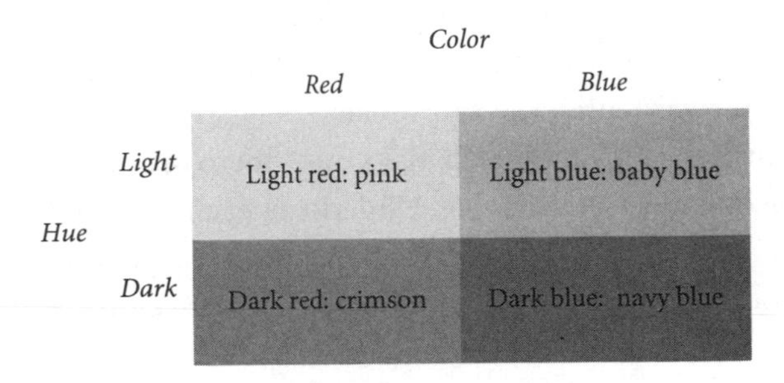

Figure 1.2 Logic of a 2 × 2 typology

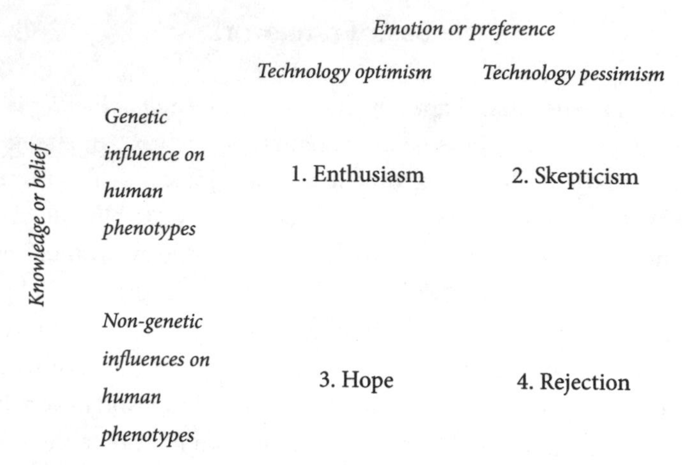

Figure 1.3 The basic framework of *Genomic Politics*

about human behaviors and traits, and they are excited about the benefits to be gained from the increasing use of genomic science. "Skepticism," on the top right, corresponds to baby blue; occupants of this quadrant agree that genetics explains a lot about human behaviors and traits, but fear that deployment of genomic science will harm society more than benefit it. "Hope" occupies the crimson quadrant on the bottom left. The Hopeful perceive genetics to be relatively or completely unimportant in explaining important human traits and behaviors, but they believe that other human or even divine actions can improve lives and society as a whole. Finally, the navy blue quadrant is represented by "Rejection," on the bottom right. Its occupants perceive genetics to have little or no influence on important traits and behaviors, and they have slight hope that any intervention will significantly improve humans' lot. Rejectors have the clearest vision of the monsters on the edges of the genomics map.

This basic framework is the vehicle for the three central purposes of *Genomic Politics*. The first goal is explanatory. Much of this book demonstrates how attitudes toward, prescriptions for, and policies about many uses of genomics technology are illuminated by locating them in one or another quadrant. The second goal is evaluative. Underlying each quadrant's perspective are moral stances, whether expressed as an ideology, a policy, a fact, or a demand. The best way to assess each moral position is to consider it in relation to the others—just as we understand pink better when we juxtapose it with both crimson and baby blue.

My third goal in using the basic framework is prescriptive (with a tinge of prediction). Although views of and goals for societal deployment of genomic science are intrinsically and inevitably political, they are not (yet?) partisan: as I show in this and later chapters, one can find liberals and conservatives, Democrats and Republicans in all four quadrants. To me, this is very good news. As a political scientist, I am enthusiastic about far-reaching disputation; how else should a democracy as diverse and assertive as ours determine how to take the right action—or any action—on new and complex public issues? But I am not a fan of strong partisan affiliations, since they tend to shut down constructive political debate. So my main prescription will be that Americans entertain an array of possibilities for working out how to govern genomics. I see suggestions of mindless partisanship creeping into this arena, like so many others—as I write, most vividly around reactions to a genomics-based Covid-19 vaccine—but I hope that the logic of *Genomic Politics* can assist readers to develop and engage with alternatives to blindered group loyalties.

BiDil: Race-Based Medicine

The basic framework will not deserve to be adopted unless it explains intense and complex battles around societal uses of genomics. My initial task, then, is to describe a few such battles and begin to show how deploying the framework makes sense of them; later chapters take up that task in fuller detail.

In 2005, the Food and Drug Administration (FDA) approved BiDil, "a drug for the treatment of heart failure in self-identified black patients." It was the first time in the United States, perhaps anywhere, that approval of a medication was linked to the race of its target recipients. A general population study of the medication had demonstrated no overall benefits of its use, but hinted at benefits for its small sample of Black participants. A follow-up study among self-identified Black patients with severe heart failure apparently showed dramatic success; on that basis, the FDA gave its approval, unusually.[6] As the FDA's news release explained:

> Today's approval of a drug to treat severe heart failure in self-identified Black population is a striking example of how a treatment can benefit some patients even if it does not help all patients. . . . The information presented to the FDA clearly showed that Blacks suffering from heart failure will now

> have an additional safe and effective option for treating their condition. In the future, we hope to discover characteristics that identify people of any race who might be helped by BiDil.

BiDil was, in short, "a step toward the promise of personalized medicine."[7]

Before approval, the FDA's Advisory Committee on Cardiovascular and Renal Drugs had spent considerable time discussing whether to recommend this novel move. One member asked if "some gene" could explain BiDil's greater efficacy among Blacks than Whites; the cardiologist who led development of the drug and its subsequent trials responded that "the working hypothesis has been that there is evidence for reduced nitric oxide bioactivity in African American populations on average compared to White populations." Another committee member queried whether Black self-identification would be an appropriate indicator for prescription. Proponents answered rather lamely that they "did not have any other biologic markers, other than self-identification," and that "it is consistent with FDA guidelines with regard to collecting ethnicity or race in clinical trials, that Black or African, self-identified African Americans is the method of doing this."[8] Despite continued concern about how close the match is between self-identification and genetic characteristics, the committee voted in favor of FDA approval.

As the Advisory Committee's discussion hinted, BiDil was controversial even before approval. At a public hearing, the president of the International Society on Hypertension in Blacks, the executive director of the National Minority Health Foundation, and the chair of the Congressional Black Caucus Health Braintrust all endorsed approval. A research associate at Howard University's National Human Genome Center, however, warned that approving a race-specific drug "smacks of a kind of influence that science has shown to be untenable.... [Such a label implies that] all members of demographics are more similar than different due to inherited biology, and that this will not/cannot change over time.... Medications work on pathophysiology and clinical phenotypes, not group labels or degrees of ancestry." Tempers rose in the aftermath of the FDA's approval. Jonathan Kahn, an attorney and scholar who has offered the most sustained critique of the FDA's action, wrote that "we were accused at one meeting of 'killing people' with our critiques of BiDil." That may have been a bit of an exaggeration—but Jay Cohn, the lead cardiologist working with BiDil, and his co-author did observe that "by railing against the idea that blacks were singled out for this study, ... Mr. Kahn has contributed to a backlash that has impeded clinical use of the drug. . . . It is tragic that thousands of patients are dying because their doctors are not prescribing the drug despite the ease of their identification." Others were rhetorically close behind. An editorialist wrote that, to some, "pooling people in race silos is akin to zoologists grouping raccoons, tigers and okapis on the basis that they are all stripey." But as sociologist Catherine Bliss summarizes the views of some of her interview subjects, "Race-free genomics is the same as the colorblind rhetoric that contributed to racism in the South. . . . Refusing to recognize the *biological* processes associated with race is seen as tantamount to scientific racism."[9]

All participants in this debate are committed to improving medical care and health outcomes for the underserved African American population, as well as for all those suffering from heart failure. All participants abhor racial injustice, endorse medical research, are knowledgeable about genomic science, accept the legitimacy of statistical analysis of clinical trials, recognize self-identified race to be a crude and inaccurate label, and endorse the promise of personalized precision medicine. Yet they come close to calling each other racists and murderers. Understanding why an obscure drug (that eventually failed commercially) can generate such intense anger among people who in other circumstances would be allies will give us some purchase in addressing Nobel laureate Jennifer Doudna's and Samuel Sternberg's question, posed in the second epigraph to *Genomic Politics*: "What will we, a fractious species whose members can't agree on much, choose to do with this awesome power?"

DNA Ancestry Testing

Genealogical research is Americans' second-favorite hobby after gardening, and the most popular target of internet searches after pornography. It lacks the life-and-death quality of seeking to alleviate heart failure, so passions do not run quite as high as in the case of BiDil. Nonetheless, people can become highly exercised about spitting in a test tube.

Various firms have offered DNA ancestry tests since the early 2000s, marketing them as a way to find particular named ancestors or possible family members, or, as the genomics testing firm 23andMe puts it, as a way to "learn about your ancestral origins, . . . to travel to the places that make you, you." After starting as a boutique game among the cognoscenti, DNA ancestry testing went mainstream; the CEO of a biometrics company judges that "the inflection point started in the summer of 2016, and from there it's gone into the stratosphere."[10]

Initially, journalists wrote clever, ironic articles about a "spit party" during New York Fashion Week ("It was a funny thing to be doing in a cocktail dress"), or more searching articles about people such as Reverend Al Sampson of Chicago, who traveled to Sierra Leone to give elders of Lunsar village the test results showing his Temne lineage. "Five hundred years ago my DNA was removed from here by slave traders and taken to America, so I'm coming back for my seat. My seat's been vacant," Sampson said. He asked for a Temne name in order "to reclaim what was taken away from me." The Scottish tourist industry developed plans to ensure that "DNA testing will be a draw for ancestral tourists who might want to 'walk in the footsteps of their ancestors.'"[11] The president of a Jewish genealogical organization in an American city reports that through DNA ancestry testing, "people are learning that they had Jewish ancestry even though they grew up as Catholic. It is very powerful. There were such ruptures in family continuity and contact [due to European emigration and World War II]—people are just thrilled to find a relative."[12]

Researchers have joined participants and advocates in finding value in DNA ancestry testing. Henry Louis Gates Jr., the prominent scholar of African American studies, points out that "for the first time since the seventeenth century, we are able, symbolically at least, to reverse the Middle Passage. Our ancestors brought something with them that not even the slave trade could take away: their own distinctive strands of DNA. . . . An exact match between an American's DNA and an African's DNA reveals a shared ancestor, and possibly a shared ethnic identity, that has been lost for centuries." Specialists in Judaic studies similarly use ancestry testing to sort out confused or thin historical narratives. Efforts to determine how contemporary Jewish groups are historically related to one another stalled for decades, until an endocrinologist's genome-wide analyses of 237 people from four distinct Jewish communities resolved the question. Sociologist Alondra Nelson envisions an even broader collective impact. Her research shows that the kinds of social engagement emerging from genealogy ancestry testing—"the aspirations for affiliation that inspire its use and the various kinds of relationships it may occasion"—are sites for creating connections among people with African heritage. Testers who discover ancestral links generate mutual "responsibilities and rights, and forms of exchange," thereby "facilitating the formation of a diasporic network."[13]

But controversy ensued. Fourteen social scientists and legal scholars wrote in *Science* that "commercialization has led to misleading practices that reinforce misconceptions." For example, if test-takers change how they report their race or ethnicity as a result of their supposed new knowledge, "this could make it more difficult to track the social experiences and effects of race and racism." Even more seriously, "because race has such profound social, political, and economic consequences, we should be wary of allowing the concept to be redefined in a way that obscures its historical roots and disconnects it from its cultural and socioeconomic context." In this polite and restrained scholarly environment, the authors declare it "unlikely that companies (and the associated scientists) deliberately choose to mislead consumers or misrepresent science." Nonetheless, they warn, "market pressures can lead to conflicts of interest, and data may be interpreted differently when financial incentives exist. . . . Unfortunately, peer review is difficult here."[14]

In the more informal medium of email listservs, scholars skeptical of DNA ancestry testing are more explicit: "A basic aspect of how this field works [is that] it draws on categories produced in particular social/cultural and institutional milieus that will always include and exclude in a manner that might help some and harm others. This is not done in any kind of random way, but in ways in which those harmed are those historically disadvantaged." Some put it bluntly: "the hocus-pocus element of arbitrary numerology" is in evidence—or even, "the racist steamroller plows ahead."[15]

The fear that DNA ancestry testers will be misled or even harmed has generated its own counterreaction. As one of my interview subjects put it, with some asperity, "People have the right to gain access to [their] own genetic information. The government should [only] control companies to ensure that tests are accurate and based on real science, and to ensure that companies explain results and give the consumer access to genetic counselors."

As with the dispute over BiDil, all disputants over racial and ethnic ancestry testing are committed to equality and sympathetic to the yearning for group identity and ancestral roots. All endorse scientific research, are knowledgeable about genomic science, accept the legitimacy of statistical analysis of genetic sequencing results, recognize that self-identified race is a crude and inaccurate label, and endorse genealogical research. Yet they too come close to calling each other racist—and perhaps corrupt, patronizing, or ignorant as well. As with BiDil, understanding what lies beneath the mutual suspicion among people on the same end of the partisan or ideological spectrum will contribute to addressing Doudna's and Sternberg's query about what we will choose to do with the awesome power of genomic science.

Forensic Biobanks

The 1994 DNA Identification Act authorized a National DNA Index System (NDIS) and specified government agencies' access requirements and the categories of data to be collected. Subsequent federal legislation expanded database laws and increased federal funding. By 2019, all states retrieved DNA from convicted felons, and many also collected it from certain misdemeanants, arrestees, probationers or parolees, juveniles, and/or undocumented immigrants and others subject to deportation. All three levels of government—local, state, and federal—have laws and regulations for collection, storage, and links to the FBI's Combined DNA Index System (CODIS), the software platform on which the FBI aggregates forensic DNA samples. As of March 2021, the NDIS contained about 14.5 million offender profiles, 4.3 million arrestee profiles, and more than 1 million forensic profiles. CODIS has assisted in over 540,000 criminal investigations, after producing close to 600,000 hits. (These data underestimate the actual results, since Covid-19 restrictions are inhibiting updates.) All three levels of government fund or run their own labs and testing. Most states also have laws and regulations for expunging samples, testing DNA after conviction, and compensating exonerated ex-felons.

Forensic databases have a higher public profile than do BiDil or DNA ancestry tests. Consider, for example, California's "Grim Sleeper," Lonnie Franklin Jr., who raped or murdered perhaps dozens of women between 1985 and 2002, and who was caught only after his DNA was linked to that of an imprisoned relative through a forensic DNA database. Franklin was the subject of several books and thousands of stories in various media (the term "Grim Sleeper" yields 130,000 hits on Google). A Google Ngram (a search engine that counts the number of times a word or phrase is used in a given year across 30 million books since 1800) shows the rise of attention to Franklin in the years after 2006 (Figure 1.4).

Most important, the distinctive techniques used to link Franklin's DNA to that of the crime scenes means that this "landmark case" will "change the way policing is done in the United States," according to Los Angeles's police chief at the time.[16]

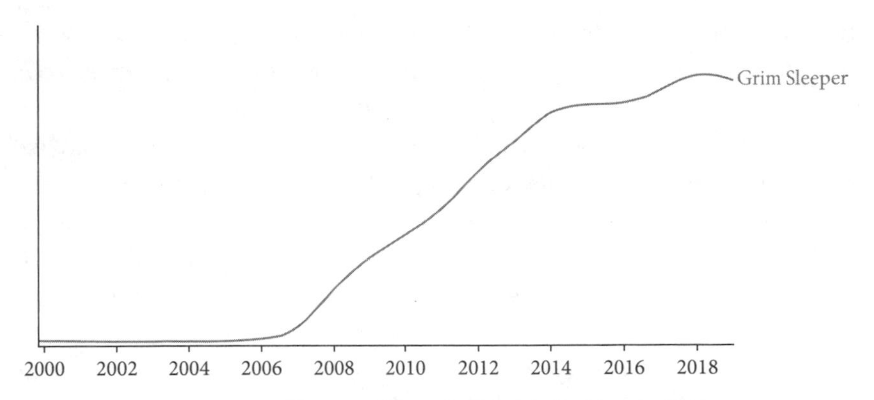

Figure 1.4 Appearance of "Grim Sleeper" in published books, 2000–2019

Fascination with horrible serial murderers is only part of what drives public attention to forensic databases. The television series *CSI: Crime Scene Investigation* ran from 2000 to 2015; it was the world's most-watched television show for six years, with its 2009 international audience estimated at 74 million. The original show spun off three parallel series, copycat shows, books, comics, museum exhibits, websites, and video games. Viewers' attention reveals a good instinct for what matters: as forensic scientist Victor Toom summarizes, "By raising issues of civil rights, i.e. being innocent until proven guilty, privacy, bodily integrity and the principle of proportionate measures, . . . forensic DNA profiling practices are at the heart of the organization of national democracies."[17]

A wide range of actors endorse expansion and vigorous use of forensic DNA databases. Some focus on their use to identify and convict criminals: President Barack Obama promoted federal "support for things like DNA testing at the state levels , . . . getting the databases set up." Pointing out that he was the father of two daughters, he observed, "There are just too many horror stories that remind us that we're not doing enough. . . . We insist on justice." As early as 2007, an otherwise skeptical analysis of the United Kingdom's national DNA database by bioethicist Mairi Levitt noted that "where DNA crime scene samples can be obtained, the detection rate for crimes increases, for example, for domestic burglary from a national detection rate of 16 to 41 percent and for theft from a vehicle from 8 to 63 percent."[18]

Another reason for support is the use of DNA databases to free people falsely convicted of a crime. Law professor Brandon Garrett, author of the most exhaustive analysis to date of mistaken convictions, wonders why any trials now occur without including DNA evidence, if it might clear a defendant. After all, "it is hard to imagine an error of greater significance in our criminal justice system than a wrongful conviction."[19] The Innocence Project reports 375 exonerations as a result of DNA testing. That may not sound like a lot, but the evidence of false convictions has led more than one governor to declare a moratorium on executions in his or her state for the reason, as California governor Gavin Newsom put it in 2019, that "most of all, the death penalty is absolute. It's irreversible and irreparable in the event of human error."[20]

But as readers will expect by now, other actors are deeply skeptical of or hostile to CODIS and its affiliates. Because of racial, ethnic, class, and gender disproportionality in arrests, convictions, and deportation, forensic DNA databases include a much higher proportion of samples from poor African American men and Latinos than those groups' share of the American population. They are "Jim Crow's database," according to the Council for Responsible Genetics. Law professor Dorothy Roberts agrees: "Forensic DNA repositories are gathered by the state without consent and are maintained *for the purpose* of implicating people in crimes. They signal the potential use of genetic technologies to reinforce the racial order not only by incorporating a biological definition of race but also by imposing genetic regulation on the basis of race." Even more broadly, Barry Steinhardt, then the director of the American Civil Liberties Union's Technology and Liberty Program, describes DNA databases as belonging to "the realm of the Brave New World." Bioethicist George Annas labels the genome a person's "future diary," by which he means that a DNA sample provides "a type of information that is revelatory in a way few things are." DNA databanks must therefore be more fully regulated and controlled in order to protect privacy and dignity; after all, "to know one's own future diary—or to know someone else's—is to call into question the very meaning and possibility of human liberty."[21]

Every actor in this system endorses a criminal justice system that convicts the guilty, releases the innocent, and distinguishes correctly between the two. No actor sanctions racism or the violation of individual liberty or privacy. All see room for considerable improvement in the United States' criminal justice system. Yet once again, some of the people who are most knowledgeable about this use of genomic science come close to calling one another racist, unjust, or totalitarian. Exploring the reasons behind this passionate disagreement will help us to determine the benefits and harms of using genomic technology in a realm that sits at the core of any polity claiming to respect the rule of law and the sanctity of individual rights.

Prenatal Genetic Testing and Genome Editing

Pregnant women have been offered prenatal testing for chromosomal other genetic disorders since the 1970s, before genomic science was well under way. Roughly half of pregnant women over age thirty-five, and a smaller but significant share of younger women, have amniocentesis, often after a prior screening has shown a possible risk to the fetus. Additional options of chorionic villus sampling and other tests ensure that prenatal genetic testing is widely available and fairly commonly used, at least in wealthy industrialized countries. After previously recommending noninvasive prenatal testing for pregnant women over age thirty-five, in August 2020 the American College of Obstetricians and Gynecologists issued new guidelines recommending that the testing be offered to "all pregnant people regardless of their age or other risk factors."

The director of the Dutch Down Syndrome Foundation, Gert de Graaf, and his co-authors estimate that on average across European countries, 54 percent of fetuses determined to have Down syndrome were terminated from 2011 through 2015. That figure varied from 0 percent in Malta to 83 percent in Spain. They report a parallel figure of 33 percent fewer live births of fetuses with this chromosomal abnormality in the United States than would have been the case without prenatal diagnosis.[22] Regional studies in the United States show variation in abortions after a finding of severe abnormalities, ranging from up to 90 percent in very liberal areas such as New York City to as low as 10 percent in highly conservative areas. As that range implies, disputes over prenatal genetic testing are entwined with the politics of abortion, parenthood, and disability. Two sociologists found "ambivalence" about prenatal testing in their analysis of a survey in the United Kingdom, since respondents were torn between endorsement of medical or individual reasons for diagnostic tests and discomfort with the societal and public health implications of possible termination of the pregnancy. As one respondent observed, "A plethora of bio-ethical issues are present."[23]

Many participants in this dispute, however, express no ambivalence. ABC News featured one mother's letter to the doctor who apparently suggested termination after an amniocentesis: "I'm not angry. . . . I'm sad you were so very wrong to say a baby with Down syndrome would decrease our quality of life. . . . But I'm mostly sad you'll never have the privilege of knowing my daughter, Emersyn."[24] If this mother chose to be more analytic, she might agree with Mary Johnson, editor of a disability rights movement journal, that the choice of an abortion because the child will have a genetic disease such as cystic fibrosis "says that those characteristics take precedence over living itself, that they are so important and so negative, that they overpower any positive qualities there might be in being alive." Gene Rudd, president of the Christian Medical and Dental Associations, describes prenatal testing for Down syndrome as simply "search and destroy."[25]

But a "writer, mom, and doula," as Jacqui Morton describes herself, did choose an abortion. She worries that "genetic testing is becoming a polarizing issue like abortion, *because* of abortion—when it's really just another moment to make sure women have access to compassionate care and comprehensive information. We could, in such moments, support each other as human beings, rather than trying to pick sides and heighten our differences." Vyckie Garrison, a blogger who is a former member of a Christian fundamentalist movement, writes more in anger than in sorrow: "When you argue that children with Down syndrome are 'special gifts' or that raising them is a 'rewarding experience' for parents, you are **appropriating their difficulties and fetishizing their difference. That is the opposite of respecting a disabled person.** . . . To refuse to act to minimize suffering (indeed, to *prevent* it) is at best selfish and at worst abusive. . . . To focus on mere 'life' to the exclusion of the quality thereof is not just stupid, it's evil." It is, she concludes, deliberately inflicting suffering on others to soothe your own conscience.[26]

The new technology of prenatal gene editing adds a sharp twist to these already agonizing debates about disability, abortion, and parenthood. Scientists can now edit an organism's genome, replacing one DNA sequence with another and thereby modifying a gene's function. That is analogous to using a word processing program to correct a typographical error, or to replace a phrase or sentence, in a document. (The core technology is known as CRISPR; Jennifer Doudna and Emmanuelle Charpentier pioneered it, for which they won the Nobel Prize in Chemistry in 2020. Doudna is one of the authors in the second epigraph to *Genomic Politics*, to which I have been referring.) CRISPR editing and its successors are now finding success in mice fetuses; pediatric surgeon William Peranteau speaks for many when he dreams that in a few years "in utero editing would provide hope for families that today have none."

In utero gene editing will be controversial both because of disputes over termination and because of the possible danger to mother and fetus of a new and invasive technology. But that controversy would fade beside the contention raised by the prospect of germline gene editing, in which all cells of an embryo are affected—including cells that will eventually become sperm or eggs. That means that, if the edits are complete and permanent, any changes to the embryo's DNA will be passed on genetically in perpetuity. In 2017, researchers successfully corrected a genetic mutation in nonviable human embryos without introducing damage to nontargeted DNA, and with complete coverage of the DNA sequences that they intended to edit. That is, they removed the typos from the manuscript without introducing new errors. Geneticist Hong Ma and his colleagues' published report in the prestigious journal *Nature* concluded, with barely veiled pride, that their new technique for successfully fixing defective genes "can potentially rescue a substantial portion of mutant human embryos," thus making preimplantation genetic diagnosis and treatment a much more viable option among women seeking to become pregnant.[27]

Germline gene editing will not be legal in the United States without a sea change in federal regulations, and perhaps laws. As of this writing, the National Institutes of Health (NIH) "will not fund any use of gene-editing technologies on human embryos" because "altering the human germline in embryos for clinical purposes . . . [is] a line that should not be crossed." Geneticist Francis Collins, director of NIH, is firm in that commitment, and he reinforces the view of probably all legislators, most advocacy organizations, most residents of the United States, and a large share of scientists who have spoken on the subject.[28]

But even though it will probably be decades before embryonic gene therapy could become widely available in the United States, emotions and rhetoric are already heated. Ma and his colleagues' research is "a pretty exciting piece of science. It's a technical tour de force. It's really remarkable," says George Daley, dean of the Harvard Medical School. But historian Marcy Darnovsky, director of the Center for Genetics and Society, describes the early research on germline gene editing as "extraordinarily disturbing. It's a flagrant disregard of calls for a broad societal consensus in decisions about a really momentous technology that could be used for good, but in this case is being used in preparation for an extraordinarily risky application." David King, head of the "independent, secular watchdog group," Human Genetics Alert, is less polite: "Irresponsible scientists [must be] stopped. . . . We call on governments and international organizations to wake up and pass an immediate global ban on creating cloned or GM [genetically modified] babies, before it is too late."[29]

All participants in these disputes endorse the rights of people with disabilities, respect the distinctive status of pregnancy, strive to live morally, and are passionately committed to decreasing human suffering and improving human society. But they come close to accusing each other of murder, sadism, or at least inexcusable treatment of totally defenseless and vulnerable entities. The intensity is easy to understand: more than all of the other issues I have discussed so far, the questions around prenatal genetic testing and inheritance point to our "unprecedented . . . ability to . . . direct the evolution of our own species." Doudna's and Sternberg's inquiry once again reverberates: "What will we, a fractious species whose members can't agree on much, choose to do with this awesome power?" The societal use of genomic science is not like other scientific innovations.

Using the Basic Framework to Make Sense of Genomics Disputes

These four topics—BiDil, DNA ancestry testing, forensic DNA databases, and prenatal genetic testing and editing—clearly differ. They range across medical research and treatment, genealogy and group ancestry, criminal justice, and abortion and disability. Actors in one arena are seldom involved in any of the other three. One arena remains in the private realm, one is entirely public, and two involve public regulation over private action. The first two involve a clear profit motive; the latter two do not. Race is central to two arenas, not far from the center in a third, and less visible in or irrelevant to the fourth. Biologically defined sex is at the core of one arena, and less prominent in or absent from the other three.

Nonetheless, these topics have a great deal in common. All revolve around recent technologies growing out of genomics; like the science itself, the technologies are evolving rapidly and have indeterminate trajectories with many plausible outcomes. All evoke passionate commitment and opposition from experts and a robust community of advocacy groups and sometimes corporations.

Most striking to a political scientist such as myself is that even in the current era of polarization among identity groups masquerading as political parties, none of these disputes is organized along partisan lines or into ideological camps of liberalism and conservatism. As Chapter 3 shows, all elected officials, of any political party, who have spoken on forensic DNA databases endorse them; few elected officials have said anything at all publicly about DNA ancestry testing, race-based medicine such as BiDil, or gene editing. As Chapter 7 shows, the American public holds different views on these issues, but unlike in other scientific arenas such as climate change, peaceful use of nuclear power, or evolution, the divisions do not follow Republican/Democratic lines. Instead, we see battles among people who usually see themselves as allies in public discourse. Most proponents and opponents of the FDA's approval of BiDil were probably liberal Democrats. The same probability holds for proponents and opponents of DNA ancestry testing. Putatively conservative FBI officials and prosecutors work with liberal Innocence Project attorneys to investigate cases proposed for exoneration; while serving as California's attorney general, the liberal Democrat Kamala Harris endorsed expanding the use of forensic DNA databases to compare crime scene or suspects' DNA samples with samples from incarcerated felons (a technique called familial matching—see Chapters 4 and 5). Doudna and Charpentier, along with genomicists Feng Zhang and George Church who also played key roles in developing the gene editing technique CRISPR, are among the most cautious about its eventual use in humans.

Yet disputants nonetheless throw thinly veiled charges at one another. What are they fighting over? The basic framework both organizes these disputes and reveals what is at stake in them. The first two disputes, about BiDil and DNA ancestry testing, mainly address the question on the vertical axis of the basic framework shown in Figure 1.3: *how significant are genetic factors in explaining human traits and behaviors?* The most explosive arena for asking that question addresses the meaning of "race." At the top of the vertical axis is the perception that being a member of a particular race, or at least sharing biogeographic ancestry with identifiable others, includes an important though not dispositive inherited genetic component. At the bottom of the vertical axis lies the perception that race is a social construction, invented in different ways by different societies in different historical periods, usually for purposes of domination and control. But race is only part of the debate over the importance of genetics in human behavior. Sociologists Melinda Mills and Felix Tropf provide one of the most ambitious lists: genetic inheritance helps to explain "a myriad of topics such as fertility, educational attainment, . . . social mobility, well-being, addiction, risky behavior, and longevity"—in short, much of life.[30] Others find this analysis "troubling," to quote one email comment.

The other two disputes, over forensic DNA databases and prenatal genetic testing or gene editing, mainly address the question on the horizontal axis of the basic framework in Figure 1.3: *how should an individual or society balance projected benefits and risks of genomics innovations?* At the left of the horizontal axis is eagerness to attain the proffered gains of better health, fairer trials, deeper knowledge of one's family, and thriving offspring, while giving due attention to the ways in which these promises might fail or backfire. At the right of the horizontal axis is concern about harm from medical interventions, genomic surveillance by police, a reborn eugenics, and designer babies, while giving due attention to the ways in which genomic science might benefit some individuals or solve some problems.

It is in the intersections of these two dimensions—the four quadrants corresponding to the four colors of the 2 × 2 grid in Figure 1.2—where the real work is done. Enthusiasts (the pink quadrant) argue that advances in genomics will save lives, invigorate diasporic networks, promote justice in courts, and help parents cope with a genetically problematic fetus. Liberal and Democratic Enthusiasts can celebrate particular aspects of these advances—more racially equal treatment of heart failure and more exoneration—while conservative and Republican Enthusiasts can celebrate other aspects, such as clearer standards for conviction and greater preparedness for a child with a genetic disability. Skeptics (the baby blue quadrant) worry that genomic science offers false hope, junk science, new surveillance, and competitive fetal perfectionism. Here too, views easily cut across party and ideological lines: skeptical liberals and Democrats can worry about racial essentialism and economic disparities in access, while skeptical conservatives and Republicans fear group-based identity politics and denial of God's plan.

The two quadrants in the lower part of the basic framework also inscribe distinctive worldviews. The Hopeful (in the crimson quadrant) turn to interventions that are not genetically based to solve societal and individual problems. Hopeful liberals and Democrats look toward social policies or political accountability; Hopeful conservatives and Republicans rely on individual will, group self-help, or divine assistance. And Rejectors (navy blue) turn to neither genetics nor other forces for betterment; in their eyes, the risks of intervention outweigh the likely gains regardless of what the intervention consists of. Rejectors are relatively unlikely even to be liberals or Democrats, conservatives or Republicans, since such appellations imply some variety of defined engagement with the public arena.

The rest of this book develops the analysis I have just sketched out, with due attention to the ideological, political, policy, and moral implications of each stance. But even at this point, I hope the stakes of *Genomic Politics* are clear. If the United States follows the path of Enthusiasm, we will be betting heavily on this wondrous and awe-inspiring technology. As *The Economist* reminds us, that is what Western nations did with regard to physics at the turn of the twentieth century—and physics brought us both space exploration and the atomic bomb. If the United States follows the path of Skepticism, we will enable genomic science to develop but will erect high fences around its use, as the European Union does for genetically modified food and the NIH does for germline gene editing. If we follow the path of Hope, genomic science will remain mostly in the labs and Americans will continue traditional debates over social welfare policies versus individual responsibility. And if we follow the path of Rejection, we will do little—so, for example, Americans will neither risk an anaphylactic reaction to a genomics-based Covid-19 vaccine nor support action by public agencies to slow the coronavirus's spread.

The right choice is not obvious; as Al Gore put it in 2013, genomics is "taking us beyond the outer edges of the moral, ethical, and religious maps bequeathed to us by previous generations. We are in terra incognita." The proximate goal of *Genomic Politics* is to create a new map of genomics technology, so that we can move beyond the paralysis of "There Be Monsters." My ultimate goal is to inform Americans' choices about where we might move and why we might go in that direction, whichever it turns out to be.

Preview of Coming Attractions

The next four chapters explore *what* shapes views in the four quadrants about societal uses of genomics. Chapter 2 further develops the basic framework, exploring the vertical and horizontal axes and then delving into each quadrant in turn. Chapter 3 explores public leaders' partisan and ideological stances regarding these four positions—or, rather, the absence of partisan divides among leaders and the complexity of their ideological positions. Chapter 4 explores disputes about genomics technology along the horizontal dimension of optimism and pessimism, and Chapter 5 does the same for disputes along the vertical dimension of genetic influence or its absence.

Chapters 6 through 8 explore *who* fits into the four quadrants of the basic framework, and *why* they do so. Chapter 6 examines the views of two sets of experts—social scientists who conduct research on and write about genomics and its uses, and geneticists, observers, and public officials who are directly involved in bringing genomics into the societal arena. It also analyzes a database of thousands of scholarly articles. Chapter 7 introduces two national surveys from 2011 and 2017 and uses them to show how Americans array themselves within the basic framework. It then shows why people occupy one or another quadrant. In addition to analyzing survey items, it analyzes thousands of open-ended responses to key questions in the surveys.

The final two chapters address *how* societal uses of genomic science should be managed and governed. Chapter 8 examines considerations used by social science experts, interview subjects, and the American public to address governance of genomics technologies. Chapter 9 concludes *Genomic Politics* with my own views, and observations intended to guide discussions of governance options and politics.

As this brief outline suggests, I approach *The Economist*'s "exciting and mysterious future"—or Doudna's "impossible but essential question," or Gore's "terra incognita"—from the cautious vantage point of the professor. But we must never lose sight of what is at stake. Genomics technology is not yet tangled up in partisan identity groups, although I see hints that such a division might be developing. I hope I am wrong about those hints; in my view, societal use of genomic science is exactly the right subject for democratic political discourse but too important to be dominated by pre-made allegiances. If I do my job, by the end of *Genomic Politics* readers will understand what our options are, what is at stake, how decisions might be made, and why all of us should be active in this arena. We can get past the blank terror of "There Be Monsters."

2
The Basic Framework
Nature and Nurture, Risks and Gains

Everything that affects humans has political undertones.
—Interview subject, in response to question about politics in the use of forensic DNA databases

Genomics is unique among the sciences in the depth of its probe into the literal and metaphorical meaning of life. It requires us to examine associations between brain and mind, genotype and phenotype, genetic inheritance and free will or God's will, the natural order versus human control. And (or therefore) it is political.

The basic framework introduced in Chapter 1 is my attempt to impose order on these existential questions and then to link them to practicable, communicable political discourse—to move us from the chaos of "There Be Monsters" to the orderliness of mapmaking. After reminding readers of the simple, almost simplistic framework, this chapter scrutinizes each dimension, each quadrant, and finally the overall structure, so that the rest of *Genomic Politics* can use it.

Is Genomic Science a Thing?

A preliminary question: is "genomics" sufficiently bounded and coherent to be an object of political maneuvering, ideological engagement, and substantive analysis? The National Human Genome Research Institute (NHGRI) definition of genomics, "the study of all of a person's genes (the genome), including interactions of those genes with each other and with the person's environment," is not much help in determining whether genomics is enough of an entity for me to write intelligibly about "genomic politics."

There are two main answers to the question of whether genomics is a thing. One can describe genomic science as a neutral, apolitical tool with no implications separate from its user's purpose—a very fancy hammer or screwdriver. A similar view is the perception of one or another aspect of genomics as an additional weapon in long-standing political disputes around, for example, the meaning of race, the moral standing of a fetus with a genetic mutation, or the planting of modified or "heritage" crops. Genomics is, in these logics, disaggregated and instrumental; actors in the arena of prenatal genetic testing will have no reason to connect with actors in the arenas of forensic DNA databases or climate change. As one person I interviewed put it, "Genomics is an umbrella term covering lots of issues. I doubt that something as abstract as genomics will attract contestation. The politics will be an intersection of a particular technology or issue with beliefs or money." A reader of this book in manuscript made the same point: "Given the disorderly, fragmented political processes that exist in the United States, and the diversity and fluidity of actors involved in debates across multiple arenas, why shouldn't the usual American pluralism be our baseline expectation for how the politics of genomic science will play out?" Most simply, "genomics is just too many things," according to legal scholar and bioethicist Hank Greely.[1]

The usual American pluralism may, of course, be the right expectation—but there is a contrasting view. Technological innovations such as movable type, the steam engine, splitting the atom, human flight, and computers transformed, over decades or centuries, how humans engage with each other and with the natural world, and even what ideas and values they were able to conceive with which to engage. Genomics plausibly belongs on that short list. As humans start to eliminate mosquitoes, enhance musical talent before birth, maintain a record of every American's "future diary," create bacteria to combat climate change, and bring the wooly mammoth (or Neanderthals?) back to life—that is, as we start "to direct the evolution of our own species"—features of public discourse and activity will change in ways that we cannot now anticipate or even imagine. That, I think, is what Doudna and Sternberg, Gore, and *The Economist* are all trying to say, as was one of the people I interviewed: "We have a genomic *revolution* in knowledge, in what it means to be human—there is deep substance to it. At some point things get so big or consequential that we have to then think about them differently." Celebrating the beginning of the Human Genome Project, George Cahill of the Howard Hughes Medical Institute was more succinct: at least in the realm of biology, "it's going to tell us everything."[2] If that is true, genomics is much more than a fancy hammer or new weapon in old battles.

The final chapter of *Genomic Politics* returns to the question of whether genomics is mainly a new tool or weapon, or a coherent, profound transformation in the way that humans understand their place in the world and engage with it. I believe genomics is the second of these, although I warn readers that much in the next few chapters examines particular genomic technologies separately from each other. Reconstruction comes only after deconstruction, at least in this case.

Genetic Influence

> We have in our beginnings some bent of mind, some shade of character. . . . In earlier times men sought its trace in the conjunction of the stars or perhaps in the momentary combination of the elements at nativity. Today, instead, we know to look within. We seek not in the stars but in our genes for the herald of our fate.
>
> —Robert Sinsheimer, 1969

> If you want to predict whether someone believes in God, it's more useful to know that they live in Texas than what their genes are.
>
> —Julian Baggini, 2015

Even those who believe that our fate is largely in our genes, rather than in the stars or Texas, do not see humans as automatons. But they do argue that humans act within a wider or narrower range of genetically inflected outcomes, for which I use the term *genetic influence*. Perceptions of the width or narrowness of that range—whether environment and behavior shape the impact of genetics, are themselves shaped by genetics, or are simply irrelevant—remain topics of intense research and debate among biologists and social scientists (as well as among theologians and parents).

Strong genetic influence lies at the top end of the vertical dimension of the basic framework. At the bottom end lies skepticism about or denial of the importance of genetics in shaping human traits, behaviors, and physical attributes. Instead of DNA, in this view, most human characteristics are caused or at least shaped by forces such as socialization, environmental context, individual choice, luck, or divine law. Despite differences and even contradictions among these causal attributions, views or policies at the lower end of the vertical dimension unite in concurring with philosopher Julian Baggini that if you want to predict whether someone believes in God—or has some other quality—genetics lies somewhere between fairly and completely unimportant.

Does Genetics Explain Important Outcomes?

Genetic Causes

Most Americans accept that genetic inheritance largely determines some phenotypes, even if few go as far as Prospero's lament about "a born devil on whose nature Nurture can never stick, on whom my pains, humanely taken, all, all lost, quite lost." Almost all people are understood to be, and see themselves as, male or female because they inherit a particular pair of chromosomes from their biological parents.[3] Genes strongly determine some diseases, even if their expression varies almost as much as the expression of gender varies within a biological sex. Perhaps 10,000 monogenic diseases usually or always result from changes in or variations of a particular gene. Examples on a typical list include Huntington's disease, sickle cell anemia, Tay-Sachs disease, cystic fibrosis, neurofibromatosis type 1, and xeroderma pigmentosum. Most people with knowledge of genetics also agree that genetic inheritance contributes importantly to phenotypes such as schizophrenia, arthritis, diabetes, musical ability, height, skin color—and perhaps Covid-19 and influenza.

Controversy spills from scientists' labs into public debate when considering the role of genetics in socially fraught traits, behaviors, or ailments such as sexual orientation, weight, aggression, intelligence, drug addiction—or racially disproportional susceptibility to Covid-19. Consider weight. A 2018 analysis combined results from studies that had previously examined associations between people's weight and their genomes. This meta-analysis included over 700,000 individuals. It revealed over 900 locations among the 3.2 billion base pairs of DNA in the genome that are associated with body mass index (BMI), accounting for roughly 6 percent of BMI variation.[4] Scientists and social scientists disagree on whether to interpret these results as an impressive scientific advance ("fully 6 percent!") or a damp squib ("only 6 percent?").

The public has mixed views on the amount of weight variation that they attribute to genetic variants. In twelve survey items between 1995 and 2009 (all of the available items in iPoll, the database of public opinion surveys), between 4 and 80 percent of Americans identified genetics as an important or the most important cause of being overweight. Much of this variation in survey results is explained by sample selection, question wording, and choices of response categories—but the very fact that these methodological choices could have so much impact suggests how volatile the issue of genetic influence is in the mind of the public.

Both experts and laypeople have similarly varying views of the importance of genetic influence for other phenotypes as well, ranging from intelligence through what pollsters call "success in life."[5] Activists argue fervently about causes, interactions, and appropriate interventions. Rising above the noise, a majority of Americans endorse action based on an assumption of genetic influence, at least in the medical arena. In my 2011 survey of American adults, more than 70 percent support testing for the likelihood of getting an inherited disease, and in the 2017 survey over three-fifths endorse research on gene therapies for disorders or diseases. (I discuss the surveys in Chapters 7 and 8.)[6]

Non-Genetic Causes

If phenotypes are not controlled mainly or at all by the individual's genetics or if genetic expression varies enormously, what *does* cause human traits, behaviors, and diseases? Unfortunately for analysts seeking parsimony, the "nurture" side of the old nature-versus-nurture dichotomy is poorly, or rather multiply, defined. Some believe that a deity creates variation among persons or even determines each human's destiny; as the psalmist put it, "All the days planned for me were written in your book before I was one day old" (Psalm 139:16 NCV). Or Hamlet: "There's a divinity that shapes our ends, / Rough-hew them how we will." (Act 5, Scene 2). In my 2017 survey, 17 percent oppose gene therapy on the grounds that "it violates the laws of God or nature."

Others believe that individuals can and do exercise choice; as Helena puts it in *All's Well That Ends Well*, "our remedies oft in ourselves do lie, / Which we ascribe to heaven" (Act 1, Scene 1).[7] A question in the gold-standard national adult survey, the General Social Survey (GSS), repeated twenty-one times from 1977 through 2018, asks whether Blacks are materially disadvantaged compared with Whites because they lack the motivation or willpower to escape poverty. Averaging responses from roughly 29,000 Americans across this period, over half agree (53 percent of Whites, 41 percent of Blacks). In contrast, just over 10 percent, with no racial differences, agree that genetic inheritance explains Blacks' relative poverty.[8]

A third explanation for differences in human phenotypes looks to neither God nor the individual, and in fact accepts the importance of genetics. The central point here, however, is that the raw material of genetic inheritance is shaped by social or environmental factors into the outcomes that we see in a person's traits or behaviors. In this understanding, even "a world where 'everything is heritable' is a world with no less work for social scientists interested in individual outcomes," points out the sociologist Jeremy Freese. Psychologist Eric Turkheimer and his co-authors provide a classic demonstration of how the impact of genetics flows through social channels. They examined IQ scores of seven-year-old twins, some of whom lived in families with incomes below the poverty line. "In impoverished families, 60 percent of the variance in IQ is accounted for by the shared environment, and the contribution of genes is close to zero; in affluent families, the result is almost exactly the reverse."[9] Most Americans intuitively agree with these researchers that context can have causal force: although a majority of respondents in my 2017 survey endorse research on gene therapy, 70 percent also agree that scientists should give equal or greater priority to "seeking to change social and environmental causes of serious diseases or disorders" than to "research on gene therapy."

The strongest version of the bottom end of the basic framework's vertical dimension is that explanations involving genetics are mistaken. Environmental analyses encompass everything from prenatal nutrition or stress to lead-bearing paint chips, family upbringing, neighborhood characteristics, societal norms, cultural or religious tropes, and government policies. Public health physician David Carpenter and his co-author argue that "social factors, including poverty, poor education, and family instability [as well as] a number of environmental exposures are documented to result in a common pattern of neurobehavioral effects, including lowered IQ, shortened attention span, and increased frequency of antisocial behavior."[10] Many Americans concur. Two-fifths of respondents (a third of Whites and two-thirds of Blacks) agree across three decades of the GSS that Blacks' disadvantage is due to discrimination, while half say that "lack of education" is the cause. Expressing the same sentiment, two-fifths of respondents in a different survey agreed in 1997 that criminal behavior was "not at all . . . determined by heredity and genes."[11]

Rejection of any role for genetic inheritance is especially prominent in debates over the meaning and importance of "race" or "ethnicity" or the assignment of a given individual to a "race."[12] Denial of a role for genetics is known as social constructivism, which sociologist Ann Morning carefully defines as "the belief that racial categories are the intellectual product of a particular (albeit enduring) cultural moment and setting, and that human biological variation does not naturally and unquestionably sort itself into. . . groups." Social constructivists understand race to be "not an individual trait but rather a characteristic of a relationship or social setting. In other words, individuals do not carry race within them." Psychologist Steven Heine injects more emotion into his definition: "The genetic variation among humans fails each biological test of being a race. . . . [R]ather, race is . . . a product of what we learn growing up. Societies decide who to classify as what, and what they decide varies from one place to another. In contrast to a biological account of race, . . . our *experiences*, not genes, differ across the races." Organizations weigh in; race, according to the American Anthropological Association, is "a worldview, a body of prejudgments that distorts our ideas about human differences and group behavior." And law professor Jonathan Kahn throws caution to the winds in pointing to "the scientifically unjustified and socially dangerous recasting of race as a social and historical construct into a reified genetic category."[13]

Dismissal of genetics may be less explicit. Four edited handbooks on crime or criminology published by Oxford University Press, with several dozen chapters by multiple authors in each, range in length from 600 to almost 1,000 pages (with small font and minimal margins). The detailed indices in all four have many subentries under the main entry "race," ranging from "hate crime" to specific groups to "racism." Indices in two handbooks do not include "genetics" or "DNA" at all; the third index has no entry for "genetics" but does cite "DNA databanks" on four pages; in the fourth, "genetics" appears on nine pages and "DNA" not at all.[14] None links genetics to race.

A final non-genetic explanation for human variation is luck, fate, or astrology— as poor Pistol describes it, "Giddy Fortune's furious fickle wheel, / That goddess blind, / That stands upon the rolling restless stone" (*Henry V*, Act 3, Scene 3). In the mid-2000s, across seven iterations of the GSS involving 8,500 respondents, 37 percent agreed that astrology is "very" or "sort of" scientific.[15]

Nature *Versus* Nurture?

Of course, explanations for human phenotypes can be combined, whether interactively (as in the Turkheimer IQ study) or in a list (as in the GSS questions about Black Americans' disadvantage). Few members of the public, in fact, impose the perhaps foolish self-discipline of choosing only one explanation for human traits, behaviors, or diseases; as communications scholar Celeste Condit elegantly portrays, they readily combine genetic and non-genetic causal attributions.[16] Consider a pair of survey questions offering respondents possible "causes of problems with a person's health" or, in a separate question, possible causes of adult health problems resulting from events in childhood. Respondents could choose as many as they wanted; as a consequence, all but one possible answer received considerable support. The combined responses to both questions are in Table 2.1.

As Table 2.1 shows, Americans accept environmental factors *and* family or social influences *and* economic factors *and* individual choice *and* medical conditions *and* God's will *and* genetics as explanations for health outcomes. A majority reject only "bad luck" as an important cause—ironically, perhaps, since medical professionals may rank it high on the list of explanations for being struck with terrible diseases such as childhood leukemia or ALS.[17]

In the more sensitive arena of defining racial groups or assigning people to a "race," some analysts point to connections between genetic inheritance and something like race. All mainstream researchers reject genetic essentialism. But many share, for example, the worry articulated by professor of nursing Elizabeth Cohn and her colleagues: insufficient representation from people around the world in scientific biobanks distorts research because "diversity in genomic studies must account for genetic variation both within and across racial categories."[18] As of 2010, 96 percent of the genomic data in the 1.7 million samples available for research came from people "of European ancestry . . . . This inequity, if it is not fixed, will turn into tremendous health inequality," according to Stephanie Devaney, deputy director of NIH's All of Us research program. For example, she continues, "African-Americans and Latinos have the highest rates of asthma in the U.S., but studies show that common drugs used in inhalers do not help them as well as they help Whites. Asians who take the antiseizure drug carbamazepine have a higher risk of severe, sometimes fatal, reaction. . . . If DNA is one important factor in our quest for more effective medical treatment, we need to address the lack of diversity in genetic data."

Table 2.1 Causes of problems with a person's health (percent saying "extremely important" or "very important")

Childhood abuse or neglect	89%
Personal behavior	84
Poor childhood diet	83
High stress*	81
Viruses or bacteria	80
Environmental pollution in childhood	80
Poor medical care	79
No friends and family to talk to and rely on*	79
No childhood vaccinations	77
Air, water, or chemical pollution*	73
Bad working conditions*	76
Poor neighborhood and housing conditions*	72
Abuse in adulthood*	70
Insufficient education*	68
Childhood poverty	66
Not graduating from high school	62
God's will*	57
Low income*	57
Born premature or underweight	51
Bad genes	49
Bad luck*	16

Note: Items without asterisk were given to half of the sample; items with asterisk were given to one-quarter of the sample. $N = 2{,}423$ U.S. adults; interviews in Spanish and English. The table combines results for two questions, one about adulthood causes and the other about childhood causes of poor adult health.

Source: Harvard School of Public Heath/Robert Wood Johnson Foundation/NPR, September 2014.

Devaney is clear that "many factors beyond our genes are at play when it comes to disease," and she provides a rich list of them. Nonetheless, the diversity sought for the All of Us biobank "will help us discover more about how DNA affects health across different communities."[19]

Others similarly work hard to avoid hints of essentialism while still recognizing genetic influences associated with what is conventionally understood as race. Neurobiologist Mary Jeanne Kreeck begins an American Museum of Natural History book on the genomic revolution with this: "Scientists now recognize that our genetic composition does not represent predestination or predetermination, but rather sets the stage for the interaction between genes and an array of factors in our individual and shared environments." The book's authors, biologist Rob DeSalle and public health researcher Michael Yudell, in a chapter pointedly titled "99.9 percent" (the proportion of the genome shared by all humans), assure the reader that "despite obvious phenotypic (observable) differences, humans are remarkably alike with respect to their genomes. Our skin color, eye color, hair color and texture, sex, height, weight, and body shape may vary, but underneath these surface characteristics our genomes are all essentially the same." Still, unease shines through the careful prose: "We know that there are grades of difference between people and populations of people that do reflect biology, and these differences may be important for understanding human evolution and to develop treatments for disease. How do we deal with the biology of human variation without giving it unwarranted social significance? . . . [A]n important component of genomics is identifying genetic differences between individuals, and also within and between human population groups."[20]

Geneticist David Reich offers a different approach to linking race and genetic inheritance while avoiding essentialism: he aims to redefine the divide between racial determinism and social constructivism. Current "orthodoxy"—"the idea that human populations are all too closely related to each other for there to be substantial average biological differences among them—is no longer sustainable." However, "racist pictures of the world that have long been offered as alternatives are even more in conflict with the lessons of the genetic data." Reich argues instead that analysis of ancient DNA samples shows that the history of human "races" consists of endless subdivisions within old groups, novel combinations of slivers of old groups in ways that create new ones, migrations and interbreeding among groups about whom we know almost nothing yet, disappearance of humanoid populations, and perhaps other forms of human mitosis and meiosis. Engaging with genetic inheritance is essential for understanding differences among what we call races, but the conventional categories of the past millennium must be understood as slippery, contingent, and essentially arbitrary.[21]

The ferocious drive through much of 2020 to determine if Covid-19 vaccines are safe and effective ran headlong into this complex interplay of genetic influence and social constructivism. Compared with European Americans, people of African, Native American, and to some degree Pacific Island or Asian descent are disproportionately likely to be infected with the coronavirus, to become seriously ill or hospitalized, and to die. (The same disparity holds for Latinos, but they are generally understood to comprise a mixture of the other races.) Researchers point to a long list of explanations for these differences. The Centers for Disease Control (CDC), for example, identifies five "key topic areas of social determinants of health" that affect Covid-19 infection, all influenced by discrimination: neighborhood and physical environment, health and healthcare, occupation and job conditions, income and wealth, and education.

If social determinants of health fully explain racial disparity, then logically there is no need for descriptive representation in vaccine trials because differences in response to the mRNA vaccine will be explained by social or health conditions, not by any biological component of what we understand to be race. But researchers and public officials are deeply uneasy at the prospect of an almost-all-White Stage 3 vaccine trial, to be followed by population-wide vaccination; the NIH and pharmaceutical companies have gone to great lengths to obtain a diverse sample of trial participants, and they face considerable criticism if they fail at this. In my view, this is correct: in addition to the imperative of increasing trust among disproportionately harmed Americans of the vaccine's efficacy and safety, no one wants to risk the possibility of a group-inflected genetic difference in vaccine response that could do serious harm. Epidemiologist Daniel Chastain and his colleagues are unusually explicit about this on-the-one-hand-on-the-other-hand view when writing in the *New England Journal of Medicine*, "Despite disproportionately higher rates of Covid-19 infection, hospitalization, and death in racial and ethnic minority groups, the direct effects of genetic or biologic host factors remain unknown. As we strive to overcome the social and structural causes of health care disparities, we must recognize the underrepresentation of minority groups in Covid-19 clinical trials." These authors conclude that laws, regulations, and "the principle of justice demand equitable selection and enrollment of participants and detailed presentation of demographic data and outcomes."[22]

There are strong historical and contemporary reasons for the asymmetry between social constructivists' certitude and assertiveness and the caution of geneticists who perceive, along with social explanations, a disproportionate proportion of particular genes in some population groups often understood as races. (That sentence is one example of such caution!) As one of my interviewees put it, "The long tail of genomics is eugenics." The possibility of a resurgent movement to promote human betterment through controlling reproduction and thus genetic inheritance, with its attendant racism and threat of genocide, haunts this field of research. Even setting aside fears of a revived eugenics movement, Morning reminds her readers that as social constructionists understand it, "racial classification is at its root an instrument of power, meant to establish social hierarchy." That is, humans find it extremely difficult to conceive of two or more groups of humans without imposing a ranked structure onto the groups. Experts who perceive considerable genetic influence struggle, scientifically and morally, to divorce their work from that history and psychology—hence the insistence that even if genetic influence for a particular phenotype is high within a population, it is narrowly defined and seldom deterministic.

Those who seek "to talk about race and biology in the same sentence without blowing up," as I describe this topic to my students, lack a usable history, and so they are on the defensive. As the authors of the American Museum of Natural History volume summarize in a classic of understatement, "The use of human population groups for scientific study is a contentious subject."[23]

Connection between sex and gender or, put differently, between genetic influence on male and female physiology and the social construction of men and women, is an almost equally contentious subject. Although scientists and laypeople are learning that there are more nonbinary individuals than has usually been recognized in modern Western cultures, few disagree that there are inherited differences between people defined at birth as male or female. But exactly *which* attributes and behaviors are shaped by biological sex is not resolved. One example of the societal importance of that question suffices: Speaking to a 2005 economics conference on diversity, Lawrence Summers, then president of Harvard University, offered "three broad hypotheses" to explain underrepresentation of female scientists and mathematicians at elite universities. First, the most prestigious occupations "expect of people who are going to rise to leadership positions in their forties near total commitments to their work," which a higher proportion of married men than women are able to provide in their twenties and thirties. Although true, Summers pointed out, that explanation does not distinguish gender disparities in science from other fields. He therefore proposed a second hypothesis: "On many, many different human attributes—height, weight, propensity for criminality, overall IQ, mathematical ability, scientific ability—there is relatively clear evidence that whatever the difference in means—which can be debated—there is a difference in the standard deviation, and variability of a male and a female population." He was talking about only "people who are three and a half, four standard deviations above the mean" and stated (later in the talk) that "I would like nothing better than to be proved wrong." But he also pointed to research evidence to support this hypothesis.[24]

Summers's third hypothesis focused on childhood socialization, discrimination, and stereotyping. He discussed remedies and then engaged cordially with the audience. But it was too late. Some participants were outraged by his second hypothesis (one reported later that she came close to vomiting), media attention was strong, and the incident contributed to the vote by the Harvard Faculty of Arts and Sciences of no confidence in President Summers a year later.

Some scholars nonetheless offer evidence of genetic differences in particular traits between human males and females. Cognitive psychologist Diane Halpern summarizes this huge body of work by pointing first to the "inconsistent findings, contradictory theories, and emotional claims that are unsupported by the research." However, "clear and consistent messages could be heard" above "all the noise in the data. There are real, and in some cases sizable, sex differences with respect to some cognitive abilities." Although "socialization practices are undoubtedly important," Halpern points to "good evidence that biological sex differences play a role in establishing and maintaining cognitive sex differences." In a similar vein, psychologist Steven Pinker describes "ten kinds of evidence that the contribution of biology [to differences between the sexes] is greater than zero, though of course it is nowhere near 100 percent."[25]

As with race, there are strong historical, political, and moral reasons for the asymmetry between arguments that vehemently reject or cautiously admit a role for genetic inheritance with regard to cognitive differences between the sexes. No reputable scholar promotes male supremacy any more than White supremacy, so geneticists balance on a tightrope between evidence showing a genetic distinction between biological men and women and repudiation of millennia of male domination. As genomic science becomes more societally prominent, the basic framework's vertical dimension of genetic influence may become even more radioactive than President Summers discovered it to be.

To put the dispute around race and sex more analytically, the framework's dichotomy of genetic influence—yes or no—is readily turned into a continuum, in which the central issue is not contention between nature and nurture but rather the relationship between them. In a further step, the midpoint of a continuum may be understood not as an equal mix of two amounts but as interaction between the two sides of the dichotomy. The idea of genetic interaction points to epigenetics. Epigenetics is the process by which behaviors or the environment (within or outside the body) affect expression or silencing of particular genes, perhaps even across generations. Scientists agree on little about epigenetics—its definition, mechanisms, importance for a given phenotype, generational reach, and even whether future research will clarify these disputes and uncertainties.[26] For my purposes, however, I can usually set epigenetics aside: despite its importance within the expert community, the interactive framing is only now diffusing out of science into the broader society, and as yet it has little impact on popular politics on policies regarding societal use of genomic science. So, with a few exceptions, I will continue to use the dichotomy of whether genetics does or does not significantly influence human traits and behaviors, adding complications and refinements only as needed.

Most Americans probably care little about genetic influence per se; their attention will be drawn only if the question of genetic inheritance is quickly shown to have an impact on something that matters to them. But most Americans presumably do care about how societal uses of genomic science will affect their lives. That question points us toward the second, horizontal dimension of the basic framework.

Technology Optimism or Pessimism

> There are people who are wired to be skeptics and there are people who are wired to be optimists. . . . At least from the last 20 years, if you bet on the side of the optimists, generally you're right.
>
> —Venture capitalist Marc Andreessen, 2014

> When you see something that is technically sweet, you go ahead and do it and you argue about what to do about it only after you have had your technical success. That is the way it was with the atomic bomb. . . . [But] I cannot very well imagine if we had known in late 1949 what we got to know by early 1951 that the tone of our report would have been the same.
>
> —Physicist J. Robert Oppenheimer, 1954[27]

Statements about the importance of genetics in shaping human phenotypes depend on one's understanding of the facts; technology optimism, in contrast, lies more in the realm of emotions. Choices about how to balance potential benefits and risks certainly invoke facts, but they emerge largely from hopes, fears, anxieties, projections, wishes, or degrees of confidence in one's own predictive capacity. One can claim to prove genetic influence or its absence; one does not prove, but rather persuades and directs, when seeking to shape the use of technological innovation. Marc Andreessen is urging his hearers to be optimists rather than skeptics. J. Robert Oppenheimer is cautioning people to attend to risks even in the face of "sweet" technical success.

People, institutions, and even societies vary in their propensity to focus on whether genomic science's move into the unknown is "for good" or "for ill." (*The Economist*, as we have noted, straddles the fence, predicting "both good and ill.") The distinction has various names: risk acceptance versus risk aversion, preference for type 1 or type 2 errors, technology optimism or pessimism, proactive versus precautionary. Its core question is the degree to which an individual, group, institution, policy or law, or society should, at the margin, choose to take risks in the hope that associated benefits will outweigh potential costs, or choose to take steps to prevent severe harms even at the expense of likely benefits. Actors differ in what they anticipate to be risks and benefits, how they see the balance, and their willingness to accept greater risks in the hope of greater gains.

More formally, technology optimism is what public health scholar Stefán Hjörleifsson and co-authors define as "underestimation and neglect of uncertainty" in favor of "widely shared speculative promise." Psychologist Abigail Hazlett and her colleagues further develop the concept: the scientific optimist "is centered on advancement concerns. . . . [He or she is driven] by motivations for attaining growth and supports *eager strategies* of seeking possible gains even at the risk of committing errors or accepting some loss." Technology pessimism, in contrast, is the overestimation of risk and harmful impact, with insufficient attention to benefits or to people's ability to respond appropriately to risk. A pessimist "is centered on security concerns . . . [and] supports *vigilant strategies* of protecting against possible losses even at the risk of missing opportunities of potential gains."[28] Andreessen's cheer exemplifies the former; Oppenheimer's disenchantment points in the direction of the latter.

Experts range along the horizontal dimension from strong optimism to ambivalence to strong pessimism when they consider genomics. At the 2013

World Economic Forum, scientists and industry leaders rated "unforeseen consequences of new life science technologies" as one of the most likely and most consequential risks to humankind. But two years later, almost all of a (different) large sample of eminent scientists endorsed genetically modified foods, as well as genetically engineered plants as an eventual fuel alternative to gasoline.[29]

The American public is similarly split between optimism and pessimism about the likely effects of genomic science. After reading matched statements about risks and benefits of synthetic biology, more respondents in one 2009 study agreed that risks outweighed benefits than vice versa. But a few years later, asked if research on DNA of plants, animals, and people "could lead to major scientific breakthroughs" or "poses unforeseen dangers," more than three times as many Americans were "excited" as were "worried" (38 to 11 percent).[30] And, of course, people can be ambivalent or change their minds.

The concept of technology optimism or pessimism can be understood as an individual stance, political platform, policy regime, ideological commitment, religious conviction, historical zeitgeist, national culture, and more. Perhaps fortunately, only a subset of that array is relevant to *Genomic Politics*.

Individuals

People vary in their level of technology optimism for many reasons. As psychologists and sociologists have shown people may be poor at or biased in estimating risk.[31] Individuals lack relevant information or analytic frameworks; media and trusted opinion leaders convey contradictory messages; the public faces novel circumstances for which risks are unknown or unknowable; people exaggerate or underestimate a few prominent features of a complicated situation; recent events loom too large. Experts can be more dug in on their position than nonexperts, so the most knowledgeable may actually be the least able to change their estimates of risk in the face of new information or circumstances.[32]

A different cluster of explanations focuses on emotions rather than on cognitive inadequacies. Anthropologist Lionel Tiger defines optimism as "a mood or attitude associated with an expectation about the social or material future—one which the evaluator regards as socially desirable, to his [or her] advantage, or for his [or her] pleasure." In this framing, to ask whether a person is too optimistic or pessimistic makes no sense, since there is no right answer to "How do you feel?" Rather, one analyzes triggers, related affects or moods, and their impacts. For example, positive denial or illusions can encourage creativity and determination or sustain one after a disaster, but delusional optimism can instead endanger oneself or others or provoke an evaluation of the optimist as a "conscienceless jerk."[33] Pessimism can be associated with unproductive fear or paralyzing helplessness, or provide the reality test needed to eschew "the psychological equivalent of junk food" embodied in "pointless" optimism."[34]

Instead of thinking or feeling, people may invoke optimism for instrumental or strategic reasons. Psychologists Ying Zhang and Ayelet Fishbach ran a series of experiments in which they gave participants tasks described as difficult or easy. They found what they labeled counteractive optimism, "a self-control strategy of generating optimistic predictions of future goal attainment in order to overcome anticipated obstacles in goal pursuit." That is, subjects "predicted better performance, more time invested in goal activities, and lower health risks when they anticipated high (vs. low) obstacles in pursuing their goals."[35] Counteractive optimism might be a scholar's way of defining the capacity that so forcefully struck observers of Abraham Lincoln: even after years of defeat and horrible mortality rates, Lincoln refused to waver in his expressed conviction that the North would win the Civil War, the Union with its promise of freedom and democracy would prevail, and (later in his thinking) slavery would be abolished and Negroes free to retain the fruits of their labor.

As the example of Lincoln suggests, explicit optimism may be a political strategy. "Partisanship is likely to trump reality. . . when the issue, however vital, is the least bit arcane or divorced from personal experience," note political scientists Christopher Achen and Larry Bartels. They point to partisans' opposite statements about whether the federal budget deficit or inflation levels are rising or falling depending on whether survey respondents are defending or attacking current fiscal policy. My co-author Katherine Einstein and I found the same pattern in views of whether Iraq possessed weapons of mass destruction in the early 2000s or was linked to the attacks of September 11, 2001.[36] In each case, survey respondents' level of optimism was strongly associated with whether "their" president held office or not.

In a series of dozens of surveys that provide enough detail to reveal two kinds of risk calculation, levels of optimism about Covid-19 show similar partisan inflections. One question is how much risk a given situation entails.

Throughout 2020, Republicans and Democrats usually differed by ten or fewer percentage points in reports of how many people they knew who had tested positive for, or had died from, the coronavirus. But three or sometimes four times as many Democrats as Republicans regularly predicted that the pandemic trajectory would soon get worse in their community or in the United States. Partisan differences in virus-related predictions obtain even within northeastern, largely Democratic states; in both the most and least densely populated ZIP codes; and in counties that have suffered high, moderate, or low effects of Covid-19.[37]

A second question is what kind of risk is most to be avoided. Consistently throughout 2020, more Democrats than Republicans reported that they know someone (perhaps including themselves) who has lost a job because of the pandemic. But in the same surveys, more Republicans than Democrats perceive a greater risk in restricting the economy, while more Democrats than Republicans see a greater risk in the virus itself.

Expressions of technology optimism or pessimism sometimes evade predictable partisan valence. In 2014, the invaluable Pew Research Center conducted almost simultaneous surveys of 2,000 American adults and about 4,000 scientists who were members of the American Association for the Advancement of Science. Nine questions implicitly or explicitly invoked an assessment of risk. The public is more risk-averse than are scientists on seven issues. In two of the nine issues, about the safety of genetically modified food and the merits of bioengineered fuel, respondents agreed across party lines; on five, Democrats or liberals came closer to the views of scientists; on the remaining two, Republicans or conservatives resembled scientists more.[38]

Policies, Laws, and Institutions

Not only people, but also societal institutions can be technologically optimistic (or proactive), rather than pessimistic (or precautionary). A proactive institution or policy promotes research and development, provides funding and other support, minimizes regulatory constraints, and creates incentives for innovation. It focuses, in short, on "eager strategies." A precautionary institution or policy encourages oversight, sets high barriers to approval of new products, moves cautiously in reconfiguring governmental agencies, and sets rules to limit misuse of information or power. It focuses, in short, on "vigilant strategies."

The policy actor's crucial decision is where to place the burden of proof. A policy entrepreneur might argue to permit or promote activity until there is evidence of unacceptable risks, while a cautious policymaker might agree with Robert Coleman, the European Commission's director general for health and consumer protection, that "those in public office have a duty not to wait until their worst fears are realized [before they act]."[39]

A classic example of risk-averse policy is an international agreement, the Cartagena Protocol on Biosafety to the Convention on Biological Diversity. Article 1 states:

> In accordance with the precautionary approach contained in Principle 15 of the Rio Declaration on Environment and Development, the objective of this Protocol is to contribute to ensuring an adequate level of protection in the field of the safe transfer, handling and use of living modified organisms resulting from modern biotechnology that may have adverse effects on the conservation and sustainable use of biological diversity, taking also into account risks to human health, and specifically focusing on transboundary movements.

As of 2020, the Protocol has 173 signatories, including the European Union but not the United States, Australia, or Russia.[40] Article 10(6) spells out the precautionary approach, albeit in opaque language: "Lack of scientific certainty due to insufficient relevant scientific information and knowledge regarding the extent of the potential adverse effects of a living modified organism on the conservation and sustainable use of biological diversity in the Party of import, taking also into account risks to human health, shall not prevent that Party from taking a decision, as appropriate, with regard to the import of the living modified organism in question . . . , in order to avoid or minimize such potential adverse effects."

Contrast the Cartagena Protocol's burden of proof to the United States' regulatory structure for genetically modified organisms. It follows the logic of a 1987 National Academy of Sciences report, which found no evidence of "unique hazards. . . in the transfer of genes between unrelated organisms." The National Academy report observed that the risks associated with introducing genetically modified organisms into the environment are no different from the risks associated with introducing any other new organism; thus by implication (and later explicitly), extant rules and regulatory agencies suffice to manage genetically modified organisms, with no extra precautions.[41]

According to Pew Research Center surveys, Russia is the country in which public opinion is most out of line with official, proactive, policy. Fully 70 percent of Russians agree that genetically modified foods are unsafe to eat, more than in any of the other twenty countries included in Pew's research. Between 30 and 60 percent of residents in the largest EU countries agree with the Cartagena Protocol and the Russian public's precautionary stance. About two-fifths of Americans see genetically modified food as unsafe, compared with 27 percent who share the National Academy of Sciences' view that it is safe until proven otherwise. (The rest do not venture an opinion; across all twenty countries, 37 percent do not know enough to have a view.) Women tend to be more risk-averse than men, while people who have taken science courses are less opposed to genetically modified food.[42]

SARS-CoV-19 also illustrates different policy treatments of the balance between risk and gain. Several state governors eased restrictions on social interaction in May 2020 before public health officials deemed it prudent to do so—as Governor Brian Kemp of Georgia pointed out, "We've got two wars we're fighting now. We're still fighting the Covid-19 virus, but we're also fighting a war to get our people back to work and reopen our economy." He authorized gyms, churches, hair and nail salons, and tattoo parlors to open, followed quickly by restaurants and movie theaters. In contrast, a week earlier, New York's Governor Andrew Cuomo had told the press that "the good news, on the overall, [is] we're finally ahead of this virus," but he insisted that New York would move slowly to reopen: "I don't wanna do a 'whoops we made a mistake' and I don't wanna have hundreds of more people go to the hospital."[43] Those divergent choices in balancing economic and health benefits and risks continued through 2020 and 2021, with only minimal apparent correlation across states between decisions about opening or closing and the actual situation of the pandemic in that state.

Societies and Issues

Calestous Juma, executive secretary of the UN Convention on Biological Diversity, observed in the late 1990s:

> In the United States products are safe until proven risky.
> In France products are risky until proven safe.
> In the United Kingdom products are risky even when proven safe.

> In India products are safe even when proven risky. . . .
> In Brazil products are both safe and risky.[44]

As Juma's "caricature of carefully considered positions by diplomats" is pointing out, technology optimism or pessimism can characterize whole societies or even epochs. As so often, Alexis de Tocqueville offers the essential starting point with regard to analyzing the United States:

> When castes disappear and classes are brought together, when men are jumbled together and habits, customs, and laws are changing, when new facts impinge and new truths are discovered, when old conceptions vanish and new ones take their place, then the human mind imagines the possibility of an ideal but always fugitive perfection. . . .
>
> I meet an American sailor, and I ask him why the vessels of his country are constituted so as not to last for long, and he answers me without hesitation that the art of navigation makes such rapid progress each day, that the most beautiful ship would soon become nearly useless if it lasted beyond a few years.
>
> In these chance words said by a coarse man and in regard to a particular fact, I see the general and systematic idea by which a great people conducts all things.
>
> Aristocratic nations are naturally led to compress the limits of human perfectibility too much, and democratic nations to extend them sometimes beyond measure.[45]

Excepting Juma and his celebrated ironic sensibility, contemporary scholars settle for less artistry and more precision than Tocqueville offers in measuring national cultures' risk acceptance or aversion. The canonical work is *The Civic Culture*, whose authors examined public opinion in five countries in the early 1960s to determine whether "human nature is fundamentally cooperative." This mattered on the theory that believing in cooperation was "in effect 'a free ride' toward optimism and faith." They found Germans' and Italians' belief in humans' cooperative instinct to be "substantially lower" than that of Americans, British, and Mexicans.[46] Subsequent research elaborated on *The Civic Culture*'s methods and frameworks; Figure 2.1 shows one example, in this case evaluations of biotechnology across European countries.[47] To demonstrate the range of views, I include not only the largest European states but also countries with particularly high or low levels of biotechnology optimism:

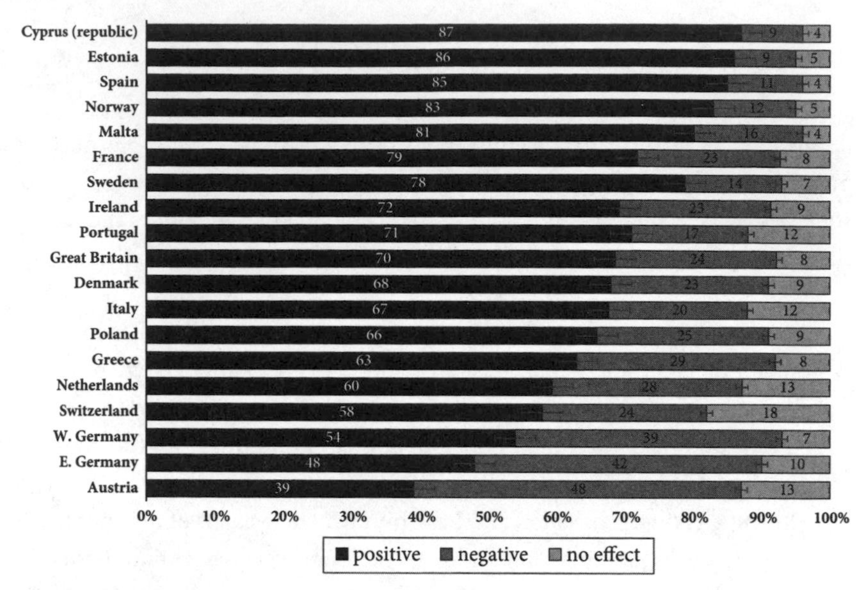

Figure 2.1 Europeans' views of the impact of "biotechnology and genetic engineering" on "our way of life" over the next twenty years
Source: Eurobarometer 2010

More Austrians are pessimistic than optimistic about the societal effects of biotechnology, whereas eight times as many residents of Norway, Spain, Estonia, and Cyprus predict positive effects as negative ones. Like most social scientists, I am not sure there is such a thing as national culture, but at least in this arena entities that we call nation-states certainly demonstrate differences in risk acceptance and risk aversion.

Other scholars—science and technology specialists Sheila Jasanoff and her co-author, political scientist David Vogel, and ethicists on England's Nuffield Council of Bioethics—all probe why some cultures or populations are more risk-averse or risk-accepting, and how those views vary across scientific arenas, time, population characteristics, or political regimes.[48] Given their distinctive theories and findings, there is no agreed-upon order of countries or logic to explain national or cultural differences in technology optimism. Luckily for my purposes, since *Genomic Politics* focuses on only one nation it suffices to note that country-level variations can be considerable, may be stable across issues, and might influence governance of genomics technology.

The Basic Framework

Bringing together the two dimensions—factual claims about the importance of genetic influence, and emotional or strategic claims about the likely impact of genomic science—enables construction of a typology that makes sense of the disputes around BiDil, DNA ancestry testing, forensic DNA databanks, gene editing, and other uses of genomic science. Figure 2.2 replicates the basic framework presented in the Introduction, now filled in with a richer understanding of how the vertical and horizontal dimensions intersect to create the quadrants.

Before we examine each quadrant, note a few features of the whole framework. One can populate each quadrant with a variety of content: ideas or philosophies, the public as a whole or segments of it, specified individuals or groups (such as the wonderfully named Society of Technophilic Engineers in quadrants 1 and 3), psychological traits or stances abstracted from concrete referents, policies, political programs, cultures, societies, or eras. Each quadrant can be understood as an ideal type or as an approximate summary of a cluster of actual empirical findings. Each quadrant can be understood to

		Emotion or preference	
		Technology optimism	*Technology pessimism*
knowledge or belief	*Genetic influence on human phenotypes*	1. Enthusiasm Genetics explains a lot about humans; new genomics knowledge is exciting and can be primarily beneficial	2. Skepticism Genetics explains a lot about humans, but use of genomics knowledge risks being harmful or inappropriate
	Non-genetic influences on human phenotypes	3. Hope Genetics explains little about humans, but understanding the actual causes of phenotypes is valuable and may enable human betterment	4. Rejection Genetics explains little about humans, and seeking to understand the causes of phenotypes may be useless or harmful

Figure 2.2 The basic framework of *Genomic Politics*, with outline filled in

have bright-line boundaries and be posed against its opposite, or to be part of a continuum in which the boundaries are softened and lowered.

Despite complexities and cautions, the basic framework enables us to make sense of the disputes described in Chapter 1, and a good deal more. We will use it as far as it takes us, which is most of the way through *Genomic Politics*, then jettison it at the end when it fails to explain a crucial finding.

Quadrants

Enthusiasm

This is the realm of committed genomic scientists who believe that their research can relieve suffering, fight climate change, and enhance human lives and societies. Geneticist Mary-Claire King once described the researchers in her lab as "banging down the doors at 7 a.m., they are so excited about making progress on their work" in discovering genetic components of breast cancer. They are in good, or at least crowded, company. In 2007, *Science* magazine editors judiciously observed "a great deal of excitement, because everyone realizes the field is changing so fast." Less sober commentators such as Keith Kleiner, an associate founder of Singularity University, describe American society as "in the midst of a revolution, . . . an explosion in ability to analyze the entire DNA sequence of individuals. . . . The medical and technological breakthroughs that will accompany widespread, cheap, and fast human genome sequencing will be far reaching and stunning." Stacey Gabriel, a senior director at the Broad Institute, proudly reports that "Broadies . . . have worked nearly non-stop since the beginning of the pandemic. We've built a completely novel automation line [for Covid-19 testing] . . . that is scalable, modular, and able to absorb anticipated future demand. The team that runs this line is here in the middle of the night, on weekends and holidays, turning tests around far more quickly than private labs, and at much lower cost. There's no end in sight." Genomic science aims at nothing less, according to philosopher Hub Zwart, than "the equitable and intelligent use of complex information" across society.[49] Most expansively, says one genomics expert whom I interviewed, genomics is "coming to push us into new stages of understanding our identity. Parts of the genome are designed to push us into larger life, stimulating another stage [of human development]. Humankind is on the verge of transformation." The genome is "forcing the question of how we define ourselves . . . , why do we do what we do? The solution is in a larger definition of ourselves."

Enthusiasm ran highest in the first decade of the twenty-first century, in the aftermath of the Human Genome Project's creation of what President Clinton called on June 26, 2000, "the most important, most wondrous map ever produced by humankind." Rhetoric of the second decade of genomics research is usually more tempered, as Enthusiasts settle in for the long haul. Former Broad Institute director Eric Lander likens the trajectory of genomic medicine to that of medicine based on the germ theory of disease, whose development "took about 75 years. With genomics, we're maybe halfway through that cycle." Even so, "the rate of progress is just stunning. As costs continue to come down, we are entering a period where we are going to be able to get the complete catalogue of disease genes." A prime target, cancer, is almost in sight: "If you understand that this is a game of probability, and there is only a finite number of cancer cells and each has only a certain chance of mutating, and if we can put together two or three independent attacks on the cancer cell, we win. If we invest vigorously in this and we attract the best young people into this field, we get it done in a generation. If we don't, it takes two generations." Lander is "not Pollyanna. . . . [I]t's not for next year. We play for the long game. I don't want to overpromise in the short term, but it is incredibly exciting if you take the 25-year view."[50]

Biotechnologist J. Craig Venter has a long list of genomics accomplishments, including leading the private venture that sequenced the human genome alongside the public Human Genome Project. Like Lander, Venter is both breathtakingly ambitious and verbally careful. He aims not merely to extinguish cancer but to create viable life-forms, use synthetic biology to solve the world's energy problems, and extend human life by slowing the aging process. No one better personifies Enthusiasm. But Venter too calls for a long view: "We are making discoveries every day that surprise us because nobody has had these data sets before. Nobody has been able to ask these questions before. With our first 20,000 genomes we have just completed the first lap in what will be a very long race."[51]

Policies, like individuals, can be Enthusiastic. Some states or countries encourage expansion of forensic DNA databases through familial searches, to widen the range of the database by exploring partial matches between a crime scene sample and the DNA of a person convicted of a felony. In a different arena, the influential British Nuffield Council on Bioethics issued a 2018 report that carefully but clearly opened a pathway to "the prospect of genome editing becoming available as a reproductive option for prospective parents." Although the United States prohibits it, Great Britain has approved conditions for conducting mitochondrial replacement therapy, in which the nucleus of a would-be mother's egg is inserted into the enucleated cell of a donor with healthy mitochondrial DNA before the egg is fertilized and implanted in the mother's uterus. Dubbed the creation of "three-parent babies," this technique is a close relative of germline gene editing. In yet another arena, the U.S. Department of Agriculture works on "introducing new traits and improving existing traits in livestock, crops, and microorganisms; . . . and assessing and enhancing the safety of biotechnology products." The U.S. Department of Energy bolsters its budget request by pointing out that "encoded in the genomes of plants, microbes, and their communities are principles that offer a wealth of potential for biobased solutions to national energy and environmental challenges." The department's genomic science programs seek to "harness this potential" by funding "fundamental research to understand the systems biology of plants and microbes."[52]

Even nations and international agencies can be Enthusiasts, betting enormous resources on the conviction that genomics research will improve the lot of humankind and, not incidentally, enhance the power of that state. BGI, the world's largest genetics research center, began as a multibillion-dollar joint public-private venture in the People's Republic of China. It has sought the genetic basis of human intelligence, along with offering customers a "genomic revolution" to "leverage genomic data to address some of the world's most acute problems in human health, food supply, and biodiversity." The multistate coalition Interpol—"Connecting Police for a Safer World"— maintains an international DNA database and search mechanism in which eighty-four national criminal justice agencies participate. Medical biobanks can be patriotic trophies: "The Saudi Biobank is a national project. . . . It will provide valuable resources for conducting cutting edge medical research with a focus on the Saudi health priorities and improving health outcomes. . . . The Saudi Biobank is one of the important medical research projects in the Kingdom of Saudi Arabia to study the impact of healthy lifestyle, genes and the environment in the Saudi population."[53]

Enthusiasm risks tipping into coercion. Since 2017, the Chinese government has been collecting DNA samples with what is thought to be the goal of 35 to 70 million men and boys across the country. It aims to "comprehensively improve public security organs' ability to solve cases, and manage and control society," according to the publication *Renmin Net*. Although officially voluntary, the project includes young children, and China analysts such as Emile Dirks and James Leibold fear that refusal to participate will lead to denial of rights to health care and travel. "China's DNA dragnet," as Dirks and Liebold label it, enables surveillance of millions of men and their male relatives.[54]

But if one can separate a value from its dangerous excesses, Enthusiasm exemplifies the ideals of the European Enlightenment, with its passion for exploring, classifying, experimenting, directing, and taking action. As with the *philosophes*, powerful moral commitment drives at least some of these rationalist ideals: geneticist Kevin Esvelt asks, "Given the power to alter the workings of the natural world, are we morally obligated to use it?" After all, if CRISPR gene editing technology can eliminate, with "no obvious ecological effects," a fly whose bite causes its human victims excruciating pain, should we not undertake a screwworm elimination campaign?[55]

In this case, although not in all discussions of gene drives, Esvelt says yes. (An intentional gene drive is a gene editing technique that increases the likelihood that the offspring of an edited organism will inherit a specific genetic variant—such as sterility in the case of screwworms or malaria-causing mosquitoes. If a large enough proportion of the organisms are appropriately edited, that species may be driven into extinction.) Enthusiasts are not naive; they concur with Skeptics that genomic science can be dangerous and misused. But they differ in having confidence that control mechanisms exist or will be created and in being convinced that the benefits of action can be made to outweigh its costs. A frequent analogy is banking: just as the invention of banks created opportunities for inventing bank robbery, which needs to be controlled by policing when norms fail (and then the police need to be controlled . . .), so genomic science may create not only cures for terrible diseases and amelioration of climate change but also opportunities for bioterrorism and designer babies for the rich—at which point self-regulation and public controls must be created to reduce risks to acceptable levels without relinquishing the benefits.

Thus Enthusiasm has a second philosophical base entwined with Enlightenment faith in experimentation and knowledge: the combined moral and utilitarian calculation that the benefits of learning more about genetic influence can be made to outweigh its perils. Humans have sufficient ability, determination, wisdom, and benevolence to be trusted to explore "the most wondrous map ever produced by humankind" even though its impact is currently beyond our comprehension. The possible Monsters that lurk at its edges can be pushed into oblivion, or at least pushed further away.

Skepticism

The European Enlightenment was not, of course, an unalloyed blessing to humankind; it was a disaster for some people. So its analogy to the genomics revolution points us toward not only Enthusiasm but also Skepticism, the second quadrant of the basic framework. Entities in this quadrant concur with Enthusiasts that genetic influence explains a lot about human phenotypes—but that fact fills individuals with foreboding, induces states to adopt precautionary postures, and leads to policies more focused on averting danger than on promoting action. A Chartgeist graphic (Figure 2.3) says it all:

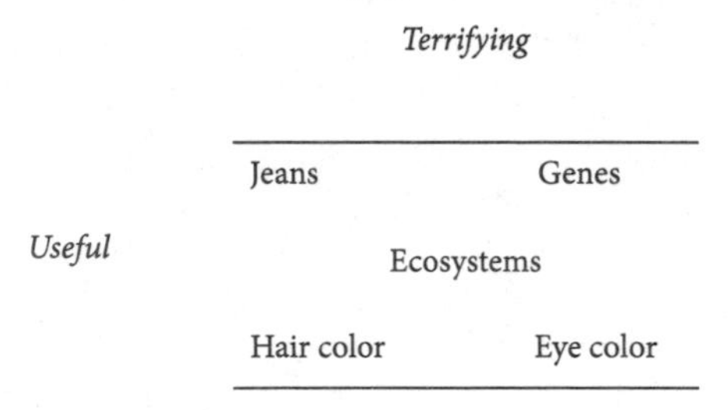

Figure 2.3 Chartgeist: Things we can alter
Source: Eilenberg 2015

The terrifying prospect of unintended consequences of genomics encompasses varied forms. Eugenics is high on the list. Writing about genetic screening for disease, sociologist Troy Duster warns that "the direct route to eugenics is not the issue. . . . It is a more insidious situation about which I would issue a warning and venture a prediction. . . . With this machinery [genetic sequencing] developing and expanding, . . . it is only a matter of time before elliptical eugenic uses are made of these new technologies." Duster warns his readers that "the hour is late, the technology is closer, and the public debate has not been vigorous." Generally an Enthusiast, George Church and his partners have explored development of a dating app called digiD8 that would invite individuals to submit their DNA for whole-genome sequencing, along with more conventional information. The goal is to "prevent two carriers of the same gene for a rare genetic disease from even meeting in the first place, by making sure they can't view each other's dating profiles. That way, on the off chance two people meet on the app, fall in love, and have children, they'll know the baby wouldn't be at risk of having a hereditary disease." Church sees the app as an extension of genetic counseling, especially Orthodox rabbis' system for genetic screening to discourage marriages that might result in children with Tay-Sachs disease. But to critics, "Church is proposing eugenics, and the project is wildly irresponsible." One commented that this app "is probably going to be ~super~ racist."[56]

The Frankenstein metaphor is almost as powerful as the invocation of eugenics in motivating Skepticism. The metaphor has itself become a metaphor; if we do not die from eating Frankenfoods, we may be killed by Frankenbugs. And concern about genetically modified Frankenfood is invigorated by Frankenjournalism, at least according to the journalism professor Keith Kloor.[57] Behind the word games is a serious concern, best exemplified by yet another metaphor, Hiroshima. Oppenheimer's somber warning in one of the quotes that start this section—"I cannot very well imagine if we had known in late 1949 what we got to know by early 1951 that the tone of our report would have been the same"—is a scientist's way of saying "We made a terrible mistake." Nobel laureate biologist David Baltimore invokes Hiroshima in explaining the temporary ban that Cambridge, Massachusetts, instituted in 1977 on experiments using recombinant DNA within the city limits. "You've got to see it in the shadow of the atomic bomb. Because that was the singular event in which the public became aware that science had done things that they never conceived of. And they kept worrying that there was an atomic bomb hidden away in modern biology."[58]

What the bomb, Frankenstein, and their spin-offs all point to is the fact that scientists devoted to the public good as well as to their craft can create a monster that spirals out of their, or anyone else's, control. It need not happen often—it need not yet have happened at all—for the precautionary principle to feel urgent. Genomic science, with its chimeras of a sheep-goat or human-rabbit fusion and prospects of xenotransplantation of animal cells, tissues, or organs into humans, is not only creepy: "Many people feel that blending even a little amount of human cellular material with that of an animal is unnatural, immoral and unacceptable."[59]

Cambridge's moratorium on recombinant DNA research revolved in large part around a Skeptical concern about process rather than substance: nonscientists deserve to be, but too seldom are, sufficiently informed and involved in making decisions about societal uses of genomics. Decisions are instead made mostly by scientists themselves, or possibly by unelected regulators. This concern may simply be an expression of the old principle that people cannot be the judge in their own case, but it can be more sharply focused. A guiding principle of the advocacy organization GeneWatch UK starts from the claim that "commitments to particular assumptions about science, technology, nature and society are [too] often made behind closed doors, with insufficient public scrutiny." Instead, "consideration of the impacts of genetic technologies on the environment, health, animal welfare and human rights should be at the heart of decision-making"—and by implication, laypeople, not experts, may be the locus for evaluating and managing those impacts.[60]

A final Skeptical concern is genomics' threat to privacy. Perhaps hundreds of articles and books warn that DNA analysts can, intentionally or inadvertently, reveal personal and family secrets, enable the hobby of genealogy to be hijacked by the criminal justice system, violate historical rights of Native American tribes, persecute ethnic minorities, or make a mockery of informed consent. Movies, memoirs, and novels—*GATTACA*, *The Lost Family*, *A Broken Tree*—amplify this apprehension; a Google search produces almost 600 million hits on "DNA + privacy" (with 2.3 million hits in the more restrained Google Scholar).[61] Overriding the many specific concerns is recognition that access to our "future diary"—reading, storing, using, revising it—is a revolutionary power that Skeptics do not believe humans can exercise without risk of devastating impacts.

Like individuals, policy choices can express Skepticism. Maryland and the District of Columbia ban expanding the reach of criminal justice biobanks into familial searching, fearing that it will drag too many innocent family members of people with felony convictions into the maw of the state. Other states employ familial searching gingerly, "without much publicity and without having passed any official law," or they "draw . . . up guidelines that assure it will be used infrequently."[62] In a different arena, two California counties passed ordinances before and during 2015 forbidding cultivation of genetically modified organisms. For a decade California banned the GloFish, a bioengineered tropical aquarium fish, and it is still prohibited in the European Union. Under a "safeguard clause," EU member states may impose even more stringent rules on cultivation of genetically modified organisms than the strict regulations imposed by the European Union itself; rules for the United Kingdom's withdrawal from the EU under Brexit include elaborate provisions for managing shipments into continental Europe of food and feed, depending on whether it has been genetically modified.

California reversed the ban on GloFish, and Cambridge, Massachusetts, rescinded its ban on recombinant DNA research—small examples of the larger point that regulatory policy may shift over time from a more to a less precautionary stance, or vice versa. In fact, "if a new risk regulation was enacted on either side of the Atlantic during the three decades prior to 1990, then it is *more likely* that the American standard was . . . more risk averse. However, if it was adopted on either side of the Atlantic after 1990, then it is *more likely* that the regulation adopted by the European Union was . . . more risk averse."[63] Politics is as central to Skepticism as it is to Enthusiasm.

If Enthusiasts' implicit motto is "the truth shall set you free," Skeptics' implicit response is lawyer and bioethicist Hank Greely's observation that "information is powerful, but misunderstood information can be powerfully bad."[64] As with Enthusiasm, Skepticism's first premise is deontological: even though genetic influence is strong, humans have the right and perhaps responsibility to resist allowing their genetic makeup to define them or direct their behaviors and practices. Also like Enthusiasm, Skepticism's second premise is utilitarianism resting on a moral foundation. Enthusiasts' ratio of acceptable risks to benefits, vision of gains, definition and calculation of harms—all seem irresponsible to the Skeptical. Instead, in the face of danger we "need to use the *precautionary* principle: hold things close to the chest till you know what you are doing. Once it's out there, you can't call it back," in the words of a person whom I interviewed.

Hope

Despite huge disagreements, Enthusiasts and Skeptics concur on a crucial point—genetics explains a great deal. People in "Hope," the third quadrant of *Genomic Politics'* basic framework, reject that claim. Even if some human phenotypes such as eye color, or some diseases caused by a mutation in only one or a few DNA base pairs are genetically inherited, Hope rests on the assertion that most human traits and behaviors are caused by something else, whether that be luck, God's will, choice, socialization, or environmental pressures.

The second defining feature of Hope is, however, shared with Enthusiasts; that is technology optimism. Like those expressing Enthusiasm, Hopeful entities operate from the belief that greater knowledge is obtainable and desirable, that people can be trusted or their destructive impulses controlled, and that initiatives based on new knowledge and new energy can be expected to generate more benefit than cost.

Genomic Politics focuses on two Hopeful explanations for behavior: individual choice and environmental pressures. These are common nongenetic explanations, as we saw in Table 2.1, and politically the most salient. They are also linked to one another in the mind of the public, while neither is closely related to genetic accounts or other explanations for human phenotypes. For example, when the GSS asked respondents in 2018 to identify the most important items in a list of possible causes of mental illness, people who chose "a genetic or inherited problem" tended also to choose "a brain disease or disorder" but seldom chose any of the other proffered explanations. Conversely, those who blamed mental illness on a person's "own bad character"—an individual attribution—tended also to choose the environmental explanations of "the way s/he was raised" and "the normal ups-and-downs of life."[65] Put together, those results suggest that Americans see individual and environmental explanations as linked, and also genetic and biological explanations as linked, but less often associate the "nurture" explanations in the bottom half of the vertical dimension with the "nature" explanations in the top half. That is what the basic framework would lead us to expect.

Another example of Hope lies in some responses to the problem of being unhealthily overweight. The Hopeful might accept Enthusiasts' aspiration for genetically based therapies linked to body mass index to help manage a potentially debilitating and dangerous condition. But they are more inclined to emphasize that even this large set of genes explains much less than a tenth of a person's BMI, and that genetic explanations encourage fatalism. (To quote that noted authority *The Onion*: "Obesity caused entirely by genes, obese researchers find.")[66] In contrast, say the Hopeful, correcting environmental causes and providing support for good choices offer much more leverage for change. Strategies ranging from exercise and appropriate food to gym classes in school, grocery stores in inner cities, bike paths through communities, control of advertising to children, and policies offering appropriate incentives to medical professionals might prove effective; that is where research and resources should be concentrated.

This logic about body weight can be generalized. As epidemiologist Nancy Krieger puts it, "We literally biologically embody exposures arising from our societal and ecological context." If disease and disability are responses to destructive contexts, then medical and social policies should focus on the "pathogenic pathways" worn through lives by continued encounters with racism, pollution, inequality, and other harms "that affect the development, growth, regulation, and death of our body's biological systems." The rhetorical emphasis here is negative, but the substantive message is positive: as Krieger notes in another article, since "this new body of scientific work is making gains and is beginning to affect policy," health disparities "can be changed by human action."[67]

Hopeful strategies can be sorted into three buckets: including genetics, tolerant of claims about genetics, and excluding genetics. The NIH's 2002 *Strategic Research Plan and Budget to Reduce and Ultimately Eliminate Health Disparities* personifies the first. It endorses "research to understand biological, socioeconomic, cultural, environmental, institutional, and behavioral factors affecting health disparities," and offers a graphic demonstration of the "multifactorial . . . genesis of health disparities," as shown in Figure 2.4.

The NIH accepts the role of genetic influence; the explanatory box of "Biological Factors" includes genetics. But biology is accompanied by two explanatory boxes that point toward individual choices ("Health Practices"

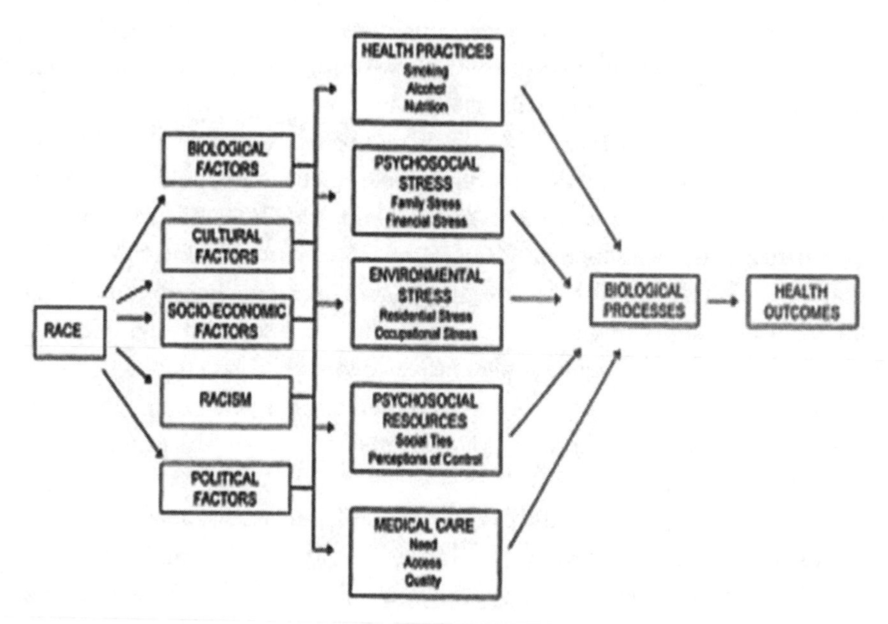

Figure 2.4 NIH's "Framework for Understanding the Relationship Between Race and Health"

Source: National Institutes of Health 2002: 15

and "Psychosocial Resources") and an additional seven of environmental factors.[68] In the text, every time the NIH strategic plan mentions genetics, it points to interactions with or the additional significance of other explanatory factors. It also, despite the focus on health disparities, is determinedly optimistic; every depiction of a problem is followed by evidence of success or strategies to promote greater equality. In sum, "medical and scientific advances have introduced new opportunities for the continued improvement of health for all Americans." NIH's mandate of "*uncovering new knowledge that will lead to better health for everyone*" will continue to promote "innovative diagnostics, treatments, and preventative strategies to reduce, and eventually eliminate, health disparities."[69]

In Hope's less leaky second bucket, genetic factors are largely irrelevant even if included in a list along with environmental or personal explanations for human phenotypes. Psychologist Steven Heine elegantly summarizes the evidence for this stance. He worries that "scientific funding is very much a zero-sum game, and by continuing to devote so much of our available resources into genetics research, we're diverting funds away from efforts to reveal the very important ways that our experiences and upbringing can also lead to harmful health outcomes." He backs this claim with an array of examples demonstrated by social scientists: "Lifespans among the elderly can be improved by providing them with more control over decisions in their lives, coronary heart disease can be effectively reduced by specific lifestyle interventions, Type 2 diabetes is more effectively treated by lifestyle interventions than it is by the leading diabetes medication, guided exercises in writing about traumatic life events can improve lung function among asthmatics and can reduce disease severity for those suffering from rheumatoid arthritis, mindfulness meditation can improve symptoms among HIV-infected adults, and group therapy can increase survival rates of breast cancer."[70] If Heine's evidence is persuasive, then even if genetics contributes to disease or disorder, scarce resources should be channeled into behavioral and environmental interventions, not diverted into would-be genetic cures that are at best unhelpful.

In the third and most tightly caulked Hopeful bucket, any attention to genetics is not merely subordinate or irrelevant but mistaken and even dangerous. Consider developmental psychologist Elizabeth Spelke's debate with her colleague Steven Pinker over women's capacity for extraordinary achievement in mathematics and science. Spelke provides evidence that "the big forces causing this gap [between the proportion of male and female scientists in elite universities] are social factors. There are no differences in overall intrinsic aptitude for science and mathematics between women and men." Instead, socialization from infancy onward, stereotyping, the snowballing effect of personal connections, and assumptions about how work ought to be done shunt girls and then women out of some professions and into others. Scholar that she is, Spelke proposes an experiment: "We should allow all of the evidence that men and women have equal cognitive capacity to permeate through society. We should allow people to evaluate children in relation to their actual capacities, rather than one's sense of what their capacities ought to be, given their gender. Then we can see, as those boys and girls grow up, whether different inner voices pull them in different directions." Spelke does not know what that experiment will reveal, "but I do hope that some future generation of children gets to find out."[71]

Most relevant American social policies follow the 2002 NIH model. Their implicit theory is that while genetic factors may influence behavior and outcomes, a combination of individual and institutional actions is the appropriate form of public involvement. Political debate turns more on what the individual and institutional actions should be and how they should be balanced (and paid for) than on whether genetic factors are more or less influential. The premise of Hope, that is, is the primacy of human agency and its resultant mandate to act to solve problems. Explanations based on genetics risk a taint of inevitability or passivity—removing control from people and their societal constructions. In this view, such a denial of autonomy is empirically wrong, normatively unacceptable, politically impotent, or all of the above.

Rejection

The basic framework's fourth quadrant comprises people, policies, or societies that are dubious about claims of genetic influence and focused on the risks of technological innovation. To put it too crudely, they are "no-no's"; the purported findings of both genomicists and activist social scientists are mostly wrong and probably dangerous. Rejecters share the fear of sociologist Eugene Rosa and his colleagues that "the advanced modernity of the twentieth century witnessed a hyper-accelerated change in the number of risks produced, their magnitude, and their global spread." One element of that modernity is "the rise of science and engineering"; it contributes to the "features of systemic risks... [that] make them especially difficult to understand and to govern: *complexity, uncertainty,* and *ambiguity.*"[72]

Like Hope, categories of Rejection fall into several clusters. The first is doubt that genomic science can fulfill its promises to solve medical, forensic, identity-based, environmental, agricultural, or any other problems. Anesthesiologist Michael Joyner and his colleagues' *JAMA* shot across the bow of medical researchers is a good example. The biomedical research community proposes that "ever-deeper knowledge of subcellular biology, especially genetics, coupled with information technology," will transform health care. But "this approach has largely failed," say Joyner and colleagues. None of these advances has shown "any measurable effect on population mortality, morbidity, or life expectancy in the United States." Instead, improvements in health outcomes result from better "prevention efforts . . . that are undervalued as outmoded and old-fashioned by the narrative." The authors' culminating challenge: "What historical precedent is there that adoption of vast new oversophisticated technology reduces costs?"[73]

Joyner and his co-authors do go on to propose ways to improve the quality and value of genomics research, so they are not full Rejecters. The second Rejection cluster is more stringent, asserting that genomic science and other new technologies are not merely ineffective or wasteful, but actually or potentially harmful. Economist Edward Perry and his colleagues find that farmers who plant insect-resistant maize seeds have been using significantly less insecticide and slightly less herbicide than nonadopters—a central goal of the genomic modifications. However, farmers planting genetically modified soybeans are using significantly *more* herbicide, as weeds in soybean fields are becoming resistant to the herbicides typically used in the new soybean regime. Worst of all, the need for herbicides for maize is also growing, as weeds in maize fields are developing resistance. The authors conclude that "continued growth in herbicide use poses a significant environmental problem"; one of them, Federico Ciliberto, laments that "I did not expect to see such a strong pattern."[74]

Unlike these economists, analysts who repudiate genomics and its societal use are occasionally angry enough to skirt norms of professional objectivity. Compare responses to analyses of disparities by race and by sex in the coronavirus pandemic. As I write, scientists are striving to understand why men have more severe cases of Covid-19 hospitalization and higher rates of death than do women. One representative article in *Nature* by five (female) physicians, microbiologists, and biochemists first reminds readers that "sex . . . is a multidimensional biological characteristic that shapes infectious disease pathogenesis." After exploring the epidemiology and biology of SARS-CoV-2, Eileen Scully and her co-authors conclude, "Although gender-related social factors, including smoking, health care-seeking behaviours and some co-morbid conditions, may impact the outcomes of Covid-19 and contribute to male-female differences in disease severity, the cross-cultural emergence of increased risk of death for males points to biological risk determinants."[75] This article is one of several drawing the same inference, none of which has evoked intense opposition.

In roughly parallel fashion, epidemiologist Kristen Azar and colleagues examine "several possible explanations" for substantial racial disparities in Covid-19 hospitalization even after adjusting for age, sex, comorbidities, and income. Their ahead-of-print article stated, "One hypothesis is that there may be some unknown or unmeasured genetic or biological factors that increase the severity of this illness for African Americans." One prominent response invoked the tropes of murder and racism that we saw in Chapter 1. Pediatrician Rhea Boyd and her co-authors addressed Azar and colleagues' draft article in a *Health Affairs* blog post, which became one of the journal's top ten posts of 2020. They begin by quoting a poem about Black lynching, and continue by listing the names of Blacks who have recently been victims of police violence. They then argue that Azar and her co-authors' "analytical framing ignores racism as the mechanism by which racial categorizations have biological consequences." Raising the possibility of genetic differences in susceptibility to Covid-19 "resurrect[s] long-refuted and disproven theories about biological race. . . . Science, through the guise of objectivity, has abetted the indignities forced upon non-White populations by probing their innate propensity for disease and thus their biologic inferiority. In 2020, such unsubstantiated claims have no place in scholarship on racial health inequities." Only when "scholars position racism as a potential driver of the inequity" between Black and White patients can "exploitation, sanctioned undertreatment, inadequate financial reparation, and increased burden of familial disease. . . emerge" into analytic and public view.[76]

Boyd and her colleagues go on to propose reforms (mainly to the system of peer review in scholarly publications). So despite their anger, these authors do not take the final step of Rejection, a wholesale dismissal of broad swaths of technology. The nonprofit organization Slow Food criticizes genetic manipulation of agriculture because the technologies "challenge a whole way of life. . . . The role of small-scale agriculture in food sovereignty and security, protection of local areas and economies, the preservation of landscape and the sustainability is becoming increasingly clear to consumers, governments and scientists." Governments should repudiate agricultural industry in favor of supporting rural communities that produce "good, clean and fair food, food whose quality is determined by attention towards sensory aspects, by respect for the environment and labor and by the cultural diversities and traditions of its producers."[77] Other Rejecters are less gentle. "The genomic revolution has induced a kind of moral vertigo," writes philosopher Michael Sandel, after serving four years on the President's Council on Bioethics. Efforts at enhancement and genetic engineering of unborn children, especially if available only to parents who can pay "hefty sums," would "represent a kind of hyperagency, a Promethean aspiration to remake nature, including human nature, to serve our purposes and satisfy our desires. The problem is not the drift to mechanism but the drive to mastery."[78]

The core philosophical premise of Rejection is "first, do no harm" in a context where, as Rejecters see it, doing harm seems all too closely linked to doing anything. Rejecters suspect or believe that much of science, technology, and even modern rationalism threatens humans' capacity to thrive with dignity and autonomy. Modernity, instantiated in this book by genomic science, is blindly pushing forward, oblivious to its current costs and future harms. If Enthusiasm leads one to bang down the doors of the lab to get to work in the morning, Rejection leads one to mount the barricades against invaders who are naive about genomics' danger, self-interested, or racist and elitist.

Putting the Framework to Work

The proof of a pudding is in the eating; a framework is only as valuable as the assistance it gives for understanding disputes surrounding such issues as BiDil, forensic DNA databases, ancestry testing, and prenatal gene editing. The next three chapters offer the proof that this pudding is indeed worth the eating. Chapter 3 shows why the more conventional frameworks of ideology or partisan politics do little to help disentangle genomics disputes, after which Chapters 4 and 5 show why *Genomic Politics*'s basic framework, with its four quadrants of Enthusiasm, Skepticism, Hope, and Rejection, does.

3

Disputes over Genomic Science Are *Not* Partisan

> [When I talk with elected officials about genomics,] there are heavy ethical, legal, social issues but not partisan dynamics. I rarely have partisan conversations. . . . [My assistant] has to remind me if the Congress member that I'm going to talk to is Democrat or Republican.
>
> —Interview with federal official

> There's always partisan politics one way or another.
>
> —Interview with federal official

Conventional wisdom holds that liberals or Democrats are pro-science, while conservatives or Republicans engage in what journalist Chris Mooney labels a "war on science." Both indirect and direct indicators support that perception. Science writer Shawn Otto's book on "the war on science," for example, includes five index entries for Democrats and fifteen for Republicans. Trend lines in the GSS show a clear break in the early 1990s between liberal and conservative respondents' confidence in scientists (Figure 3.1).

Even after subjecting these results to a variety of controls, sociologist Gordon Gauchat shows that conservatives show "a fairly steep decline in their trust in science" from the 1970s through the first decade of the 2000s, whereas moderates and liberals show a mild or no decline.[1]

Similar results obtain with regard to views on climate change, human evolution, research on embryonic stem cells, and the SARS-CoV-19 pandemic. In mid-June 2020, three months into the spread of Covid-19 across the United States, three-quarters of Democrats but only half of Republicans agreed that the Centers for Disease Control and other public health organizations get the facts on the coronavirus right almost all or most of the

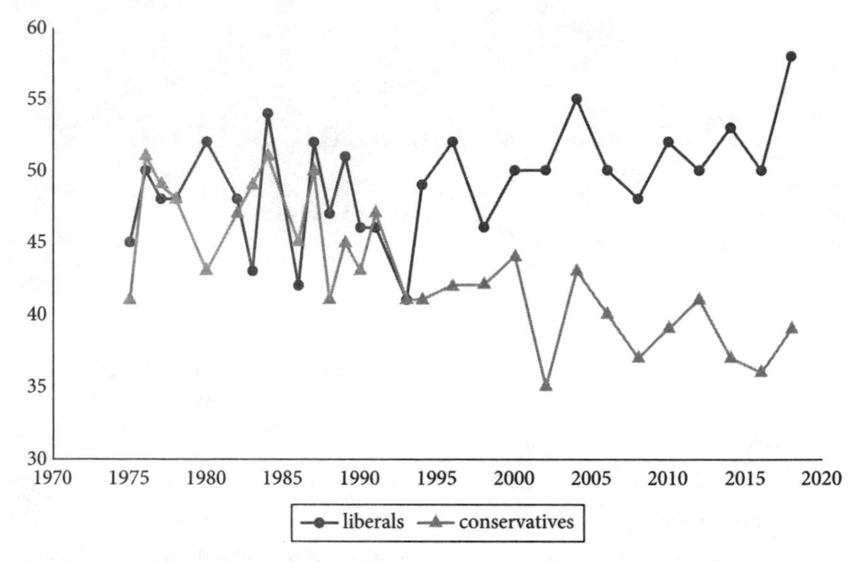

Figure 3.1 "A great deal of confidence in the scientific community," by ideology, GSS, 1974–2018 (percent agreeing)

time.[2] Since no one was capable of full knowledge of the facts at that point in the pandemic, this is in part a declaration of faith in or disbelief in science, inflected or even caused by partisan identity.

Nonetheless, conventional wisdom about the conservative or Republican war on science is too simple. Liberals (or Democrats) are more likely than conservatives (or Republicans) to oppose fracking, peaceful use of nuclear power, offshore oil drilling, and genetically modified foods.[3] Although Republicans are less likely to endorse public health organizations' statements about Covid-19 than are Democrats, nonetheless at least half agree that these organizations get the facts right.

Once we relax the trope of a Republican war on science, several fascinating patterns emerge. First, some partisan differences are simply odd or symbolic, sometimes regarding scientific questions that have no answer. Three in ten Democrats, but only a fifth of Republicans, agree that life has existed on Mars.[4] Many more conservatives than liberals were sure that Terri Schiavo, whom doctors had declared to be brain-dead and who had no capacity for communication, felt "pain and discomfort" when her feeding tube was withdrawn. With some variation across diseases, higher proportions of Democrats than Republicans agreed in 2001 that multiple sclerosis, depression, brain injuries, deafness, tuberculosis, HIV/AIDS, heart disease, child abuse, and prostate cancer are "very serious" health problems for Americans. Only concern about baldness is equally shared by Democrats and Republicans. Even with controls for education, in the GSS and other surveys Democrats are significantly more likely than Republicans to see scientific merit in astrology.[5]

Second, some scientific controversies generate alliances between people usually identified as right-wing libertarians and people usually identified as left-wing progressives, against the mainstream left and right. An important case is the merits of childhood vaccinations; school districts with the lowest rates of vaccinated children include wealthy progressive communities such as Malibu, California, and Boulder, Colorado, as well as poor "off the grid" towns in southern Oregon. In May 2020—a point at which the issue was still hypothetical, with no evidence of safety or efficacy—a fifth of Democrats and a third of Republicans (and their respective leaners), would not or probably would not get a Covid-19 vaccine if it were immediately available.[6]

Third, patterns shift or emerge over time. Sometimes a nonpartisan issue becomes politicized. Left and right shared a new concern for the environment in the otherwise contentious 1960s. President Richard Nixon and the high-heeled First Lady Patricia Nixon planted a tree at the White House to honor the first Earth Day (Figure 3.2), and President Nixon created the Environmental Protection Agency by executive order.By 1990, however, more Democrats than Republicans (78 to 64 percent) worried a great deal or a fair amount about global warming, and by 2018, the partisan gap had widened to 91 percent of Democrats but only 33 percent of Republicans expressing concern about climate change.[7]

Environmentalism exemplifies what law professor Dan Kahan calls "identity-protective cognition," as distinguished from straightforward empirical evaluation. It is "an unconscious tendency to conform information processing to . . . protection of one's status within an affinity group whose members share defining cultural commitments."[8] Though most likely unaware of the concept, politicians may gleefully intensify identity-protective cognition. One classic case is Senator Jim Inhofe (R-OK), who in 2015, as chair of the Committee on Environment and Public Works, waved a snowball on the Senate floor to prove that "manmade global warming is the greatest hoax ever perpetrated on the American people" (Figure 3.3). After all, "God's still up there. The arrogance of people to think that we, human beings, would be able to change what He is doing in the climate is to me outrageous."

Figure 3.2 Earth Day celebration at the Nixon White House, 1970
Source: White House Photo Office

Figure 3.3 Senator Inhofe's snowball, intended to prove the falsity of claims of human-made climate change, 2015
Source: C-SPAN, February 26, 2015

Throughout 2020, President Trump similarly worked strenuously to associate wearing a mask during the Covid-19 pandemic with, as *The Economist* summarized, emasculation or—worse—being a Democrat. Some Republican governors and mayors denigrate and reject mask mandates regardless of the state of the pandemic in their arena of authority; for a while, Billy Woods, sheriff of Marion County, Florida, forbade his deputies in most circumstances and visitors to his office building to wear masks.[9]

Nonetheless, despite the temptations of identity-protective cognition, some scientific controversies do in fact defuse over time. A relevant instance is in-vitro fertilization (IVF). It was highly controversial in 1978, when the first child was born as a result of this technology; the respected magazine *Nova* published an article "suggesting that test-tube babies were 'the biggest threat since the atom bomb' and demanding that the public rein in the unpredictable scientists." Even in 2006, after thousands of births, at least two-fifths of Americans thought the procedure was or might be unethical. That proportion, however, declined by half as of 2013, after millions of IVF babies had been born.[10]

In the fourth and most relevant pattern for *Genomic Politics*, science may evade the partisan orbit. Political partisans do not disagree on whether smoking is dangerous, traditional medicine should be taken seriously, or space exploration warrants public support. About two-fifths of both Republicans and Democrats agree (in contrast to most scientists) that genetically modified food risks harming people's health.[11] Historically, coalitions to promote, or to regulate, IVF in the United States were not organized along partisan lines. On one hand, encouraging the development of IVF procedures was "a constellation of scientists, start-ups, and women's and patients' groups favoring freedom of inquiry for scientists, increased state funding for competitiveness-enhancing research, expansive reproductive and privacy rights for individuals, and improved access to therapeutic innovations." On the other hand, promoting greater regulation, was "a less predictable coalition of the right and part of the left of the political spectrum demanding caution, oversight, and some prohibitions in order to protect embryonic life and, more generally, the sanctity and integrity of the human."[12]

Defusing or remaining outside the typical liberal-conservative political controversy are unusual paths for policy issues to take at a time when polarization in the United States is reaching levels unprecedented for more than a century.[13] Genomic science is one dog that (mostly) doesn't bark in the otherwise raucous arena of political discourse. Thus only when the realm of party politics is set aside can we return to explaining the disputes involving charges of murder and racism that are embodied by the basic framework's groupings of Enthusiasm, Skepticism, Hope, and Rejection.

Public Officials Are Not Partisan Regarding Genomics

Forensic biobanks can generate ferocious debate; they are "Jim Crow's database" or a response to the perennial concern for maintaining social order. They are a lever to pry open prison doors for the falsely convicted or a hammer coming down on poor men of color and their families. Intense as they are, these positions are not aligned with political parties or conventional political ideology. To my knowledge, *every* public official who has taken a public stance on the subject has endorsed the creation and use of DNA databases in the criminal justice system. I have already quoted President Obama's demand for justice. Consider several others:

- Obama administration attorney general Eric Holder issued memoranda "to ensure the Department of Justice uses DNA evidence to the greatest extent possible to convict the guilty and exonerate the innocent. . . . DNA evidence is one of the most powerful tools available to the criminal justice system, and these new steps will ensure the department can use DNA to the greatest extent possible to solve crimes and ensure the guilty are convicted. . . . [That] will help law enforcement and prosecutors keep communities safe."
- After California's arrest of the Grim Sleeper in 2011, the Democratic, Black state attorney general Kamala Harris provided funds to double the number of familial DNA searches for horrendous crimes and to reduce the DNA backlog for other criminal investigations. She proclaimed that "California is on the cutting edge of this in many ways. I think we are going to be a model for the country."
- In 2012, New York's Democratic governor Andrew Cuomo worked with legislative leaders of both parties to expand the law for collecting DNA samples from criminal suspects. Maryland's bill to permit DNA collection from some arrestees "was pushed through vigorously by [Democratic] Governor O'Malley with the support of the Democratic Party."[14]

- A 2012 Massachusetts law permitting prisoners to request DNA evidence for forensic testing that might lead to a new trial passed unanimously in both the state's Senate and House.

Elected officials have said little about other societal uses of genomic science. Two exceptions have had an impact. Donna Christensen (D-VI), a member of the Congressional Black Caucus, urged approval of BiDil at the FDA's 2005 advisory committee hearing: "You have before you an unprecedented opportunity to significantly reduce one of the major health disparities in the African American community and, in doing so, to begin a process that will bring some degree of equity and justice to the American healthcare system. . . . Knowing that diseases are expressed differently in different racial and ethnic groups, the challenge is not to avoid research but to act appropriately when this research is conducted and reported." A decade later, President Obama started a Precision Medicine Initiative by observing that "doctors have always tried to tailor their treatments as best they can to individuals. You can match a blood transfusion to a blood type—that was an important discovery. What if matching a cancer cure to our genetic code was just as easy, just as standard?"[15]

Direct-to-consumer genetic testing has received a little more attention from elected officials, but with neither a partisan inflection nor an impact. In 2010, a House of Representatives subcommittee held a hearing on "Direct-to-Consumer Genetic Testing and the Consequences to the Public Health." The subcommittee chair, Bart Stupak (D-MI), thanked everyone for "working together on this important, bipartisan inquiry," which grew from an earlier, equally bipartisan request to the Government Accountability Office to investigate the rapid growth of direct-to-consumer testing and the possibility of consumer fraud.[16] There is no evidence of party-based or ideological dispute in or resulting from this hearing (or any other outcome).

Seven years later, Senator Charles Schumer (D-NY) expressed concern that genetic tests sold directly to consumers risk violating consumer privacy, and he asked the Federal Trade Commission to "ensure that these companies have clear, fair privacy policies and standards."[17] The FTC duly posted a warning on its website about not violating privacy and revealed an "ongoing investigation" of DNA testing companies; the companies themselves updated their terms of agreement.[18] So far as I know, these are the only public statements by federal elected officials about medical or scientific uses of genomic science.

Finally, from the consequential to the ridiculous: in 2018, President Trump called on Senator Elizabeth Warren (D-MA), whom he nicknamed "Pocahontas," to engage in genetic testing to prove her claims of Native American ancestry. "I will give you a million dollars to your favorite charity, paid for by Trump, if you take the test and it shows you're an Indian. I have a feeling she will say no."[19] Warren accepted the dare and took a DNA ancestry test, with the predictable result of a highly charged and arguably racist public debate about the appropriateness of doing so. Although her march into Trump's trap received criticism from both left and right, this interaction was, unlike all others involving federal officeholders and genomics, indubitably partisan. But it too had no discernable impact.

Laws Are Not Partisan Regarding Genomics

Genomics-inflected legislation is no more partisan than are elected officials' statements. The only federal law focused directly on societal use of genomic science is the Genetic Information Nondiscrimination Act (GINA). It was passed in 2008, after a vote of 95–0 in the Senate and 414–1 in the House; President George W. Bush signed the bill into law. The law prohibits health insurers from using information about a healthy person's genetic predisposition to a disease in order to deny him or her coverage or to charge higher premiums. It also forbids employers to use genetic information in decisions about hiring, firing, promotions, or job placement. The law was preemptive—"the first civil rights act of the twenty-first century," according to Senator Ted Kennedy (D-MA)—and "effective at comforting people," as one person I interviewed described it. Another interview subject concurred: "GINA was a non-issue for either party. There weren't any people that were discriminated against. Was not a wedge for either party."[20]

States have followed, and in some cases extended, the federal GINA's precautionary lead. The Republican Speaker-designate of Florida's House of Representatives, for example, sponsored a bill in 2020 to close a gap in GINA's privacy protections: "Given the continued rise in popularity of DNA testing kits, it was imperative that we take action, in order to protect Floridians' DNA data from falling into the hands of an insurer who could potentially weaponize that information against current or prospective policyholders in the form of rate increases or exclusionary policies." The bill passed the House unanimously, and it passed the Senate with three dissenters (from both parties) arrayed against thirty-five supporters. Although at this writing, very few states mandate DNA privacy for long-term care, life, and disability insurance, most do limit health insurers' and employers' use of genetic information, prohibit employment discrimination based on genetic test results, and restrict requests for genetic information from job applicants. As in Florida, I can discern little or no partisan distinction between states with and without such laws.[21]

According to the NHGRI, nine states have taken the proactive step of requiring health insurers to provide coverage for genetic testing for at least some disorders or for a medical procedure based on genetic testing results. Some of these states are relatively liberal, well-off, and urban—Connecticut, Illinois, Massachusetts, and Rhode Island—but California and Minnesota are not on the list, whereas the relatively conservative states of Kentucky, Louisiana, Utah, and Arkansas are. So there is no discernible partisan or ideological slant to this more expansive law.

States vary with regard to other genomics-related laws. Forty-one have established one or more types of privacy protection for DNA samples; about half restrict or regulate direct-to-consumer genetic testing companies; fourteen restrict the use of residual newborn screening specimens for research. Some states are more activist than others, but again, this is not a publicly salient or politically polarized arena.

Legislation on forensic DNA databanks is the locus of the most consequential and extensive federal involvement with genomic science. Laws in the 1990s created and funded NDIS and its associated CODIS system. Subsequent laws extend the reach of forensic biobanks, support local and state DNA collection and analysis, and bring more categories of crimes or alleged perpetrators under federal jurisdiction. Congress has passed eleven laws, starting with the DNA Identification Act of 1994, that focus on or include substantial provisions regarding forensic DNA databases. Despite the intensity of advocates' and experts' disputes, these laws have generated almost no partisan controversy among legislators, as Table 3.1 shows.

The Violent Crime Act of 1994 was deeply partisan and tapped into Americans' charged views on race, crime, and criminal justice. I include it here because of the importance of its DNA Identification Act, but votes on the bill as a whole surely had little to do with forensic biobanks. The only other evidence of partisanship is the 2013 Violence Against Women Reauthorization Act, which was also saturated with partisan disputes having little to do with databases. One or (usually) both houses of Congress

Table 3.1 Federal legislation on forensic DNA databases, by party of sponsors, vote on passage, and presidential signing, 1994–2017

	Sponsors		*Vote on final passage—House of Representatives*				*Vote on final passage—Senate*				*President's party*
	Rep	*Dem*	*Rep*	*Rep*	*Dem*	*Dem*	*Rep*	*Rep*	*Dem*	*Dem*	
	yes	*yes*	*no*	*yes*	*no*	*yes*	*no*	*yes*	*no*	*yes*	
1994: Violent Crime Control and Law Enforcement Act (DNA Identification Act is one subtitle)	0	3	4	131	188	64	7	36	54	2	Dem
2000: DNA Analysis Backlog Elimination Act	5	6	Voice vote				Unanimous consent				Dem
2000 Paul Coverdell National Forensic Sciences Improvement Act	24	9	Passed without objection				Unanimous consent				Dem
2004: Justice for All Act	10	9	198	14	194	0	Unanimous consent				Rep
2006: Adam Walsh Child Protection and Safety Act	33	5	Voice vote				Voice vote				Rep
2008: Debbie Smith Reauthorization Act	10	39	Voice vote				Unanimous consent				Rep

2012: Katie Sepich Enhanced DNA Collection Act	3	3	Voice vote				Unanimous consent				Dem
2013: Violence Against Women Reauthorization Act	7	53	87	138	199	0	23	22	53	0	Dem
2014: Debbie Smith Reauthorization Act	5	12	Voice vote				Unanimous consent				Dem
2017: Rapid DNA Act	18	7	Voice vote				Unanimous consent				Rep
2019: Debbie Smith Reauthorization Act	13	11	178	I	224	0	Unanimous consent				Rep

did not bother with a roll call for the other nine laws in which DNA plays more than a trivial role; the message from legislators is that all right-thinking politicians concur in using genomic technologies to protect Americans from crime, punish the guilty, and preserve the innocent from false conviction.

States have passed many more laws on forensic DNA databases than has the federal government, but once again not in any clearly partisan way. As of 2014 (the most recent year for which data are available), all states require at least some convicted offenders to submit a DNA sample to their forensic biobank; most states' mandate includes those convicted of any felony and at least some misdemeanors. Some states include juveniles. New York and Wisconsin—whose partisan majorities have in recent years tended in opposite directions—have the broadest catchment area; the politically and socially odd combination of two New England states, two Deep South states, and three upper Midwest states have the narrowest.[22] As of 2018 (the most recent data compendium available), thirty-one states and the federal government authorized the collection of DNA samples from people arrested but not convicted of some crimes.[23] Thirty-five states and the federal government have some sort of statute to compensate exonerated people for false imprisonment; those that do not include a few liberal eastern states—Pennsylvania, Delaware, and Rhode Island—as well as more usually conservative states such as Georgia, the Dakotas, and Arizona. Confounding stereotypes, President George W. Bush concurred with Congress about what was understood to be a generous payment of $50,000 per year of wrongful incarceration, with more for each year that a person was wrongly held on death row. Texas has not only engaged in by far the most executions of any state, but also has the most generous compensation framework for false imprisonment.[24]

Since state laws on forensic biobanks change, two points matter more here than the exact number of states or exact terms of a law at a given time. First, the three-fifths of states that collect DNA samples from arrestees—that is, the most aggressive users of forensic DNA databases—are disproportionately southern and midwestern, but the list also includes California, Delaware, and Connecticut. Second, apart from laws on expungement and compensation, to my knowledge *all* relevant legislative changes in states, whether liberal or conservative, have expanded the range of situations in which DNA is collected or used.[25]

Greater normative and political controversy follows familial searching, as I discuss in Chapter 4. The FBI defines it as "an intentional or deliberate search of the [forensic DNA] database conducted after a routine search, for the purpose of potentially identifying close biological relatives of the unknown forensic sample associated with the crime scene profile." As it repeatedly points out on its website, the FBI does not conduct familial searches or take a position on them. States vary. While Maryland and the District of Columbia forbid intentional searches for family members, as of 2018, laws in about a dozen states call for labs to conduct them under certain circumstances. Twenty-four states (including, oddly, Maryland) disclose or proceed with partial matches if they are found by accident, but do not authorize looking for them. This peculiar, ambivalent policy points to the technical and political complexity of familial searching—it is a "last resort method when all other investigative leads have been exhausted and a case has gone cold"—but endorsement does not show a partisan pattern.[26] Semi-permissive states may be heavily Democratic (California) or Republican (Utah); they are spread over the country and vary in size, wealth, urbanization, and race or ethnic composition.

Political opposition can be strong, but again it is not clearly partisan. California's robust 2004 proposition, the DNA Fingerprint, Unsolved Crime and Innocence Protection Act, passed with more than a 60 percent majority. Opponents included the American Conservative Union, as well as liberal advocacy groups such as the American Civil Liberties Union of California and the Children's Defense Fund.[27]

Like forensic databases, granting the right to post-conviction DNA testing as part of a fight for exoneration has strong supporters among public actors, and no overt opponents whom I could find. All states have statutes regulating post-conviction DNA testing, but they vary in scope and many have considerable limitations. Restrictive states are scattered across the geographic and political map; most are relatively rural, and some are disproportionately White—but not all. Any plausible partisan pattern has multiple, and important, exceptions.[28]

Federal Funding of Genomics Research

"Follow the money" is advice that is as good for identifying policy commitments as for identifying Watergate malfeasants. The NIH, including but expanding beyond the NHGRI, is the main source for public funding of genomics research, translational clinical work, and some specialized genetics-related patient care. Figure 3.4 shows the budget for NIH (top panel) and NHGRI (bottom panel), from 1990 through 2020.

The message of the upward trajectories in Figure 3.4 is easy to read. Over three decades, the NIH budget rose in current dollars to $41.6 billion in 2020—almost three times larger in constant dollars than its $7.6 billion in 1990. The steep rise during Bill Clinton's Democratic presidency (first light segment) became even steeper in the early years of George W. Bush's

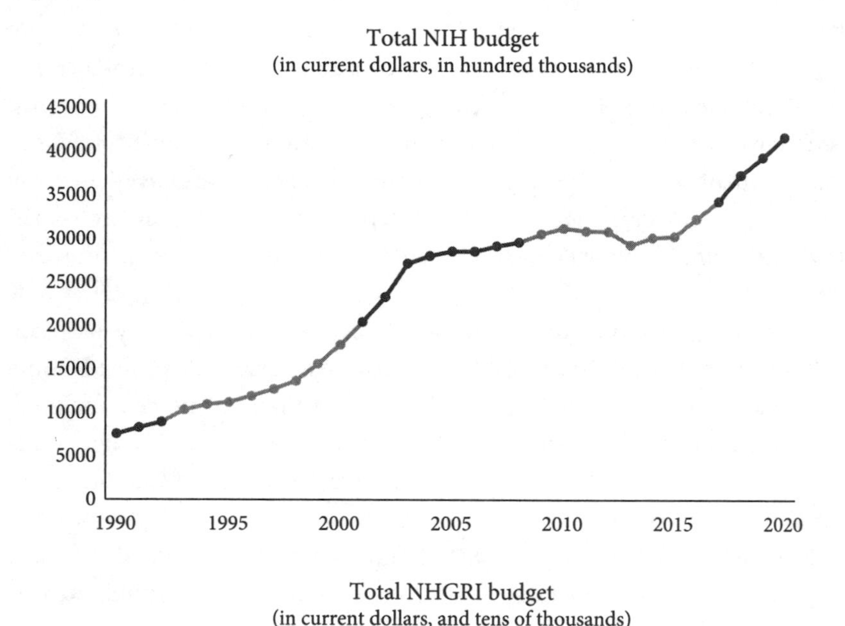

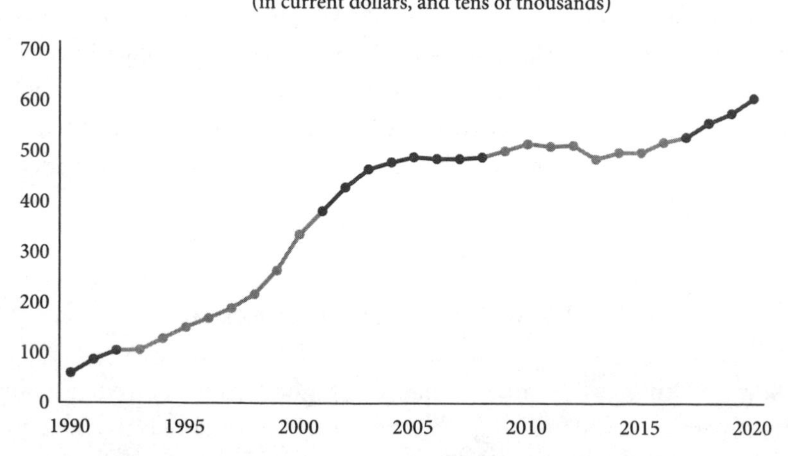

Figure 3.4 Congressional appropriations for NIH and NHGRI, 1990–2020
Source: National Institutes of Health 2020a

Republican presidency (second dark segment), and the plateau of Obama's presidency returned to a steep rise during Trump's. Appropriations for NHGRI rose even more, proportional to its starting point, also during both Democratic and Republican presidential administrations.

Genomics was spared partisan dispute even in the rancorous final years of Obama's and Trump's presidencies. After Obama's speech promising personalized medicine, the Republican-controlled Congress passed the Twenty-first Century Cures Act with little debate in the last months of his presidency. Among other provisions, the law authorized close to $5 billion over nine years for precision medicine; it involved new technologies, private-public partnerships, dissemination of genetic and clinical information, and protections for human participants. The bill passed the House of Representatives 392 to 26 and Senate 94 to 5—almost matching GINA's feat. Five years later, despite an unprecedented challenge to the 2020 presidential election results and the near-death experience of the 2020 budget bill to fund the federal government, the NIH received a 3 percent budget increase for 2021.

An onset of vitriolic partisan politics around genomics, with intense moral, personal, and institutional impact, is not hard to envision. The most likely trigger would be contestation over prenatal genetic testing or editing, or over germline gene editing of viable embryos. So far, American scientists' restraints, laws, the FDA's and NIH's ban on research involving human germline gene editing, and cross-cutting moral, ideological, and financial considerations (examined in what follows) have kept such a conflagration at bay.[29] That equilibrium may persist—but if it does not, my claim that genomic politics is not partisan politics will become a historical artifact. Later chapters consider that prospect in more detail.

Courts Are Not Partisan Regarding Genomics

One judicial arena sets the terms for courts' use of forensic DNA databases, and here too we see the dog that seldom barks with regard to partisan or ideological disagreement. The arena is the operational meaning of the Fourth Amendment's guarantee of the "right of the people to be secure . . . against unreasonable searches and seizures." May prosecutors or defense attorneys use DNA that people "abandon" by virtue of walking, breathing, smoking, or eating as evidence in a trial?

Legal scholars find this a difficult and troubling question, with implications for privacy, autonomy, justice, authority, and genetic influence.[30] But, apart from questions of exactly what constitutes "abandonment," courts find this an easy question to answer. Legal scholar Palma Paciocco summarizes: "Police officers can collect these discarded items [such as chewing gum or used tissues] and test them for DNA. Under the governing case law, they are free to do so without a warrant, and without triggering the Fourth Amendment right to be free from unreasonable searches and seizures. . . . Numerous courts have . . . denied Fourth Amendment challenges to surreptitiously collected DNA samples. Their decisions have established and entrenched the concept of 'abandoned DNA.' "[31] Federal judges, regardless of their ideology, interpretive norms, history in prior decisions, or the party of the appointing president, accept this entrenchment. Such operational consensus is not noteworthy for the federal judicial system, but it remains useful evidence to demonstrate that at least this intersection of genetics and law is not ideologically or politically fraught.

More surprising is the ideological coalition in a second arena of Fourth Amendment jurisprudence. Although Supreme Court decisions are intended to be above politics, analysts have no difficulty in characterizing justices along the standard conservative-to-liberal dimension.[32] Justice Antonin Scalia was widely recognized as highly conservative, while Justices Ruth Bader Ginsberg, Sonya Sotomayor, and Elena Kagan have been among the most liberal. These four justices all participated in 359 cases, in which they joined in dissent in only five, fewer than 2 percent.[33] One such case, *Maryland v. King*, addressed whether mandatory collection of an arrestee's DNA sample through a cheek swab violates constitutional restrictions on search and seizure.[34] The 5-4 decision, written by centrist Justice Anthony Kennedy, held that "when officers make an arrest supported by probable cause to hold for a serious offense and bring the suspect to the station to be detained in custody, taking and analyzing a cheek swab of the arrestee's DNA is, like fingerprinting and photographing, a legitimate police booking procedure that is reasonable under the Fourth Amendment." Scalia's dissent was scathing, and was joined by the three liberals:

> Today's judgment will, to be sure, have the beneficial effect of solving more crimes; then again, so would the taking of DNA samples from anyone who flies on an airplane (surely the Transportation Security Administration needs to know the "identity" of the flying public), applies for a driver's

> license, or attends a public school. Perhaps the construction of such a genetic panopticon is wise. But I doubt that the proud men who wrote the charter of our liberties would have been so eager to open their mouths for royal inspection.

However else one might dispute this decision—in particular, whether fingerprints and DNA really are comparable—it is indisputable that the usual partisan or ideological divides do not characterize the majority and minority coalitions in *Maryland v. King*.

The Public (Perhaps) Is Not Partisan Regarding Genomics

Only a few surveys before my own examine ideological or partisan differences in the American public's views on genomic science. I return to this question in much more detail in later chapters; what I wish to establish here is that the scattered extant results show little partisan or ideological pattern. Liberals tend to see more genetic influence with regard to sexual orientation; under some conditions, conservatives see more genetic influence with regard to racial, class, or individual characteristics. Some scholars' previous analyses show no clear or consistent impact of partisan or ideological differences.[35] And even where it does appear, partisanship or ideology seldom stands out as a very powerful predictor of any outcome in any analysis. And yet intense controversies abound, to which we now turn.

4
Enthusiasm and Skepticism

I think that more scientific knowledge is always better. I drink the Enlightenment Kool Aid, so to speak.

—Comment on survey of experts, 2018

[We now have] 2,400 things that are highly predictive and medically actionable.

—Interview with genomic scientist

As greater knowledge is gained there's more of a risk that it could be used for evil purposes.

—Comment on survey of experts, 2013

The framework I employ is too spare to portray reality, as is any dichotomy or ideal type. Nonetheless, if a simple typology captures real fault lines—as in the cases of "exit, voice, and loyalty"; "credit-claiming, advertising, and position-taking"; "hedgehogs and foxes"; "thinking fast and slow", to name a few of the most famous and fruitful—it can open possibilities for new insights, moral clarification, and political agenda-setting.[1] Most likely, I fall short of the inventors of these frameworks—most of us do—but they are models to which I aspire. So, having cleared away the potential competitor of partisanship as an explanation for disputes over ancestry testing, BiDil, forensic DNA databases, and prenatal gene editing, let us put the framework to use and see how well it illuminates responses to this new and complex science.

Enthusiasm

In each century since the beginning of the world wonderful things have been discovered. . . . In this new century hundreds of things still

> more astounding will be brought to light. At first people refuse to believe that a strange new thing can be done, then they begin to hope it can be done, then they see it can be done—then it is done and all the world wonders why it was not done centuries ago.
>
> —Frances Hodgson Burnett, *The Secret Garden*, 1911

> Scientists are optimists—why else would we devote so much effort to devising intricate experiments to tease out new knowledge? . . . Our world is rife with potential tragedies: rapidly dwindling resources, new diseases that spread with frightening speed, the effects of global warming. The role of science in protecting our lives and our planet is crucial and dramatic.
>
> —Frances Arnold, 2012

Frances Hodgson Burnett's technology optimism is over a century old and her language a bit quaint, but the sentiment that keeps bringing young readers to *The Secret Garden* persists. The cynic would call it a Whiggish view of history; the analyst, a preference for type 1 over type 2 errors; and the strategist, a proclivity for proactive intervention. Burnett, of course, knew nothing about genetics. But her promise that "in this new century hundreds of things still more astounding will be brought to light" can be read as a beacon to show the way for genomic Enthusiasts such as Frances Arnold.

Medical and Scientific Research

Scientific research is the central locus of Enthusiasm. Is genetic influence strong? "The sequence of the human DNA is the reality of our species, and everything that happens in the world depends on those sequences," according to Nobel laureate and virologist Renato Dulbecco. Should we be optimistic about the benefits from genomic science? "When you walk down the street [in Cambridge, Massachusetts, a center of biotechnology], there are young excited people who are racing to get someplace, busy at some piece of inventing the future," according to Susan Hockfield, neuroscientist and former president of MIT.[2]

Enthusiastic rhetoric emerged in the 1980s, with successes in identifying genes implicated in some forms of cancer or diabetes, Huntington's disease, sickle cell anemia, and other rare and devastating hereditary disorders.

The prose got even more purple as scientists persuaded Congress to fund the Human Genome Project. Geneticist Norman Anderson assured a crucial early meeting of decision-makers that "a century from now, as history books are written, the big projects that were important in this century are the genome project, and after it possibly space and then the atomic bomb." This sort of elation achieved its goal of extracting several billion dollars from Congress; the usually buttoned-down senator Pete Domenici (R-NM) observed, "As someone who is supposed to know all about the federal budget, I am rarely in a position where I can look at a program and say that it is exciting enough to keep somebody like myself energized while we are trying to reduce the deficit, but I have found one here."[3]

Perhaps inevitably, the Human Genome Project has not yet fulfilled its grandest promises, despite President Clinton's ever-repeated encomium to "the most wondrous map ever produced by humankind." The genome turns out to be vastly complex, and its interactions with the environment and individual behavior even more so; scientists agree that it will be decades, if ever, before humans can understand and control genetic influence for important phenotypes such as intelligence or bipolar disorder. So their enthusiasm is more realistic than it was several decades ago, even if no less fervent.

Scientists can, of course, point to impressive technological and substantive advances. As gene sequencing drastically decreasd in cost and increased in efficiency, the number of sequenced human genomes rose exponentially. The two panels of Figure 4.1 show the trend lines.

"Genotyping cost is asymptoting to free," so within a generation "it will be easier to know someone's genome than their name," predicted bioengineer Russ Altman early in the development of sequencing.[4] On National DNA Day, April 25, 2018, the Broad Institute announced the sequencing of its 100,000th human genome, enough to gratify those hearing Eric Lander's quip at an Institute retreat that "single cell sequencing is the answer to almost everything." As of the end of 2020, the Broad Genomics Platform had processed more than 2 million samples, an average of "one 30X human whole genome every 10 minutes."

Actual medical success, of course, matters more than any set of numbers. As of January 2020, the FDA had approved four gene therapies to treat, for example, certain forms of leukemia and retinitis pigmentosa, a rare genetically inherited blindness. It "anticipates many more approvals in the coming years, as evidenced by the more than 900 investigational new drug (IND) applications for ongoing clinical studies in this area."[5] The ClinicalTrials.

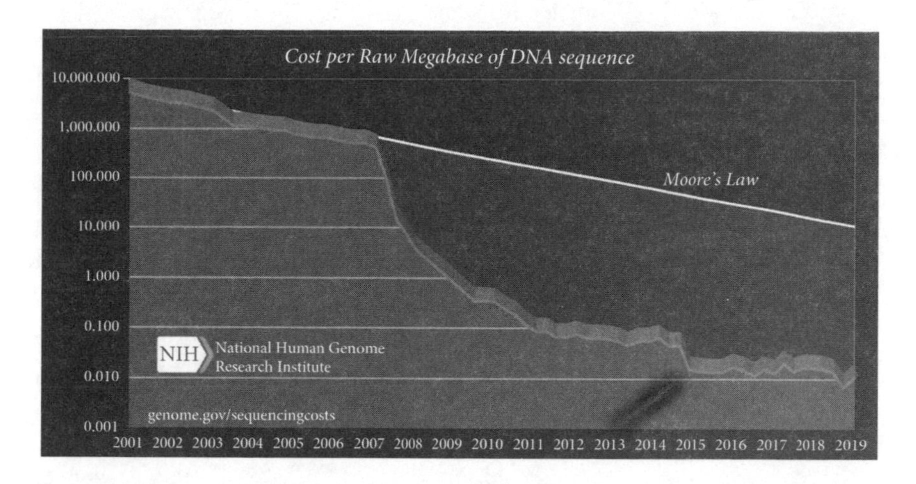

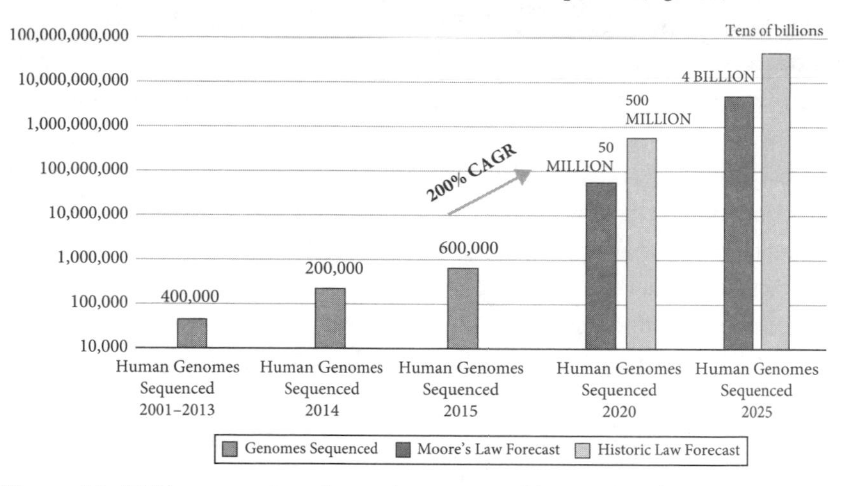

Figure 4.1 DNA sequencing: decreasing costs and increasing data
Source: National Human Genome Research Institute 2019a

gov database lists hundreds of recruiting, active, or completed studies in the United States for gene therapy of diseases such as Alzheimer's, Parkinson's, hemophilia A and B, severe combined immunodeficiency, various cancers, and others. Reputable science journalists report on experimental gene therapies for "a fatal degenerative brain disease, an achievement that some experts had thought impossible," and on "replacement skin to save a dying boy."[6] The unprecedented pace of discovering several efficacious vaccines for

SARS-CoV-2—accomplishing in months what typically takes years—results from the rapid sequencing of the coronavirus genome, development of RNA-based vaccines, and other genomics innovations that have been accumulating over the past decade.

Many vaccine trials or research on the disorders in the ClinicalTrials.gov database will come to nothing; Derek Lowe, a chemist who regularly blogs on the coronavirus for *Science*, anticipated with regard to vaccines "sudden reversals, and sudden bursts of hope, despair, and confusion." But the occasional success can induce euphoria. A geneticist who worked for decades to develop the "breakthrough therapy for cystic fibrosis" described it to me in an email as "a tremendous story of science and medicine and perseverance and teamwork. I feel blessed to be a part of it."[7]

Enthusiasm will grow exponentially if, as seems plausible, CRISPR-Cas gene editing, base editing, and prime editing become sufficiently reliable and precise to be used in humans. CRISPR is the pathbreaking innovation (it has its own online journal), but researchers are creating even more refined second-generation techniques. They are making it possible to change one base pair, or a string of base pairs, in the cell's DNA sequence so that the chemical properties that yield a particular genetic impact are modified. I return to the original metaphor: CRISPR and subsequent gene editing techniques are analogous to a word-processing program that enables a writer to correct a single-letter typo or to replace one word or phrase with another. Just as Mark Twain pointed out that the difference between the almost right word and the right word is "the difference between the lightning-bug and the lightning," so CRISPR-based gene editing and its successors can change the physiological trajectory of an organism with a small tweak.

Initial experiments with gene editing were done in vitro; that is, cells were removed from the body, small regions of the genetic code were appropriately edited, and then the cells were sometimes reinserted into the body. Scientists then looked to see what happened—whether the edits changed what they intended to change and did not change what they intended to remain untouched. (In our metaphor, did the word-processing program search and always replace "man" when the word was supposed to be "woman," but did not ever replace "man" with "women" or "men" with "woman"?) By March 2020, researchers had gained sufficient experience and confidence so that a patient with Leber's congenital amaurosis 10, a hereditary condition that causes blindness, received a CRISPR edit directly in the body—that is, in vivo. The distinction between in vitro and in vivo gene editing sounds like a technical distinction, which it is. But scientists see much more: in geneticist Fyodor Urnov's metaphor, "It is akin to space flight versus a regular plane trip."[8]

Assuming that CRISPR and successor modes of gene editing move from single-person trials for one disease to extensive medical use for several or many diseases, Enthusiasm, along with Skepticism, will reach greater heights. Echoing Jennifer Doudna and Samuel Sternberg, who I quoted in the book's epigraph, geneticist David Corn observes, "We are the only species that can control its own evolution. This stuff is incredibly profound—it can take us anywhere." Ambitions know few bounds. A Chinese researcher envisions gene editing as the solution to intractable diseases that are now prohibitively expensive to treat: "CRISPR is one of the most beautiful technologies available to us. . . . It could be the cheapest medicine ever. . . . Six thousand genetic diseases can be cured by this technology."[9]

Where big science marches, big money is close behind. Mayor Al Vellucci, who led the move to halt recombinant DNA research in Cambridge, Massachusetts, in the late 1970s, changed his mind after seeing development of this new technology in other communities. In 1982, he welcomed the pharmaceutical company Biogen to Cambridge, saying he had "no fear of recombinant DNA as long as it paid taxes."[10] The mayor's fiscal instincts were sound. On a day in July 2020 when the stock market was "treading water," reported *U.S. News & World Report*, the small pharmaceutical company Moderna, "which reported impressive developments in the first human trials of a [Covid-19] vaccine earlier this week, continued to see its stock rally into the close of trading Friday, as shares picked up another 15.9 percent on the day." Moderna's vaccine innovation "makes it a potential gold mine for investors. . . . The stock is already up more than 700 percent from 52-week lows."

Moderna's and other pharmaceutical companies' extraordinary success in developing genomics-derived Covid-19 vaccines is, rightly, the focus of headlines and head-spinning financial transactions. But less spectacular arenas of genomic medicine are also big business. In 2019, Roche bought the "gene therapy specialist" Spark Therapeutics for $4.3 billion. Spark seemed unpromising at the time of the sale; its stock price "tumbled [the previous] year after two of 12 patients showed an unfavorable immune response when treated with a higher dose of Spark's haemophilia therapy SPK-8011." Nonetheless, economist Severin Schwan, Roche's CEO, assured stockholders and others that "it's now the right time to step up."[11]

Until gene therapy and gene editing are fully mature technologies, one strategy for recouping development costs is pricing a few drugs at unprecedented levels—up to $3 million for a one-time-only proposed therapeutic at this writing. Jack Grehen, a recipient of this experimental gene therapy for hemophilia, is a quintessential Enthusiast: "It's been absolutely brilliant and life-changing for me. I can just go about my day and not have to worry."[12] Dr. John Pasi, director of the study in which Grehen participated, concurred: "Not to have to worry about hemophilia any longer—I think it's essentially transformational for many patients." BioMarin Pharmaceutical's chief commercial officer, Jeff Ajer, points out, perhaps with a touch of defensiveness, that hemophilia is "gigantically expensive to treat [so that] it's likely that our gene therapy would save a lot of money—millions, perhaps many millions."[13] As of this writing, insurance companies, labor unions, and one state legislature are seeking to develop policies and procedures to cover the cost of these drugs, which are enormously expensive but need to be taken only once (probably—no one really knows). No such policy, however, yet exists.

Battelle Institute estimated in 2013 that the economic impact of genomics on the United States was at least a trillion dollars—and the scale of the enterprise has grown exponentially since then. I know of no update on Battelle's estimate, so we must draw partial inferences from piecemeal data. In a series of reports, a prominent marketing research firm estimated in 2019 that markets for sample preparations (part of a "vast . . . global market" for genomics materials), gene editing products and services ("at the beginning stages of its growth story"), DNA sequencing ("significant market growth potential"), genome editing ("popularity is on the rise"), and proteomic technologies ("an explosion of new . . . content") totaled almost $55 billion.[14] Each report predicts a large imminent rise, with a compound annual growth rate of around 20 percent—and none of these data speak to employment of scientists or the immensely greater impact of genomics on American industries, livestock production, and agriculture.

Patent data provide another reason for financiers to extol genomics technologies. In 2018, the last year for which these data are available, about 15 percent of U.S. patent applications in the technology field were for medical technology, pharmaceuticals, and biotechnology. These were among the largest categories in that field, and one could add many genomics-heavy patent applications in agriculture, nanotechnology, or energy. U.S. bioscience-related patents more than doubled in number in the decade before 2019, and biotech reached 13 percent of total business R&D in 2017 in the United States.[15] Researchers' passion may not be financially motivated, and corporate executives may care a great deal about therapeutics—but the financial and exploratory incentives are surely reinforcing.

Genomic scientists and corporate executives do not quite claim that the more we learn about genomic science, the more the data will reveal that "everything is heritable," as suggested (tongue in cheek?) at a Broad Institute retreat. Nonetheless, underlying the arcana of scientific Enthusiasm is the classic Enlightenment conviction that "science has granted us the gifts of life, health, wealth, knowledge, and freedom," as psychologist Stephen Pinker puts it.[16]

Criminal Justice

Enthusiasm is also a driving force in the criminal justice arena. "It's very clear. This saves lives," according to attorney and forensic DNA expert Christopher Asplen.[17] Proponents of forensic DNA databases care little about genetic influences on phenotypes; what matters for them is that a properly collected and analyzed DNA sample is almost always unique to a particular individual.[18] The issue in this arena is how best to use, rather than how to analyze or treat, the particularities of a genomic profile.

Although started in the mid-1990s, forensic DNA databases acquired preeminence in what many described as an "alarming" report from the National Academy of Sciences. In a genre that is usually painstakingly modulated, its authors were scathing about failures of precision and reliability in most so-called forensic sciences: "only nuclear DNA analysis has been rigorously shown to have the capacity to consistently, and with a high degree of certainty, demonstrate a connection between an evidentiary sample and a specific individual or source."[19] Defense attorneys point out that samples get contaminated, laboratories make errors, prosecutors can be overzealous, and even a tiny overlap in DNA profiles can be dangerous for defendants. But Enthusiasts counter by pointing to the benefits. Forensic DNA databases offer a corrective to fallible or biased eyewitness reports, potential witnesses' refusal to participate, police corruption or mistakes, jurors' racial, gender, or class discrimination, incompetent or overworked lawyers, and misleading experts. Benefits from use of forensic DNA databases, points out political scientist David Lazer, include harms that do not occur and expenditures that are not made. For example, plausible suspects can be excluded before trial, saving both investigative resources and problems for those individuals; cold hits in investigations can streamline a case, saving time and resources; convicted perpetrators who would otherwise be out on the streets are not commiting new crimes. DNA databases might even deter strategically sophisticated would-be criminals.[20]

It is of course hard to determine DNA databases' impact on crimes not committed, but some have taken on the challenge. Stanford student Can Wang and management professor Lawrence Wein calculate that if all 8,307 unanalyzed sexual assault kits in Detroit were tested, almost a third of their biological evidence would match at least one DNA sample in CODIS or in another kit. They estimate that an offender commits at least seven sexual assaults annually, and on average 3.7 years pass from first offense to first conviction. Thus conviction based on a DNA match in a sexual assault kit could avert roughly twenty-six rapes. Even using a conservative estimate of thirteen rapes averted per offender, that is a lot of misery not experienced. As a further incentive, if one is needed, testing the kits would be cost effective. A kit analysis that costs an estimated $1,600 could save roughly $134,000 in victim costs for a single rape.[21]

Estimates on whether a DNA database hit actually leads to conviction range widely, from well under 5 percent probability for New York's reduction of its backlog of sexual assault kits to Cuyahoga County's 22 percent from its backlog and Denver's 50 percent from selected cold cases. But results could be strengthened: conviction rates from hits in analyzed sexual assault kits are low in most jurisdictions partly because the statute of limitations for the crime has run out—which would presumably not occur if sexual assault kits were regularly analyzed and/or the statute of limitations extended.[22]

Unlike Wang and Wein's city-by-city analyses of how sexual assault kits are treated, one study offers a wide geographic sweep for examining the crime that does not occur because of forensic DNA databases. Economist Jennifer Doleac compares the trajectory of people convicted of a felony just before, and those convicted of a felony just after, mandatory DNA submission in each of seven states. She finds that DNA databases have a substantial impact, reducing violent offenders' likelihood of another conviction within five years by 17 percent. Property offenders were less strongly affected, but the pattern is the same. Given how they are measured, even these results probably underestimate "the true deterrent effects," according to Doleac. That is, as state DNA databases grew, crime rates further declined; measurements are imprecise, but "growth in the average database from 2000 to 2010 decreased violent crime by 7–45 percent and property crime by 5–35 percent."

As with sexual assault kits, taking steps to prevent crime is cost-effective. When Doleac was writing, the marginal cost of using DNA profiling in a case was $600 (with the costs falling). In contrast, imposing longer sentences to deter crime cost more than ten times as much, and adding police or police hours cost tens of thousands more dollars.[23]

Anne Anker, a Danish criminologist, and her co-authors show broadly similar results from Denmark's 2005 expansion of its DNA database. Most notably, recidivism declined 43 percent in the year after implementing the law, and the drop persisted for at least three years. Offenders who were violent, young, fathers, or newly criminal were least likely to commit another crime after the DNA databases were introduced. Just as important, an increase in quality of social life accompanied the decline in recidivism: "First-time offenders are more likely to be married after they are added to the DNA database, and recidivists are more likely to be with the same partner and to live with their children." Although cautious scholars, Anker and colleagues conclude triumphantly: "Keeping people out of trouble (and out of prison) can put their lives on a more positive track."[24]

Familial searching sets a high, but not insuperable, threshold for Enthusiasm. "The technology is powerful . . . there's demonstrable success," summarizes Frederick Bieber, a renowned medical and forensic geneticist.[25] The usual use of a forensic DNA database requires a perfect match between the characteristics of the DNA sample taken from the crime scene, victim, or sexual assault kit and a sample included in CODIS or another database. If the DNA does not match perfectly, the two samples are not from the same individual. Taking advantage of the nature of genetic inheritance, familial searching turns that stringency on its head. If a crime scene sample is very similar but *not* a perfect match to a sample in the database, there is likely to be a close family relationship between the person who inadvertently left DNA at the crime scene and the person who submitted DNA to the database at some other time and place. Genealogical and other search methods can be used to identify plausibly germane relatives of the known person in the database, thereby generating a set of persons of interest. From that point on, police detective work takes over; that is how California's Grim Sleeper was identified after decades of assault and murder.

Police, and especially elected officials, are cautious about the ethical implications, resource expenditures, and logistical difficulties of conducting familial searches. But with media portrayals of people such as Lonnie Franklin Jr. and Dereck Sanders, a.k.a. the "Roaming Rapist," as well as a growing record of careful use, endorsement is growing. The dean of the UCLA Law School, Jennifer Mnookin, worries about the privacy and racial implications of familial searching. But she nonetheless decided after the Franklin case and a few others that "if it's helping us solve big cases, it seems like a worthwhile trade-off."[26]

In contrast to familial searching, and with the possible exception of judges and prosecutors embarrassed by their consequential mistakes, almost everyone is enthusiastic about using DNA to exonerate the falsely convicted. Most exonerees are Black men or Latinos who were young and poor at the time of conviction; many have spent years, even decades, in jail or prison.[27] A few contrarians worry that Innocence Project attorneys are self-righteous, the criteria for accepting cases are too stringent or slow, or the success of exoneration efforts risks falsely implying that the criminal justice system is effectively self-correcting. But it is hard to criticize even a partial remedy for the fact that up to 12 percent of murder or sexual assault convictions can be challenged. Although these results were obtained for Virginia in the 1970s and 1980s, comparisons with other states suggest that they are not anomalous.[28] Faced with this accumulating evidence of wrongful conviction, states are reforming criminal justice procedures by, for example, recording interrogations and improving eyewitness reporting procedures. An FBI official told me matter-of-factly that "we work with the Innocence Project constantly. We don't want people wrongly convicted. They help us with research, we give them info and leads."

Although not close to the scale of medical and scientific genomics, there is money to be made in the intersection of genomic science and criminal justice. Local police departments are buying expensive Rapid DNA machines; the New York University Policing Project envisions the "potential political and financial impacts as a global multibillion dollar industry." The Bureau of Justice Assistance reports over a thousand grants totaling more than $800 million to proposals including the keyword "DNA," and the federal government spends over $100 million annually on "DNA related and forensic programs and activities" in states and localities.[29] Adding in FBI and state or local expenditures on forensic DNA databases, and occasional legislative allocations for improving labs or reducing sexual assault kit backlogs, genomics-related expenditures for criminal justice run into the billions of dollars.

No police chief envisions profits analogous to those in the mind's eye of pharmaceutical company executives. Nor, I would guess, do most criminal justice officials retain for long the Enlightenment faith that reason and science can solve the problems they face. Instead, criminal justice Enthusiasm rests on the liberal principle of using facts and fair procedures to make society more just for rights-holders. Supporters of forensic DNA databanks point to the gender implications of analyzing sexual assault kits, the racial implications of less biased trials and more exonerations, the family implications of reduced recidivism, the budgetary implications of more targeted investigations, the moral implications of achieving closure for victims and their families—and the benefits to the whole community of less crime. As President Obama put it, "We insist on justice."

Biogeographical Ancestry

The arena of DNA ancestry testing is rich with Enthusiasts. The company African Ancestry "helps expand the way people view themselves and the way they view Africa!"—all in the service of "improv[ing] the cultural, emotional, physical, spiritual and economic well-being of people across the African Diaspora." Customers' testimonials report that the "experience moved me in ways that I cannot adequately express in words. The unveiling of my ancestral connections ranks right up there with witnessing the birth of my children, the difference was that I was the one being (re)born" or that "learning my maternal lineage has been profoundly enlightening to myself, my daughters, mother, aunts and so forth. I learned my ancestors are great artisans, royalty, and warriors of great strength and skill. I can now pass this on to my children."[30]

The testimonials could be multiplied. Maya Sen and I conducted a survey experiment in 2010 to see how a random sample of Americans would respond to the idea of DNA ancestry testing. After reading vignettes of hypothetical people taking tests with varied outcomes, 55 percent of our respondents expected that those discovering ancestry from a given region of the world would be pleased with the result, and a third thought these test results would matter a lot to that person's identity. Most other respondents expressed neutral views; few were negative.[31]

African Ancestry has imitators and competitors. The 23andMe DNA test kit was a best-selling item on Amazon.com's 2017 Prime Day.

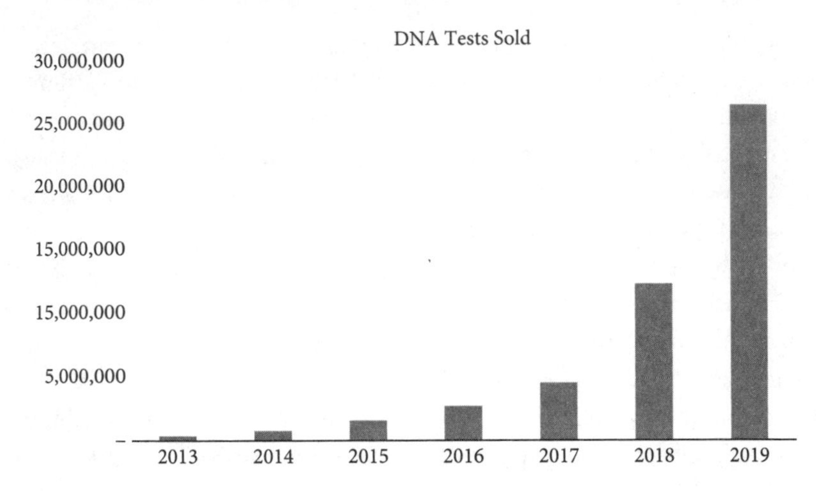

Figure 4.2 Growth of the DNA ancestry test market
Source: Genetics Digest 2019

ConsumersAdvocate.org analyzes the "10 best DNA testing of 2020"; the website DNAWeekly analyzes the seven best; the website PCmag.com examines its five top picks for humans and three top picks for dogs. And no wonder: "The DNA market doubled in 2019—for the third year in a row."[32] Figure 4.2 shows the trajectory.

Firms selling ancestry tests show great ingenuity in generating new Enthusiasm. Moving beyond testimonials, 23andMe urged soccer fans to "Root for Your Roots" in the 2018 World Cup. Lest this promotion seem crassly commercial, 23andMe's vice president of consumer marketing offered an elevated appeal: "We hope this collaboration becomes a new way for people to experience global events like the World Cup. We are creating an authentic experience that can open our minds and bring people together to celebrate diversity, and our global teams rooted in our DNA."

Even the satirical publication *The Onion* promotes genetic ancestry testing, although it cannot resist leavening fake Enthusiasm with genuine Skepticism:

PRO:

- A definitive way to finally prove to everyone you have $79.
- Can trace proud lineage back to ancestors who decimated America's indigenous population.
- Something to talk about for a couple days.

- Get to mail spit!
- Handy reference for government to use when eugenics program kicks off.

CON:
- Takes the fun out of making up your heritage.
- Risk living with truth that you're actually Latvian instead of Lithuanian.
- Finally getting caught for all the murders you committed in the '70s.
- Finding out via email that you probably have cystic fibrosis.
- Gives false hope to anyone who believes their results may be remotely interesting to anyone else.[33]

Genuine Enthusiasm for DNA ancestry testing stems from many motivations. Adolescents participate in the hope of shocking their parents by finding an NPE (non-paternity event— when someone's presumed father is not in fact the biological father) somewhere in the family tree, or a great-grandparent of the "wrong" race or ethnicity. More mature test-takers, writes sociologist Alondra Nelson, engage in "affiliative self-fashioning." The self is in fact fashioned in two ways: participants are self-selected, and they select preferred responses to the test results. Thus Nelson finds that people use DNA testing for various purposes—to demonstrate racial sincerity, strengthen family links, deepen cultural ties, or generate community locally or across a diaspora. Sociologists Wendy Roth and Biorn Ivemark describe a similar dynamic: test-takers "choose selectively" from the results, emphasizing or disregarding elements of their ancestry profile depending on their "identity aspirations" (which race or ethnicity they want to be a part of) and "social appraisals" (how they think others will respond to their claims of a particular identity). Like Nelson's self-fashioners, Roth and Ivemark's selective choosers become "consumers picking and choosing the genetic ancestries they want to embrace."[34] Most end up, not surprisingly, enthused about the persona they have created and the community they have joined.

DNA ancestry testing is situated in a profit-oriented market. More than 26 million people have taken DNA tests, at a current cost of anywhere from $50 to $400, depending on how much ancestral and/or health information the customer seeks. As of 2017, the company Ancestry had more than 3 million paying subscribers, had tested over 1 million individuals with its AncestryDNA test, hosted over a billion searches monthly, and had 17 million app downloads. It boasted more than 27 billion genealogical records, with 2 million more being added to its website daily. This activity yielded revenue of over $1 billion in 2017—30 percent higher than the previous year. The private equity firm Blackstone Group announced in 2020 that it would acquire Ancestry for $4.7 billion. Perhaps it was undervalued, if co-founder Paul Allen is correct in his claim that "the value of the [genealogical] records *increases* over time; they don't decrease over time."[35] The canonical direct-to-consumer company 23andMe was acquired in January 2021 by a firm that will take it public, valuing the company at $3.5 billion. 23andMe announced that it plans to use its new financial resources to conduct research on and develop drugs for genetic disorders, mining its huge customer database for (voluntarily provided) genomic information.

23andMe's diversification may be wise. The genome sequencing corporation Illumina reported "a weakness" in 2019, in contrast to the previous exponential growth of the direct-to-consumer testing market. "Experts speculate that the natural expiration of the early adopter wave and growing privacy concerns are to blame."[36] Companies laid off staff.

But even if custom is waning and Paul Allen is incorrect in predicting the rising worth of genealogical records, a reason for Enthusiasm beyond individual curiosity remains. As one person I interviewed put it, the millions of tests already completed are socially beneficial: "As more of these stories make public the interconnections, flow of genes across races, the whole idea of racial mixture eventually becomes ho-hum."[37] African American scholar Henry Louis Gates Jr. celebrates not only people's ability to recover their African ancestry but also his own discovery of his "Yiddishe mama" and his visit to his long-lost Irish cousins. Gates learned that "my mitochondrial DNA, my mother's mother's mother's lineage . . . [was not] Yoruba, as I fervently hoped. . . . A number of exact matches turned up, leading straight back to that African Kingdom called Northern Europe, to the genes of (among others) a female Ashkenazi Jew."[38] His reaction: "I have the blues. Can I still have the blues?"

In a more serious vein, Gates points out that DNA ancestry testing is a democratizing force, enabling groups whose ancestry is lost to history—descendants of enslaved Africans, White indentured servants, Jews fleeing pogroms, Chinese contract railroad workers, exiled Native Americans—to access the kind of history previously available mainly to Daughters of the American Revolution or Sons of the Confederacy. Sociologist Alondra Nelson even sees DNA ancestry testing as a "curious instrument of racial reconciliation," through, for example, Georgetown University's pledge to compensate descendants of enslaved workers.[39]

A third basis for Enthusiasm about DNA ancestry testing has nothing to do with identity and connection. I include it here because the concept of clustering people in or across races through knowledge of their genetic heritage opens the door to thinking about that messy topic in societal venues beyond genealogy. Even after the controversy around BiDil, some researchers tiptoe through. As geneticist Esteban Burchard puts it, "Race is a complex construct. It includes social factors; it includes self-identity factors; it includes third-party factors of how you view me. But it also includes biological factors."[40] In research on disease and therapeutics, Burchard and his colleagues identify "marked differences in drug response to asthma therapies between racial and ethnic groups, which contribute to health disparities in asthma morbidity and mortality." He offers a striking conclusion in undramatic language: "Racial/ethnic differences in drug response are partially explained by genetic differences."[41]

Burchard has company. For example, neurologist and epidemiologist Nita Limdi and her co-authors find that statistical analyses make better predictions about the impact of a given dose of warfarin on the recipient if they are conducted separately for European Americans and African Americans, rather than if the analyses combine the two groups. That is because the gene variant influencing Blacks' response to the drug is different from the gene variant influencing Whites' response to the same drug. In another example, among lung cancer patients from Latin America, Native American ancestry is associated with an increased likelihood of mutations in the EGFR gene. That finding is independent of whether the person smokes. Geneticist Matthew Meyerson also finds that East Asians with lung cancer are about five times more likely to have the EGFR mutation than are Europeans or North Americans with lung cancer, although he points out that "it is not clear whether the ethnic difference in EGFR-mutant lung cancer is due to environmental or genetic factors." Using genome-wide association studies of the Million Veteran Program biobank, psychiatrist Murray Stein and his co-authors find that veterans of European ancestry show different genetic links to post-traumatic stress disorder than do veterans with African ancestry.[42]

Reasons vary for accepting the legitimacy of linking these three elements—a person's ancestry, a racial designation, and adjustments to medical diagnoses and medications. Representative Christiansen sees those associations as an important step toward compensatory justice: "The health justice framework that promotes race-based medicine as a necessary shortcut to leveling the playing field via scientific redress is now securely an upstream part of research conceptualization and a downstream trope of social justice." Others see it as a (hopefully brief) way station on the path toward individualized, precision medicine. One geneticist whom I interviewed described his research on racial differences in drug response as "a space holder till we can actually look at individuals." A practicing clinician told me, with some edge in her voice, that attention to patients' race is simple common sense:

> Ethnicity is the poor man's substitute for genetic information—we should use it only if it is validated in predictive studies that a patient is more likely to have drug response if they have a given skin color. There is wonderful research to show who will respond to interferon for hepatitis C. We know from clinical practice that African Americans don't respond as well [and we now have] three studies that found a polymorphism that conferred interferon resistance. Was that racist? No—African Americans have a higher risk but we can test to see if anyone has that polymorphism. I would test everyone. So [race serves as an] intermediate surrogate. I understand concern regarding the "biology of race," but ethnicity per se is a genetic variant.[43]

Moving away from medicine as well as from identity and social connections, the study of deep ancestry is a final arena for Enthusiastic examination of biogeographic inheritance. Genomic research on human migration out of Africa roughly 70,000 years ago has moved in waves, perhaps mimicking in vastly briefer compass the migration patterns themselves. The first wave, corresponding roughly to the 1990s, emphasized the inconsequence of nominal racial groups. Geneticist Luca Cavalli-Sforza and his colleagues' iconic, 1,100-page *History and Geography of Human Genes* concludes the section "Scientific Failure of the Concept of Human Races" with the flat declaration that "classification into races has proved to be a futile exercise." This is not a humanist manifesto against false precision; the 550 pages of charts and maps prove that one can indeed classify populations into races. But doing so is "completely arbitrary"; ancestral groups blur and merge as people move and mate. Thus there is no way to determine that a classification into, say, five races (as defined in 1775 by naturalist Johann Friedrich Blumenbach, the five were Caucasoid, Mongoloid, Ethiopian [later Negroid], American [Indian], and Malayan) is better or worse than zoologist Ernst Haeckel's ten races of the 1870s.[44] "There is no biological reason to prefer a particular one" among these typologies, insist Cavalli-Sforza and colleagues. Their book demonstrates the intrinsic and irreducible incoherence of conventional racial boundaries in a series of magnificently colored illustrations showing the swirls and blends of group lines (Figure 4.3):

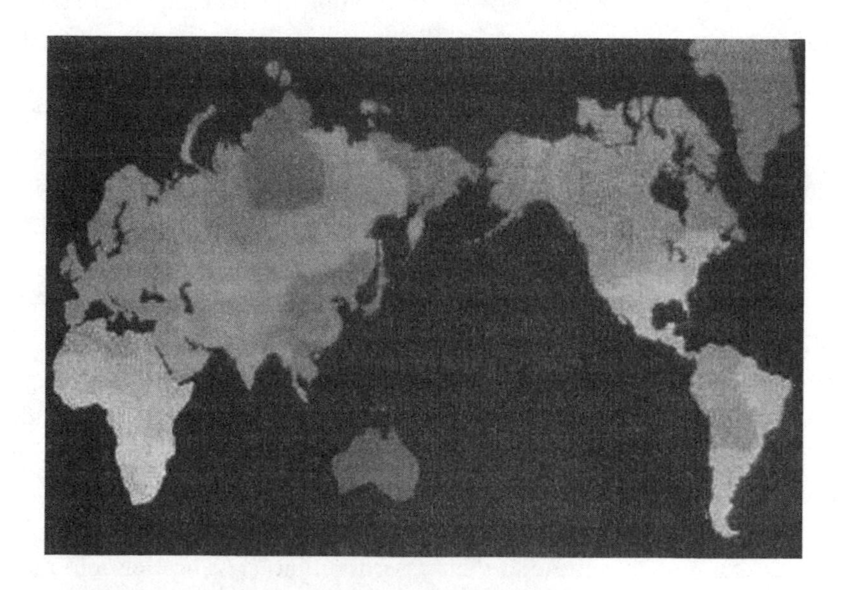

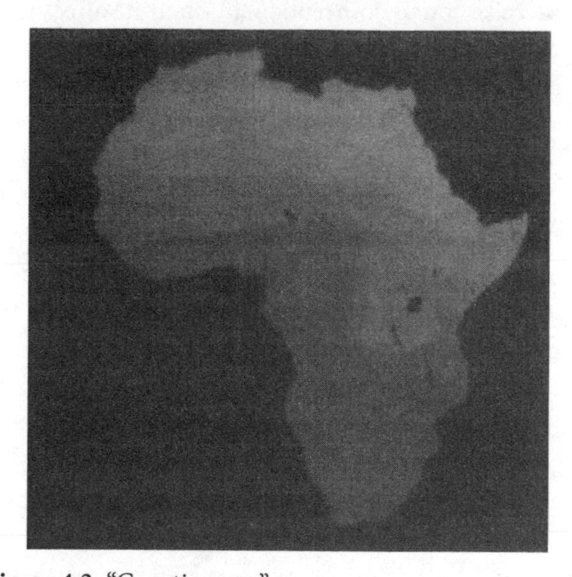

Figure 4.3 "Genetic maps"
Source: Cavalli-Sforza et al. 1994: following p. 541

In case that argument is not strong enough, Cavalli-Sforza and co-authors add two more to underscore the scientific failure of the concept of defined human races. First, since the boundaries in any human classification system are blurred, "minor changes in the genes or methods used shift some populations from one cluster to the other." And finally, following Richard Lewontin and many others,[45] they show that due to migration histories, "the difference *between* groups is . . . small when compared with that *within* the major groups, or even within a single population."[46]

As happens most of the time, however, this ringing conclusion was not the end of the story. In a second wave of research, roughly in the early 2000s, some researchers once again found that they need a concept like race even if they shy away from it in favor of terms such as ethnicity, biogeographic ancestry, or region of the world. Regardless of how it is labeled, these Enthusiasts contend that the concept of a reasonably coherent genetically characterized group does essential work; it turns out to be very difficult to report genomic analyses of populations without using group categories. Even Cavalli-Sforza and his colleagues write that "the color map of the world [Figure 4.3] shows very distinctly the differences that we know exist among the continents: Africans (yellow), Caucasoids (green), Mongoloids, including American Indians (purple), and Australian Aborigines (red)."[47] This does not quite contradict their assertion that "classification into races has proved to be a futile exercise," but it does tend in the opposite direction.

As we saw earlier, one can make a parallel analysis with regard to biological sex. The long-standing debate over whether racial boundaries are in some sense real or are purely social constructs is analogous to the more recent debate over how to understand being transgender or having intermediate sexual characteristics. Statements such as Harvard president Summers's speculation about sex-linked genetic differences in extraordinary mathematical and scientific ability generate the same kind of controversy as do statements about racially inflected levels of intelligence or athletic ability. Regardless of the right answer, if there is one, what matters for my purposes is that for sex as for race, geneticists find it difficult to do their work without using conventional group terms, despite knowing the pitfalls of being regarded as a genetic determinist or even essentialist.

One difference, however, is intriguing: as we saw with the Covid-19 example earlier, researchers are often much less anxious about linking sex to genetic inheritance than they are about linking race. Consider this Enthusiastic announcement: research at or funded by NIH "has resulted in a dramatic increase in our appreciation of sex differences across many disease states. Neurological and psychiatric disorders are among those most impacted. . . . Our understanding of the biological underpinnings of such differences has advanced only recently with the advent of unbiased genome-wide data." There are, in addition, "evolving data showing that genome-sequence variations influence disease risk in a sex-specific manner." As a result of this new knowledge, the National Academy of Sciences organized a workshop called "Sex Differences in Brain Disorders" in September 2020.[48] It is hard to imagine substituting "race" for "sex" in these statements; NIH seems very unlikely to do so.

Even Enthusiasts appropriately cautious about associating racial groups with genetics may be explicit when studying "the new science of the human past." As we have seen, according to David Reich, analyses of ancient DNA show that for thousands of years human populations have differed genetically from other human populations in ways related to the likelihood of having traits we now value. However, he demonstrates, today's supposed races "are mixtures of highly divergent populations that no longer exist in unmixed form." For example, to everyone's astonishment, ancient North Eurasians "contributed a large amount of the ancestry of present-day Europeans as well as of Native Americans."[49] No one has traced the migration patterns that produce such results—but "race" in some sense is genetically real even though what we now conventionally label as races are genomic hodgepodges.

Given this mysterious history of population mitosis and meiosis, Reich rejects arguments claiming genetic distinctiveness of contemporary groups labeled as races as much as he rejects claims that race is only a social construction. Instead, he makes a classic Enthusiast's plea: "Even if we do not yet know what the differences are, we should prepare our science and our society to be able to deal with the reality of differences instead of sticking our heads in the sand and pretending that differences cannot be discovered." At base, Reich is urging the Enlightenment creed: "The data provided by the genome revolution are potentially liberating, providing an opportunity for intellectual progress."[50]

"Potentially liberating" is the key phrase. By precisely and carefully linking group categorization to genomics, Enthusiasts seek to pry classification away from denigration, and detoxify it. Speaking as sociologists rather than geneticists, Dalton Conley and Jason Fletcher deny that "consideration of genetics in racial analysis is always pernicious. The ability to control for genotype actually places the effects of social processes, like discrimination, in starker relief. . . . Controlling for genetic differences de-naturalizes the [inequality of group] outcomes."[51] In short, Enthusiasts hope, just as denigrated groups recapture for their own use terms like gay, queer, pussy, or slant, the study of group ancestry through genomics will help to free "race" from Harvard biologist Louis Agassiz's nineteenth-century racial science and psychologist Henry Goddard's twentieth-century eugenics.

Advancing into the Unknown

Some Enthusiasts are simply fascinated by their research. "A pink-haired grad student named Abbie Groff," who discovered a distinctive RNA molecule in a mouse's genome, says, "You come in in the morning, and it's like Christmas." Others foresee solutions to seemingly intractable social problems: a rural American health care system is incorporating genomic screening into its primary medical care to assess disease risk, eventually for millions of underserved patients. Fast, cheap genomic sequencing of pathogens such as Ebola or Zika enables a worldwide surveillance system that may eventually prevent the spread of infectious diseases in poor countries. Covid-19 vaccines, based on genomic sequences posted on Twitter within weeks of the first human outbreak, are typically described as miraculous. Passenger pigeons or the wooly mammoth may be (sort of) brought back from extinction; genetic technology may preserve endangered species threatened by climate change. Genetically engineered bacteria can generate electricity and turn sewage into clean water.[52] The agonies of sickle cell anemia and cystic fibrosis may be conquered. Mosquito gene drives may eradicate malaria, dengue fever, and screwworm; CRISPR gene editing may enable elimination of thousands of Mendelian (single-gene) diseases. The decrepitude of old age may be slowed or even reversed. Genetically compatible partners could reduce divorce rates. Serial rapists and killers might be convicted before dozens of crimes and years of anguish ensue; poor men of color can be freed after decades of wrongful imprisonment.

In short, genomics Enthusiasm rests on the conviction that taking advantage of rather than fighting or ignoring genetic influence opens the door to understanding and advancement. "The beauty of scientific progress is that it never locks us into a technology but can develop new ones with fewer problems than the old ones," writes Steven Pinker.[53] Christmas every morning, indeed.

Skepticism

> We face danger whenever information growth outpaces our understanding of how to process it. The last forty years of human history imply that it can still take a long time to translate information into useful knowledge, and that if we are not careful, we may take a step back in the meantime.
>
> —Statistician Nate Silver, 2012

> Research in genome editing conducted by countries with different regulatory or ethical standards than those of Western countries probably increases the risk of the creation of potentially harmful biological agents or products. Given the broad distribution, low cost, and accelerated pace of development of this dual-use technology, its deliberate or unintentional misuse might lead to far-reaching economic and national security implications.
>
> —Director of national intelligence James Clapper, 2016[54]

Skeptics agree with Enthusiasts about the facts—genetics shapes human phenotypes, more or less in different circumstances and for different traits—but their emotional or strategic reaction to that presumed fact is the opposite. At best, more engagement with genetic influences cannot achieve Enthusiasts' dazzling promises; at worst, it is a danger to humankind. Skeptics share the view of former director of national intelligence James Clapper that genomic science is a "dual-use technology"—in diplomacy-speak, something that can be used both to promote peace and to wage aggressive war.

Medical and Scientific Research

Some Skeptics are disillusioned by genomic science's failure to live up to early promises. One person I interviewed remembers with exasperation a decades-old *Science* editorial that seemed to guarantee that with the Human Genome Project, "homelessness will never [again] be a problem because we'll know the genetic basis for schizophrenia etc." Geneticists "absolutely hyped the project"—but they were joined by other scientists, the U.S. president, the British prime minister, and journalists. Science writer Nicholas Wade described sequencing the human genome as "an achievement that represents a pinnacle of human self-knowledge" (although he later became disillusioned about genomics progress). Even the World Socialist Web Site joined the chorus, depicting the Human Genome Project not only as "an extraordinary scientific achievement" but, reassuringly, as a "refutation of the prevailing nostrums that private profit, personal advantage and competition are the only driving forces for scientific research."[55]

"Deflating the genomic bubble" was probably inevitable.[56] Although scientists remind critics that decades often pass before a fundamental discovery makes its way into technological use, concern persists that genomic science will prove largely futile in the face of daunting physiological and societal complexity. But a deeper cause of Skepticism is the opposite concern—that knowledge gained through genomic science will become all too powerful. One fear is that evidence of partial genetic influence will be treated as dispositive and slide into genetic essentialism. The *locus classicus* of this anxiety is the 1997 movie *GATTACA*, anointed by NASA in 2011 as the most scientifically plausible science fiction movie ever made.[57] *GATTACA* (the letters in the title refer to the bases that constitute DNA) portrays a society in which fetal "valids" are edited to accord with a genetic registry that selects for particular characteristics enabling them to perform professional jobs. "Invalids" (accent on first syllable, as in "not valid") are created in the old-fashioned way and relegated to menial jobs. Only by excruciating attention to corralling every "abandoned" DNA cell, and usually not even then, can an invalid secretly rise above his or her allotted station.

GATTACA crystalized fear of "genoism," discrimination based on genotype rather than on actions, aspirations, or other characteristics that philosopher Michael Sandel calls "gifted," in the sense of being a gift from nature, chance, or a deity. NHGRI offers a more sober warning of genetic discrimination. As genomics becomes more and more effective at linking DNA differences to bodily disorders, health care providers will increasingly turn to individualized precision medicine to prevent, diagnose, and treat disease. That is the goal—"but" warns NHGRI, "unless this DNA information is protected, it could be used to discriminate against people."[58]

Just as the steady rise in NHGRI's budget that we saw in Figure 3.4 is an institutional manifestation of Enthusiasm, so the Genetic Information Nondiscrimination Act (GINA) described in Chapter 3 institutionalizes Skepticism. In the context of few or no validated claims of genetic discrimination, GINA is consensually understood to be preemptive—"anticipat[ing] a form of discrimination that may pose a future threat." Nonetheless, some Skeptics see the law as too weak. It does not cover disorders already manifested, small employers, or life, disability, and long-term-care insurance. More worryingly, "few, if any, applicable legal doctrines or enactments provide adequate protection or meaningful control to individuals over disclosures that may affect them."[59]

Concern about genomics' excessive power typically follows one of three paths. First, societal deployment may go too far; the phrase "designer babies" and Sandel's "case against perfection" encapsulate this fear. People with this concern may endorse personalized treatment for cancer or exoneration of the wrongly convicted, but they are more swayed by anticipation of inappropriate or dangerous uses.

Given this mixture of plausible benefits and risks, "Yes, but . . ." arguments abound. Michael Cook, editor of the bioethics newsletter *BioEdge*, agrees that "In some respects, IVF has been quite a success." Not only is there a "thriving industry," but also the estimated 8 million children born through IVF to date are "the happiness side of the ledger." Nonetheless, "how about the women who endured cycle after cycle of IVF without conceiving? Their lives have been filled with suffering as a result. And there are far more of them than women who eventually conceived." Perhaps of deeper concern is "the destruction of millions upon millions of human embryos. And how about the disturbing future of IVF—designer babies and genetically engineered children? That is the pain side of the ledger." He concludes, with some hesitation, that "the balance is negative."[60]

Religiously based Skepticism about human overreach is a second variant of "yes, but . . . " Again, one rich example suffices. Bioethicist and CEO of Christian Healthcare Centers Mark Blocher concurs with Enthusiasts that "it is good to discover a prenatal genetic condition if we can correct it." But he seeks appropriate limits: "Since we live in a fallen world populated by depraved people, all are destined to die (Hebrews 9:27; Romans 6:23)." So "maintaining health is ultimately a losing proposition. . . . [A]t what point do we stop? Perhaps the key moral question in this advancing age of biotechnology will be this: As we press forward to break existing scientific constraints, are we doing so to give glory to God or to deify man?"[61]

Skeptics' "yes, but . . ." might instead signal concern about slippery slopes. Great Britain's Nuffield Council on Bioethics concluded in 2018 that prenatal germline gene editing for selection of certain traits "could be ethically acceptable" if intervention is "intended to secure, and be consistent with, the welfare" of the edited individual, and if interventions "uphold principles of social justice and solidarity." This position is not Skepticism; it is, in fact, the most Enthusiastic organizational statement on gene editing of which I am aware. As a result, Skeptics such as biologist Paul Knoepfler responded vigorously. This "green light . . . is likely to do harm" given the report's "aspirational undertone . . . that suggests they have placed too much emphasis on hypothetical potential benefits and haven't weighted the range of risks sufficiently." Like the others I am quoting here, Knoepfler begins a further explanation with "yes" views: "I'm in favor of responsible research using CRISPR and other genetic modification technologies in human stem cells, which my own lab conducts, and potentially on a limited, careful basis of the in vitro use of CRISPR in viable human embryos limited to the lab if there is a compelling scientific rationale and ethical oversight." His "but" follows apace: "The heritable, reproductive use of genetic modification technologies like CRISPR in humans is another matter, which I oppose. It is too risky on so many levels ranging from medical to societal, including eugenics."[62]

Skepticism about germline gene editing can be boiled down to a bumper sticker: "Adopt a moratorium on heritable genome editing." That is the position of eighteen luminaries, including Eric Lander, Françoise Baylis, Feng Zhang, Emmanuelle Charpentier, and Paul Berg, among others, as summarized in the title of an article in *Nature*.[63]

A final "yes, but . . . " argument is economic. Skeptics usually celebrate the successful vaccines against SARS-CoV-2. But they also draw attention to dangers, including medical risks and especially the creation of a new, deep, inequality among individuals and across countries. Affluent individuals in affluent countries with strong health care systems are the first to receive protection, and there may be a long time between the initial and final rounds of vaccination around the world. Even worse, affluent individuals in affluent countries could be the *only* people to gain immunity if genetic mutations make this (or the next?) virus vaccine-resistant faster than people can be vaccinated in numbers great enough to stop its spread. Poor people in poor countries could then be even more vulnerable than before a vaccine was created. Thus Oxfam calls for a "people's vaccine" for Covid-19 that is "available to all, in every country, free of charge." As seen in Figure 4.4, it urges supporters to act to combat genomic inequality.

Beyond Covid, although few dispute the right of pharmaceutical companies to earn a profit, Skeptics have long protested what they describe as the common business practice of exploiting vulnerable people and even countries. Medical researcher Peter Bach's "yes" is agreement that "the clinical

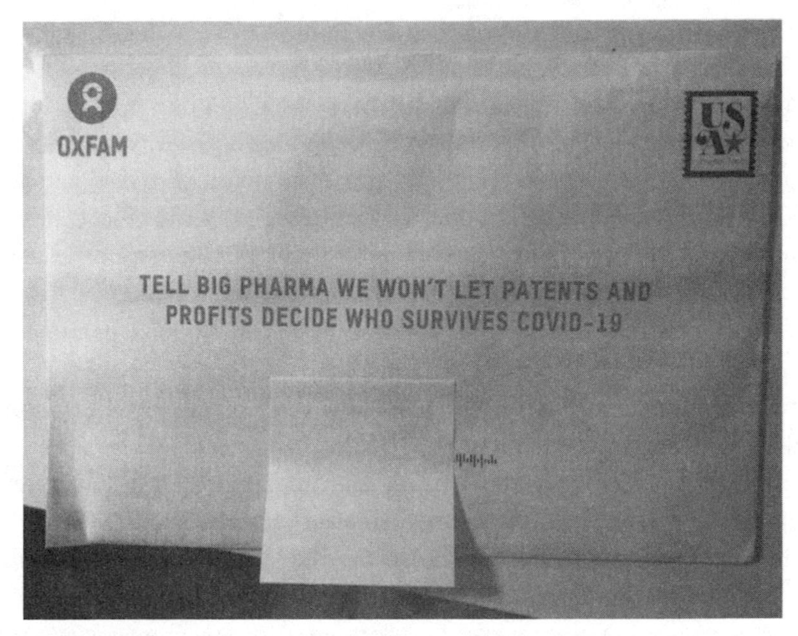

Figure 4.4 Oxfam: "The people's vaccine" must be "available to all, in every country, free of charge."
Source: photo by author

breakthrough [of BioMarin's gene therapy for hemophilia] is prodigious. We should be thrilled by it." Nonetheless, in his "but," the $3 million price is "outrageous." Pharmaceutical companies' "greatest innovation" lies not in their scientific prowess but in their "ability to extract money from society that we could put into other things—like better benefits in Medicare, lower out-of-pocket costs for poor people, dental coverage and things like that."[64]

Anger at extortionate pharmaceutical companies can generalize into resistance to the embedding of genomic science in neoliberal economies. As sociologist Jenny Reardon puts it, "In an era of biocapital—in which biotechnology and capitalism are fundamentally entwined—alienation of human beings from the value that inheres in their own bodies takes new forms." The economic engine of genomic science that Enthusiasts celebrate is in Skeptics' view a giant incubus, sucking private, personal information and money from millions of people and aggregating them not merely into surveilling biobanks but into "life . . . [as] a business plan," or a "pharmocracy," in the words of anthropologist Kaushik Rajan. Reardon's concern is whether "individuals have the right to own and control their own cells, including their

DNA"; Rajan's issue is exploitation of the poor as experimental subjects in drug trials while excluding them from medical treatments reserved for the rich.[65] Anxieties about privacy and inequality are tightly and perhaps intrinsically intertwined; pharmaceutical companies are only the tip of the iceberg in a Foucauldian nightmare of control through biopolitics and neoliberalism.

Other Skeptics warn not of individual or corporate malfeasance, or even of "dual-use" military threats, but of the possibility that genomic science will slip out of *anyone's* control. We simply do not know enough about what we are doing; genetic experimentation may escape the bonds of human intention and create a monster. A Google search in May 2021 for allintext: "Frankenstein + genetic + engineering" yielded about 1 million hits (and about 13,000 hits on Google Scholar). Exemplifying this wide concern is a 2009 debate in Key West, Florida, about whether to control the devastating disease dengue fever by releasing genetically modified male mosquitoes whose offspring die before reaching maturity. The director of the Florida Keys Mosquito Control District was cautiously optimistic: "If this actually worked we would win in every possible way. Other approaches are more costly and more environmentally challenging. The data looked solid, and certainly we need to think differently about mosquito control than we have in the past."[66] But he ran into a buzz saw at a public hearing. A researcher spelled out the concern: "Genetic modification leads to both intended and unintended effects. . . . What will fill the gap or occupy the niche should the target mosquitoes have been eliminated[?] Will other pests increase in number? Will targeted diseases be able to switch vectors?" Although not at the hearing, the director of the environmental group GeneWatch was blunter: the genetically modified *Aedes aegypti* is "Dr. Frankenstein's monster, plain and simple. To open a box and let these man-made creatures fly free is a risk with dangers we haven't even begun to contemplate." Local residents at the hearing were the bluntest of all: "It breaks my heart to think that you guys have the nerve to come here and do this to our community. Anything genetically modified should not be touched," said one. Said another: "I, for one, don't care about your scientific crap. . . . I am no guinea pig."[67]

The Defense Department's Safe Genes program, which includes leaders in the field of gene drive (in which organisms are modified so that a high proportion of their descendants, for example, die young or are sterile), agrees that at present, "such drive systems lack control mechanisms and are consequently highly invasive." Accidentally or by human intention, an uncontrolled gene drive could move beyond the borders of the nation that initiated it or the species in whom it was engineered. Geneticist Kevin Esvelt concurs with Key West residents that "the social and diplomatic consequences of an unconstrained release" could create a "profound tragedy." He differs, however, in asking whether such a fiasco would or should inhibit later use of a more refined CRISPR-based gene drive. It is the phrase "later use" that distinguishes cautious Enthusiasts from unconvinced Skeptics. The former see gene drive as "the technology most likely to help eradicate human scourges such as malaria and schistosomiasis . . . . Now is the time to be bold in our caution." The latter see it as, if not "bioterror, [then i]t's going to be 'bioerror.'"[68]

Some Skeptics eschew these finicky "yes, but..." probabilities and Frankensteinian metaphors; certain types of genomics activity are just wrong and not to be tolerated. Chapter 1 quotes NIH director Francis Collins's mandate that "NIH will not fund any use of gene-editing technologies in human embryos." This view of a "line that should not be crossed" transcends conventional political ideology. Brendan Foht, a self-defined conservative writing in the *National Review*, proclaims that the family's "basic structure—a married man and woman having children whom they love and care for unconditionally—should not be tinkered with by social or biological engineers." Although Michael Sandel might not share Foht's view of the family, he does share the view that gene editing promotes undesirable perfectionism, individualism, and consumerism. The progressive bioethicist Arthur Caplan is equally opposed, albeit for a different reason: prenatal gene editing "may lead to . . . homogenisation in society where diversity and difference disappear." Marcy Darnovsky, historian and executive director of the Center for Genetics and Society, focuses on inequality: germline gene editing is "an unneeded and societally dangerous biotechnology, one that could be leveraged by privileged elites seeking purported genetic improvements to ensure that their children are treated as superior to the rest of us." She urges battle against the racism and eugenics of "biologically defined hierarchies." Americans "can refuse to allow inequalities to be inscribed in our genomes. We can forgo a future in which class divisions harden into genetic castes. We can instead affirm the widespread rejection of heritable genetic modification, and reclaim biotechnology as an instrument for fostering solidarity and serving the common good."[69] Dual-use technology, indeed.

Criminal Justice

As in science and medicine, Skepticism about using genomics in the courts takes several forms. What unites these arguments is agreement that the risks or proven costs of engaging with genomic science to promote criminal justice outweigh actual or potential benefits.

Skepticism sometimes focuses on courtroom use of DNA databases. Human activities, especially repetitive, technical jobs calling for continual attention and nuanced judgment, are inescapably flawed. Financial, professional, or personal incentives can shape perceptions of fact. Training, supervision, and quality control inevitably vary across disparate and unconnected organizations. Funding is never sufficient. Prosecutors may seek status and reelection through aggressive drives for conviction. Judges and juries who need to believe that their evaluations are just may resist counterevidence. These banal points can add up to the conviction that police collection, laboratory analysis, courtroom use, and sentencing impact of data from forensic databases are not to be trusted.

Thus writer Matthew Shaer reported in 2016 that forensic DNA use is becoming "ever more common—and ever less reliable." Criminologist William Thompson told Shaer that "it was no longer a question of whether errors are possible. It was a question of how many, and what exactly we're going to do about it." With widening use of forensic DNA, "the answer has only become more elusive."[70] Even new proprietary software promoted as more accurate than human technicians and better able to resolve inconclusive samples poses risks, since it is also less transparent than humans so that its possible built-in biases are harder to discern.

As Shaer implies, concern about overreach multiplies concern about human error. Criminologist Jesse Jannetta and his co-author at the Urban Institute fear that DNA collection from juveniles is excessive, and "can generate devastating long-term collateral consequences." Nor is excess readily remedied; despite state laws mandating destruction of unneeded samples, legal scholar Elizabeth Joh finds that *"arrestee DNA expungement is largely a myth."* In many states, the arrestee must request expungement and often provide multiple letters, witnesses, court orders, and other impedimenta; given these barriers, expungement almost never happens.[71]

Economics plays a role in the increasing reliance on forensic evidence. Promoters of Rapid DNA machines in police departments emphasize their speed (they can "search unsolved crimes of special concern while a qualifying arrestee is in police custody during the booking process") and their efficiency ("Persons of interest that have no link to a case can be identified and removed from consideration more quickly, dead-end leads can be discarded and, in general, scarce investigative resources can be more effectively utilized"). Furthermore, since Rapid DNA machines are "*not authorized for use on crime scene samples*," according to the FBI, any technical flaws that might creep in can be corrected at relatively little cost.[72]

But technology pessimists are not reassured. The ACLU warns that despite handling highly sensitive information, Rapid DNA machines' "design and use is largely unregulated." The wider their dissemination, the more these machines will be handled by relatively untrained and inexperienced operators. Mistakes will ensue. "Local police are not listening" to FBI warnings about unauthorized use. Most seriously, the availability of flashy new technology gives incentives for expanding forensic DNA databases, with all of the attendant concerns about privacy and overreach. In short, " 'cheap and easy to use' is a perfect recipe for overuse."[73]

Databases with ever more samples create a temptation to engage in more familial searching, a third concern of criminal justice Skeptics. Although no one endorses such searches as a routine strategy, Enthusiasts value their use in especially difficult cases. But Skeptics doubt the value of familial searches, deem the costs too great, or find them to be immoral. Since racial, ethnic, or religious groups in which relatively large families are common are statistically more likely to be subject to genetic surveillance, "demographic trends ensure that innocent members of the Hispanic community will disproportionately experience privacy invasions," observes law professor Daniel Grimm. Molecular biologist Rori Rohlfs and her colleagues start from a different discipline but end up with the same warning about racial discrimination. Genetic techniques used to partially match DNA samples of family members risk identifying a fairly distant relation as a close relative. As a result, too many or the wrong family members may undergo investigation, and "this risk falls disproportionately on those ethnic groups that are currently overrepresented in state and federal databases."[74]

Further objections to familial searching include claims related to equality, accuracy, privacy, democratic accountability, legality, and "societal interest in intact families," in a representative list created by law professor Erin Murphy. Despite these objections, such searches are becoming easier. If permitted by the website proprietors, law enforcement officers can enter a profile into an online genetic genealogy website; through triangulation with partial matches among hobbyists seeking long-lost relatives, they can identify a few plausible suspects. The price of this success is too high for some. As one bioethicist summarizes, "Although we might all like to allow DNA testing to be used to catch criminals like the Golden State Killer, such testing comes with an unavoidable loss of privacy, and competing interests must be considered."[75]

The Golden State Killer, as he was known in the media, was Joseph James DeAngelo. California charged him in 2018 with eight counts of murder and related kidnapping and abduction charges. (The statute of limitations had expired on an additional alleged fifty rapes, sometimes combined with torture. He eventually pleaded guilty to thirteen murders and kidnappings.) DeAngelo was found when a law enforcement agency uploaded crime scene DNA sequences to GEDmatch, an open-source, slightly countercultural, volunteer-run genetics genealogy website. Working with a genealogist, police were able to construct family trees and derive possible suspects. One was DeAngelo.[76] Including GEDmatch, genetics genealogy companies subsequently revised their consent forms, generally to require users to opt in for their DNA samples to be used in the criminal justice system. The Department of Justice also issued guidelines in 2019 controlling how cases involving federal funds or investigators could conduct forensic genetic genealogy. (The rules do not apply to state and local law enforcement agencies, who engage in most criminal investigations.)

Regardless of stricter rules, the cat is out of the bag; we are close to the tipping point where no one can opt out of genetically driven identification. Using GEDmatch data from almost 1.3 million people, computational biologist Yaniv Erlich and his colleagues estimate that "about 60 percent of the searches for individuals of European descent will result in a third-cousin or closer match, which theoretically allows their identification using demographic identifiers. Moreover, the technique could implicate nearly any U.S. individual of European descent in the near future."[77]

To no one's surprise given this history, views on forensic genealogy are mixed. Richard Shelby, a former detective who participated in the decades-long search for the terrifying Golden State Killer, observes that "if criminals out there know they can be tracked down this way, they are going to have to try to not leave their DNA at the scene, and that's nearly impossible. It's one of the best crime-fighting tools to come in a long, long time." The implications for greater gender equity of this new strategy are powerful since it is mainly used for sexual assaults and other forms of violence typically conducted by men against women. But even the liberal *Washington Post*'s headline writer describes "the ingenious . . . DNA technique" as "dystopian." Americans as a whole are evenly split, according to the only survey item I have found for a national sample. Forty-eight percent accept "DNA companies sharing their customers' genetic data with law enforcement agencies in order to help solve crimes"; a third find it unacceptable and the rest are not sure.[78]

Money changes hands, with predictable evaluations. Start-up entrepreneurs offer "an expert team" to conduct forensic genealogical analyses able to "revolutionize criminal investigations and the identification of human remains, . . . answer family questions, and open up a world of hidden data." That boast is just what Skeptics fear. Sites such as Facebook, Twitter, and GEDmatch "are subject to incentives to make millions by selling our data to others," writes legal analyst Nila Bala. It is bad enough that they monetize our daily use of social media sites. "Now, we're talking about our DNA, and the potential buyer is the government," with its enormous power to conduct "dragnet searches—the genetic equivalent of stop-and-frisk."[79]

If possible, Skeptics' anxiety about genetic surveillance rises another notch when race is brought into consideration. In Chapter 1, I quoted legal scholar Dorothy Roberts's view that forensic DNA databases are "maintained *for the purpose of* implicating people in crimes." DNA databanks "create suspects from an ever-growing list of categories." Echoing and ratcheting up Bala's concern, Roberts further argues that the state authorizes itself "to take citizens' private property—in this case, their genetic information—without due process. Those are the features of a totalitarian state, not a liberal democracy." Racial bias throughout the criminal justice system means that young Latino and Black men are especially vulnerable to this technological totalitarianism.[80]

Biogeographical Ancestry

Given these concerns, it is no surprise to find hostility to genomics uses that explicitly invoke racial or ethnic groups. At a minimum, Skeptics perceive DNA ancestry tests to have little to no scientific value, so they are costly both emotionally and financially. Geneticists Robert Green and Adam Rutherford protest that companies such as 23andMe and African Ancestry are "asking people to pay for something that is at best trivial and at worst astrology," since there is no "agreed-upon approach to pick the right number of markers and combine them mathematically. Everyone is sort of just making it up as they go along." Just as journalists love stories about people finding their identity through a spit test, they also love stories such as "How DNA Testing Botched My Family's Heritage, and Probably Yours, Too." As we saw in Chapter 1, Skeptics even hint at illicit behavior by ancestry testing companies: "Market pressures can lead to conflicts of interest, and data may be interpreted differently when financial incentives exist."[81]

A deeper concern than misleading customers or relieving them of extra cash is that associating conventionally understood racial or ethnic groups with the purported science of genetic inheritance risks being a "backdoor to eugenics," in the words of sociologist Troy Duster's classic book. Ann Morning fears that DNA ancestry testing could revive the nineteenth century's supposed racial science, since it is "a modern version of what early scientists were doing in terms of their studies of skulls or blood type. We have a long history of turning to whatever we think is the most authoritative sense of knowledge and expecting to find race proved or demonstrated there." Especially among people who know little about genetics, DNA ancestry tests may "promote an essentialist view of race as fixed and determining innate abilities," according to yet another sociologist, Wendy Roth, and her colleagues. The tests are too fraught to be treated as "spit parties."[82]

Even if they do not extend as far as eugenics or phrenology, genetic ancestry tests reinforce contemporary group hierarchies, according to Skeptics. Self-identified White supremacists use DNA ancestry testing to prove, or if necessary invent, their racial (and non-Jewish) purity. Roth and her coauthor worry that genetic ancestry testing "reinforces their [test-takers'] White privilege" since it offers supposed proof of racial identity in a "symbolic, optional form, removed from structural inequalities."[83] This observation elegantly demonstrates the contrast between Enthusiasts and Skeptics, both contesting racial hierarchies: the former see DNA ancestry tests as an opening for multiracialism and new affinities, whereas the latter see reinforcement of White supremacy.

Skeptics' most serious concern is the impact of DNA ancestry testing on Native Americans; it raises the issue of the moral and empirical legitimacy of blood quantum, about which Native Americans have been deeply split for decades. Blood quantum is defined as the fraction of a person's ancestors, or "blood," who were documented "full-blood" Native Americans. The term is a physiological relic of federal laws that distinguished among American Indians according to how much of their ancestry was Native. The Department of the Interior still issues a Certificate of Degree of Indian or Alaska Native Blood, based on proven connection to ancestors identified on Indian censuses or tribal rolls from the decades around the turn of the twentieth century. Over half of Native nations require a given blood quantum as a necessary if not sufficient condition for tribal membership.

That is the context within which nations decide whether to include results from DNA ancestry tests as one criterion for tribal membership. They differ. Some abjure DNA testing, while others accept it as part of an application—for example, to document that a child is related to a particular set of parents. Individuals sometimes take a DNA ancestry test to determine if they have Native ancestry before starting the arduous process of proving ancestry and applying for citizenship in a particular tribe. Tribes may themselves use DNA ancestry testing. In a series of elections and court cases, a majority of the members of the Cherokee Nation sought (unsuccessfully) to disenroll Cherokee Freedmen—descendants of Blacks who were enslaved before the Civil War but granted tribal membership after it—on the grounds that DNA testing revealed little Native ancestry.

This complex and sensitive context provides rich grounds for concern, according to bioethicist Nanibaa' Garrison, about genocide and imperialism, integration and identity. Anthropologist Jessica Blanchard and her colleagues focus on a related issue, the collective impact of genomic testing on a Native community. They fear that the desire to learn about one's own genetic heritage may "usurp long-standing community values, systems of governance, and forms of relationality." Anthropologist Darryl Leroux draws the broadest conclusion: power dynamics shift as genetic scientists "intervene in complex settler-Indigenous relations by redefining indigeneity" according to their own (i.e., outsiders') logic. These shifting power dynamics, in turn, reinforce political and economic inequality as genetic scientists "mobilize genes," intentionally or not, "in ways that support . . . White settler claims to Indigenous lands."[84]

People who all agree that the genome influences human traits and behaviors disagree on whether or how societies should use the science behind that fact. They differ in defining what is a risk or benefit, in judging how risky or beneficial an innovation is likely to be, in the relative priority they give to pursuing gain and avoiding cost, and in their judgment of whether American society can, or will even try to, solve new problems that may arise by using genomic science. Disputants may differ at a more ideological or philosophical level as well; some espouse one or more moral absolutes in regard to which the logic of a risk-benefit analysis is inappropriate or even pernicious.

The depth and breadth of these conflicts make them quintessentially political, but they are not yet arrayed along a left-right or Republican-Democratic continuum. Some Enthusiasts, such as pharmaceutical company executives or most police chiefs and prosecutors, are conservative. Some Enthusiasts, such as Innocence Project lawyers or representatives who belong to the Congressional Black Caucus, are progressive. Skeptics may be conservative—fearful of the use of genome editing by "countries with different regulatory or ethical standards than those of Western countries," as James Clapper delicately put it—or liberal, such as social constructivists or disability rights advocates.

If this cross-partisan or non-partisan pattern continues, one indirect but enormous impact of genomic science could be its contribution to reconfiguring Americans' associations between ideologies and science policies. Or not. Perhaps genetic influence is weaker than both Enthusiasts and Skeptics understand it to be, and other causal factors matter much more in shaping human phenotypes and trajectories. That possibility creates its own set of political disputes: ignoring or resisting the magnetic pull of genomic technologies creates a new challenge for technology optimists in the Hope quadrant, while the siren call of genomic technologies is yet one more reason for despair among those residing in the quadrant of Rejection. The basic framework described in *Genomic Politics* thus helps to explain views even among those who downplay genomics; demonstrating that claim is the purpose of Chapter 5.

5
Hope and Rejection

> If you want to be rich, you would do better to inherit your parents' business than their genes.
>
> —Stephen Reicher, 2019

> The story of an "enhanced" humanity panders to some of the least attractive tendencies of our time: techno-scientific curiosity unbounded by care for social consequence, economic culture in which we cannot draw lines of any kind, hopes for our children wrought into consumerism, deep denial of our own mortality.
>
> —Tom Athanasiou and Marcy Darnovsky, 2002

The fall of 2020 saw universal astonishment at the speed with which vaccines against SARS-CoV-2 were developed and approved for wide use by regulatory agencies in many countries. Covid-19 vaccines did indeed appear at warp speed, as the Trump administration had promised and fostered. They evoked two diametrically opposite reactions: gratification and suspicion—or, in the terms of *Genomic Politics*, Hope and Rejection. Scientists, public officials, and most of the public in most countries proclaimed that victory against Covid-19 was at hand, or at least would be once billions of people around the world received their vaccinations before the virus mutated too much. Leaders express confidence that vaccination would win; if we can conquer smallpox and polio, we can conquer Covid-19, and in months rather than decades, as in the earlier cases. Many Americans rushed to their doctor's office (or at least a website) or sat in their car for hours in order to get what journalists insist on calling a jab.

But substantial portions of the populations of many countries—for example, up to two-fifths of Americans and half of Japanese and the French according to some polls—ran in the other direction. They reject the vaccine as unproven and potentially dangerous, and in a few cases even take direct action to prevent others from getting their jab. Signs at protests proclaim that "vaccines can cause injury and death," or "vaccines increase coronavirus risk by 36 percent."

Reasons for hope about or rejection of a new technology range from too much to too little knowledge or faith, and from bitter or rewarding experience to acceptance of a trusted leader's word. Differing levels of hope or rejection can invigorate social movements demanding change or reinforce political withdrawal and alienation; they can be a literal matter of life or death, as with the Covid-19 vaccines.

In *Genomic Politics*, Hope and Rejection have precise meanings; they refer to residence in the two quadrants sharing the view that genetic influence is minimal but disagreeing on whether "techno-scientific curiosity," in the words of Athanasiou and Darnovsky in the epigraph to this chapter, offers grounds for running toward or away from something like a new vaccine.

Hope

Medical and Scientific Research

Hopeful explanations of societal phenomena that play down the importance of genetic influence take several forms. One approach is to reject the partition of the basic framework's vertical dimension into two categories, "genetic influence" and "not genetic influence." Instead, one can turn its either/or logic into a both/and list. As a student asked after hearing me describe the basic framework's dimension of genetic influence, "What if God created or chose your genes—is that genetic influence or not?" Good question!

NIH's agenda is a classic manifestation of the both/and approach in the public arena. As we see in the flow chart shown in Figure 2.4, its 2002 *Strategic Research Plan* to tackle racial inequality in health and health care endorses "research to understand biological, socioeconomic, cultural, environmental, institutional, *and* behavioral factors affecting health disparities" (emphasis added). That list implies funding to study "predisposing genetic factors," but the phrase is surrounded by a plethora of nongenetic funding targets. NIH's 2009 *Strategic Plan* states more forthrightly that "genomic research is beginning to shed light" on why some groups "might disproportionately suffer from certain diseases and conditions"—thus edging into a position on the contentious question of whether races are in some ways genetically distinct.

It continues, however, by asserting that interactions among "complex, dynamic, and multidimensional" factors, always located in particular contexts, are what explain health disparities.[1]

Since "genome" is in its name, it is not surprising that a strategic plan of NIH's National Human Genome Research Institute (NHGRI) shines a brighter light on the importance of genomics research. It proudly points to scientific and health care advances in the field and specifies future research goals. Nevertheless, despite the plan's Enthusiasm, NHGRI is reticent on the sensitive issue of group-based health disparities. "Genomics research may still have a role in informing the understanding of population differences," but genomicists have been "appropriately cautioned" that "most documented causes of health disparities are not genetic." Sounding much like Nancy Krieger and Stephen Heine, whom I quoted in earlier chapters, NHGRI's director and his co-authors warn that genomics "must not be mistakenly used to divert attention and resources from the many non-genetic factors that contribute to health disparities, which would paradoxically exacerbate the problem."[2]

The both/and approach, implied by NIH's "interactions" and NHGRI's "role in informing," encourages the Hopeful to blur the boundary between the top and bottom halves of the framework's vertical dimension through attention to epigenetic effects. This is a second Hopeful strategy, based on interactions within the body rather than on additive lists of causes or research foci. NHGRI defines epigenetics as the process by which "chemical compounds and proteins . . . can attach to DNA and direct such actions as turning genes on or off." That is, epigenetic activity "change[s] the way cells use the DNA's instructions" over time in the body and perhaps even across generations.[3] Social scientists emphasize that epigenetic processes inside the body can be set off by factors outside it, including everything from access to prenatal health care to stress or air pollution.

As I note in Chapter 2, investigation of epigenetic influences on human traits and behaviors is only beginning. It is an arena of intense investigation; as of May 2021, "epigenetic" elicits 25 million hits on Google and 1.1 million results on the more disciplined Google Scholar. Even if only 10 percent are relevant to genomic politics, that is a lot.

The policy impact of this research is small so far, but the field reinforces my central theme in *Genomic Politics*: the social implications of epigenetic processes do not lend themselves to partisan position-taking. As the political philosopher Shea Robison observes, the science "contains elements of both dispositional and environmental influences at once. Thus, both conservative and liberal narratives can . . . be constructed simultaneously." Bioethicist Mark Rothstein and his colleagues offer the progressive narrative, pointing to "individual and societal responsibilities to prevent hazardous exposures, monitor health status and provide treatment. Epigenetics also serves to highlight the effects of inequality in living and working conditions and adds a multigenerational dimension to environmentally caused adverse health effects." But another bioethicist, Luca Chiapperino, fits epigenetic science into a more conservative frame of individual responsibility. He suggests that "we might have a moral obligation (where a choice can be made) to bring to birth the 'best' child possible." Epigenetic science reinforces that obligation "by shedding a light on the potential harmful consequences of individual unhealthy lifestyles . . . on future generations." The new science "thus compel[s] individuals to mitigate the risks and threats to the well-being of those we (will) care for over the whole life course."[4]

Robison exemplifies Hope; epigenetic science and its accompanying left-right muddle offer "potentially transformative possibilities for politics and policies which transcend the conventional ideological dichotomizations."[5] As a tiny straw in the wind, I note the strikingly cheerful, proactive tone of mass-market books about epigenetics—*Change Your Genes, Change Your Life* (by medical professor Kenneth Pelletier) or *You Are What Your Grandparents Ate: What You Need to Know About Nutrition, Experience, Epigenetics and the Origins of Chronic Disease* (by journalist Judith Finlayson). These books offer Hope and even transformation through individual or perhaps familial choice; whether a parallel message about collective responsibility for genome regulation gains a mass audience remains to be seen.

A third and final form of Hope sets aside the logics of interactive epigenetics and of both/and additive lists. It opts instead for the original depiction of the basic framework as an either/or dichotomy, and for the tougher-minded message that genetics does not influence most phenotypes—something else is the causal driver.

The political right, center, and left all offer versions of this argument, depending on how they conceive of the "something else." A conservative stance holds that traits or behaviors such as sexual orientation, weight, addiction, or a tendency toward violence are under an individual's control, even if particular contexts make it more or less difficult to exercise that control. The belief that one can overcome same-sex preference through discipline, prayer, and marriage to a person of the opposite sex is a particularly strenuous view.

Psychiatrist Charles Socarides, co-founder of the National Association for Research and Therapy of Homosexuality, reported treating about 1,000 homosexuals. "He claimed to have 'cured' more than a third through psychoanalysis that dealt with what he said was the cause: an overbearing mother and an absent father."[6] Conversion therapy now ranges from controversial to illegal. All mainstream medical associations repudiate "treatment" for same-sex preference; as of this writing, two-fifths of the states and many smaller jurisdictions ban conversion therapy for minors. In keeping with *Genomic Politics*'s theme of nonpartisanship, most of those states had liberal Democratic governors at the time of passage (California, Hawaii, New York, Oregon), but some did not (Illinois, Maine, New Jersey, Utah).

Setting aside such harsh versions of individual responsibility, Hopeful conservative explanations for behavior and prescriptions for changing it abound. Alcoholics Anonymous is perhaps the epitome; it aims to enable members to "stay sober and help other alcoholics achieve sobriety." The famous twelve steps include admitting one's helplessness with regard to alcohol, seeking aid from a higher spiritual power, working to make amends to others and remove character defects, and assisting other alcoholics. AA's more than 2 million members do not look to genetics, or even to context or circumstances, for explanations and solutions, and the association eschews visible policy or political involvement.

Liberal versions of Hope are much more likely to make policy demands. For example, while recognizing the extensive research linking multiple genes to up to a tenth of variation in body mass index, a blog post on the Harvard T. H. Chan School of Public Health's "Obesity Prevention Source" nonetheless reminds readers that "the contribution of genes to obesity risk is small, while the contribution of our toxic food and activity environment is huge.... That's why obesity prevention efforts must focus on changing our environment." RAND Corporation scientist Deborah Cohen is less polite: "The obesity epidemic has absolutely nothing to do with genetics. [The research is] a distraction and prevents us from addressing the problem in a way that will ultimately make a difference." Instead of carrying out more studies about the role of genetics in obesity, we should be "standardizing portion sizes" and shifting unhealthy food away from grocery stores' prominent displays. If commercial tactics that encourage overeating were "a bigger story, we might shift public opinion. There might be demand to change those things."[7]

Or perhaps the problem is cultural tropes about weight rather than a physical condition: if Americans stop assuming that BMI by itself is a meaningful indicator, we can solve the "problem" of over- or underweight for many people "without needing societal or medical intervention of any kind."[8] Alternatively, the Affordable Care Act, more commonly known as Obamacare, ignores grocery stores' shelving algorithms and cultural tropes in favor of formal policy mandates. It requires all individual, family, and small group health insurance plans to cover free obesity screening and nutrition counseling; as of 2019, twenty-three states require insurers to cover weight-loss surgery. In keeping with *Genomic Politics*'s theme of politics but not partisanship, these states range from conservative (Georgia, Iowa, North Dakota) to liberal (California, Massachusetts, New York).

The United Health Foundation provides an elegant visualization of Hope. A 2019 report explains health disparities between males and females as "not strictly based on biology but develop[ed] through gendered experiences over the lifecourse and . . . influenced by factors such as differences in access to resources and societal gender norms." Its model of health outcomes is intersecting, multifaceted—and leaves out genetics altogether (Figure 5.1).[9]

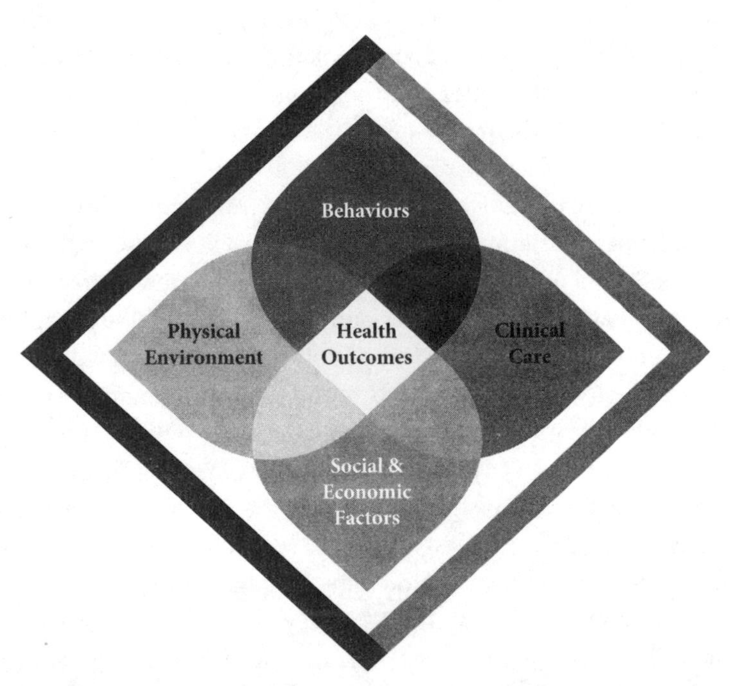

Figure 5.1 United Health Foundation's "framework for identifying . . . health drivers and outcomes"
Source: United Health Foundation 2019

Criminal Justice

Hopeful people or institutions in the criminal justice arena resemble those in the medical and scientific arena. They too sometimes but not always incorporate genetic influences into causal explanations, while always keeping the focus on nongenetic reasons for unlawful behavior.

Analyses by criminologists Kevin Beaver and Joseph Schwartz exemplify the first two Hopeful logics—both/and lists and epigenetic interactions. First, given that heritability estimates "can wax and wane depending on social context, maturational development, and other situational-specific factors," considering both genetic and environmental influences on lawless behavior will make criminal justice policies better targeted and more effective than considering only one type of influence. For example, determination of whether offenders carry genes "known to predict risk level" can be combined with factors such as neighborhood context, personal ties, or job status that are already included in risk assessment tools for deciding who will benefit most from behavioral therapy.

Epigenetics comes next. Beaver and Schwartz point out that the "logic of responsivity" (individual characteristics that enable some people to respond more effectively than others to a particular treatment program) can be expanded beyond the usual list of traits to include gene-environment interactions. Implicitly responding to critics of any attention to genetics in the criminal justice arena, they conclude that when combined with other information, "far from being oppressive, genotypic information . . . could actually open up more opportunities for treatment."[10] They are, in short, technology optimists.

Some Hopeful analysts concur. Psychologist Terrie Moffitt observes that "throughout the past 30 or 40 years most criminologists couldn't say the word 'genetics' without spitting. [But] today the most compelling modern theories of crime and violence weave social and biological themes together."[11] One person whom I interviewed points to the next step: "If we could understand pathways for extremes of aggression, that might be very helpful. As a society, we [would then] need to decide if it would be appropriate in any way to intervene."

That, however, is exactly the locus of the most intense conflict within this quadrant. Other Hopefuls perceive any analysis that includes genetic influence in explaining unlawful behavior—even if it is part of a list and especially if it suggests intervention—to fall somewhere between discriminatory and purely evil. In 1995, NIH sponsored a conference called "The Meaning and Significance of Research on Genetics and Criminal Behavior." The organizer, bioethicist David Wasserman, posited a broad arena of evidence-based consensus—"everyone agrees that any complete explanation of human behavior would have to include genetic and biological factors as well as a vast array of social and environmental factors, and the complex interactions among them"— but he was mistaken. The conference became something of a circus. Protesters, including some conference participants, denounced the event as eugenicist "racist pseudoscience," chanting, "Maryland conference, you can't hide—we know you're pushing genocide."[12]

Although most conferees ended up exchanging gracious comments about each other's careful scholarly work, the claim persists that genetic and environmental causes of crime cannot coexist. Perhaps ironically, Wasserman elegantly articulates what is not fully his view: including genetic inheritance in criminological research "will only deflect attention from more global and less tractable causes of crime. In a society pervaded by racism and inequality, whose most powerful institutions inflict widespread social harm, it is obtuse and diversionary to look for genetic idiosyncrasies that lead some of the least powerful and worst-off members of the society to commit acts of violence and predation."[13] Instead, Hope lies elsewhere—in directly confronting the social ills that impel people to commit crimes.

While the left points to global and intractable causes of crime, more conservative residents of the Hope quadrant hold that criminal behavior is ultimately under an individual's control. We have no settled knowledge about what share of Americans see criminal actions as a matter of choice rather than of context or genetics. In a 2015 survey, political scientists Stephen Schneider and his colleagues asked respondents whether genetics, environmental factors, or personal choice best explain each of 18 traits or behaviors. Respondents could choose only one explanation. On average, they attribute 55 percent of a person's criminal record to personal choice, 34 percent to environmental factors, and 11 percent to genetic factors. But in a survey twenty years earlier, in which respondents could choose multiple possible answers, only 16 percent of respondents agreed that "crime is the product of a person's free will" (although up to a third also accepted individualist explanations such as seeking the American dream, abstention from wholesome activities, indifference to family or morality, or "they figure they can get away with it").[14] Views on such a difficult and abstract question are most likely unsettled or nonexistent; much depends on a survey item's framing or wording, the respondent's broader context, or top-of-the-head considerations.

Biogeographical Ancestry

Residents of the Hope quadrant may have complex views about DNA ancestry testing. They acknowledge little to no genetic influence in individual phenotypes and reject any understanding of "race" as a biological phenomenon incorporating meaningful distinctions. However, as technology optimists, the Hopeful may find the practice of DNA ancestry testing to be an intriguing innovation, particularly when focused on individual genealogy.

This complex set of views can play out in different ways. Some endorse the both/and additive logic: geneticists writing for the charitable trust Sense About Science describe DNA ancestry testing as "credible" if used "to supplement independent, historical studies of genealogy . . . . Where we can make a connection between a tribal group and a particular section of DNA, for example, we could say that if you carry it today there is a possibility that some of your ancestors were in that group." That, in fact, is how most Native nations that engage with DNA ancestry testing treat it—as a starting point, as confirmation of family relationship, or as a piece of evidence in the more significant demonstration of blood quantum, such as direct Indian ancestry or cultural ties to the tribe. In a different geographic sphere, many of Alondra Nelson's African American self-fashioners fit the quadrant of Hope: they " 'align' DNA analysis with other evidence of their ancestry as well as with their genealogical aspirations, prior experience, or extant relationships," thereby creating an "interpretive arc of ethnic lineage."[15]

Other ancestry test-takers simply reject any implied genetic influence when results of a DNA ancestry test conflict with more meaningful personal, cultural, historical, or political affiliations. Despite taking a DNA test, for example, one participant in Roth and Ivemark's study chose a race on the 2010 census "really from a sociopolitical standpoint." Having done conventional genealogical research, Pearl Duncan told the *New York Times* editor that she "would not be stopped from claiming my heritage" by a mere genomic analysis.[16]

The clearest manifestation of Hope in this arena is the goal of turning DNA ancestry testing into a lever for destroying the whole idea of fixed racial categories. In this view, the genuine phenomenon of ancestral heritage can be used to put an end to the false concept of race. Geneticist Charles Rotimi is carefully optimistic: "The potential exists to describe simultaneously our similarities and differences without reaffirming old prejudices."[17] Talking with her students after they take ancestry tests, communications scholar Anita Foeman is also cautiously optimistic: "The conversation is complicated and jagged, and it mercifully undermines neat, simplistic stories." In one sample of eighty-five testers, Foeman and her colleagues found that 53 percent of self-identified non-African respondents (especially women and young adults) express "positive affect" after a discovery of unexpected African ancestry.[18]

Hope is the most politically complicated quadrant in the basic framework. "Not genetically influenced" can mean many things with many ideological connotations. Hopeful conservatives reach for individual choice or a deity's plan; liberals point to environmental or social impacts. In a 2015 survey conducted by the Robert Wood Johnson Foundation and Harvard's T. H. Chan School of Public Health, a higher proportion of African Americans than Whites see lack of good medical care (56 to 41 percent) and poverty (47 to 31 percent) as causes of bad health—but Blacks are also more likely than Whites to invoke God's will (47 to 29 percent).[19] Then there are nonideological, or at least unclassifiable, explanations for phenotypes that invoke fate or luck.

Even Hopeful technology optimism is politically complex. Both/and lists imply liberal activism: adding another causal factor to those already in the mix reveals a new insight to be pursued or a new lever for action. Interactive analyses might imply liberalism, or at least activism: understanding how another causal factor changes the behavior of those already in the mix enables more effective or targeted leverage. But zero-sum, either/or, models are ideologically ambiguous. Rejection of genetic explanations is sometimes associated with greater acceptance of people with conventionally defined undesirable characteristics ("people addicted to opioids are victims of their environment, or of the pharmaceutical industry") and sometimes associated with more blame ("their willpower is weak, and look what addiction is doing to their children").[20] But all of the Hopeful share the belief, by definition, that *something* can ameliorate bad conditions or make good ones even better. Rejecters do not.

Rejection

> "Cheap and easy to use" is a perfect recipe for overuse, particularly when it comes to sensitive technologies in the hands of the government. We have seen this dynamic with cell phone location tracking, face recognition, and communications eavesdropping: intrusive information collection that was once subject to "natural limits" because it was expensive gets deployed far too broadly when new technology makes it cheap. . . . Our DNA is . . . the "nuclear weapon" of identifying technologies.
>
> —ACLU attorneys Vera Edelman and Jay Stanley, 2019

> We understand life by studying life, not by picturing a vague mechanism capable of directing its course.
>
> —Science writer Stephen Talbott, 2017

Rejection, the basic framework's fourth quadrant, holds that genomic science cannot help to solve problems or even enhance good situations both because genetic influence is weak or nonexistent and because technology is fallacious or destructive. The main outcome of efforts to harness genomics for social use will be wasted resources, unacceptable risk, or normatively repugnant actions.

Medicine and Scientific Research

In repeated survey items from 2002 through 2020, consistently only a tenth of respondents to Gallup polls found human cloning to be morally acceptable. Even Enthusiasts hesitate; Skeptics see successful cloning as technically infeasible; the Hopeful deem any work on cloning to be a waste of precious resources. But Rejecters abhor it; persistent animal experiments will create, as biologist Tanja Dominko summarizes, "a gallery of horrors."[21]

The idea of human cloning plays little role in *Genomic Politics*. Its widespread condemnation means that this technology does not help to explain the sort of disputes depicted in Chapter 1 or to distinguish among the basic framework's quadrants. But rejection of other genomics technologies does illuminate the fourth quadrant's political import. Physician Michael Joyner and his colleagues ask, for example, "What historical precedent is there that adoption of vast new oversophisticated technology reduces costs?" NIH support of "underperforming initiatives" should be rerouted to "a potentially more fruitful research area," such as "work of clear public health importance and imaginative . . . biomedical research." Only then can taxpayers' support for science research be deployed "to benefit the many, not the few." Other Rejecters similarly point to wasted resources and unfilled promises of gene therapy for Mendelian (single-gene) diseases, rejuvenation through stem cells, cancer remission through precision therapies, or gene editing. Even pharmacogenomics (the study of how an individual's genetic makeup affects his or her response to therapeutic drugs), says psychiatric nurse Barbara Limandri, "soon became a disappointment to clinicians."[22]

Even setting aside cloning, Rejecters argue that use of genomics technology may go beyond ineffectiveness and waste to actual harm. Across twenty-eight experimental studies of public opinion, psychologist Erlend Kvaale and his colleagues find that telling respondents that there is evidence for genetic explanations of mental illness "reduce[s] blame . . . but induce[s] pessimism about recovery" and "increase[s] endorsement of the stereotype that people with psychological problems are dangerous." Thus invoking genetics may make people sympathize more with those suffering from mental illness, but it may also "set the stage for self-fulfilling prophecies that could hamper recovery from psychological problems."[23]

Troy Duster is a sharper critic. People exposed to the herbicide atrazine over a long period of time have an elevated chance of developing some cancers. For example, exposed factory workers at one production facility have over eight times more prostate cancer than workers not so exposed, "and it just so happens that this plant is located in a community that is more than 80 percent African American." Despite this result, some scientists study "population differences" in rates of prostate cancer "through a unidimensional genetic prism, but with no understanding of the larger context in which humans are exposed to environmental insults." They should also, or instead, analyze "the systemic pattern of African Americans living close to toxic waste dumps across the whole country." Why, demands Duster, "should the decontextualized genetic inquiry of differing prostate cancer rates between Americans of European and recent African descent be characterized as apolitical 'science'—while the rate of their increased risk to exposure to atrazine is seen as 'political' science?"[24]

Duster's challenge to genetics rests on a perception of a negative-sum game between technologies associated with different types of explanation. Since physical or mental health issues will not be satisfactorily addressed if they are attributed to genetic inheritance rather than to discriminatory treatment, inquiries focused on genetics not only overlook but even worsen societal inequities. A vicious cycle gets established: problems caused by social hierarchy are interpreted as evidence of a race, sex, or social class's genetic weakness or distinctiveness. Given underlying bias, that interpretation reinforces the social hierarchy, whose environmental consequences exacerbate the bodily problems, which strengthens the genetic attribution, which reinforces . . . and so on. Anthropologist Michael Montoya explores how the search for causes of diabetes in the American Southwest manifests this vicious cycle: Medical researchers interpret the "social conditions" of Mexicanas/os as "*biogenetic* conditions." Thus "the conditions of [an individual's] life, those events that shaped their physiological condition and the social and political history that attached itself to their lives in the form of an ethnic identity, are conscripted into the service of a biogenetic disease research enterprise." Deployment of genomics technology, intended to help find a solution to the alarming rate of diabetes among Mexicanos/as, ends up worsening the "structures of inequality" and the "racialized . . . meaning systems" within which the research subjects live.[25]

The Criminal Justice System

Rejecters anticipate only harm from technologies that enable genetic profiles to be incorporated into criminal justice policies and practices. In an essay only a few pages long, criminologist Colleen Berryessa and her co-authors catalogue a daunting array of concerns about research on heritability in "antisocial behavior." These include, among others:

- The belief that "males are more likely to exhibit criminal behavior, especially rape, due to . . . an evolutionary propensity to violence"
- "Medicalizing norm-defiance," which risks "reducing the role of personal responsibility and opening the door to pharmacological approaches"
- "Undermin[ing] the foundations of criminal responsibility and punishment"

- Alternatively, "caus[ing] judges or juries to view the defendant as so dangerous as to require indefinite incarceration"
- Violating genetic privacy
- Stereotyping traits or behaviors purportedly associated with a particular genetic profile
- Labeling and tracking children before any behavior warrants monitoring
- "Justify[ing] the increased surveillance of certain groups" or individuals with certain genetic profiles, as identified through forensic databases
- Encouraging judges, juries, attorneys, and defendants themselves to defer too much to experts

Even this list is "only the 'tip of the iceberg' arising from behavioral genetic research."[26]

Berryessa and her colleagues' list of risks in bringing genetics into criminal justice is especially powerful because it mixes a libertarian concern about insufficient attention to personal responsibility with the more typical liberal concern about impugning individuals or denigrating groups. They worry that a "biological culture" emphasizes "therapy as much as control"—which is exactly what some of the Hopeful are aiming for. But to Berryessa and coauthors, predicating crime control strategies on "a rationale of public health" may be just as dangerous to freedom as is the old eugenics because it is more insidious, equally targeted on powerless entities, and not obviously racist.

Social theorist Nikolas Rose concurs. The use of technology that purportedly finds a biological predisposition to criminality can mutate into "genetic discrimination . . . and the use of quasi-consensual `treatment' for supposed biological tendencies, as conditions for a non-custodial sentence." Searches for a genetic predisposition to aggression may also lead to preventive detention and "other pre-emptive interventions for 'the protection of the public.' "

A person I interviewed put it most succinctly: when asked about eugenics, he answered, "We're past that. But there's always a possibility of misuse of technology."[27]

Biogeographical Ancestry

In contrast to some in the Hope quadrant, Rejecters have a simple view of DNA ancestry testing: the tests mislead customers because there is little to no genetic influence on "race." Although direct-to-consumer testing companies never claim to be identifying a person's race, that inference is easily drawn. At a collective level, the temptation of "defining identity in genetic terms," as law professor Dorothy Roberts writes, "creates a biological essentialism that is antithetical to the shared political values that should form the basis for unity. . . [and that are] needed to fight racial oppression." At an individual level, Duster and legal scholar Pilar Ossorio warn that ancestry test results "at odds with a person's self-identified or attributed race" can cause real pain. Science writer Amy Harmon describes the reaction of a former Black Power activist who discovered that she had no African ancestry on her father's side: "What does this mean; who am I, then? For me to have a whole half of my identity to come back and say, 'Sorry, no African here,' it doesn't even matter what the other half says. It just negates it all. . . . It doesn't fit, it doesn't feel right."[28] On the television show *Lopez Tonight*, rap artist Snoop Dogg jokes about discovering that he is too White, at least compared with his competitor, basketball player Charles Barkley. But his body language hints at genuine discomfort (Figure 5.2). George Lopez's needling gifts to him of "things White people like," such as a size-medium T-shirt and a DVD of *Glee*, perhaps cut a little deep.

Ancestry testing may be even more harmful when it does confirm a person's supposed race. Shlomo Sand, an Israeli historian, reacts uncomfortably to stories of people using DNA ancestry tests to confirm their ethnicity: "I feel like the Nazis lost the war, but they won the victory of an ideology of essentialist identity through the blood."[29]

Evolutionary biologist David Reich posits that "the reality of [group] differences" is "potentially liberating," but critics are unpersuaded. Sixty-seven experts (mostly social scientists) published a blog post reiterating that even unintentionally making "race" seem like a real biological phenomenon, which they perceive Reich's work to do, causes damage. The insights of population genetics are too readily "misunderstood and misinterpreted." They "urge scientists to speak out when science is used inappropriately to make claims about human differences" because "science and the categories it constructs do not operate in a political vacuum." As I write, the most dangerous politicization of science is the Chinese government's use of DNA testing to determine if someone has Uighur ancestry. Uighurs are a Muslim minority group in China, some of whose members are accused of terrorism or terrorist tendencies; using DNA ancestry testing to determine who is "really" a Uighur all too often results in surveillance, placement in detention centers, and forced "re-education."[30]

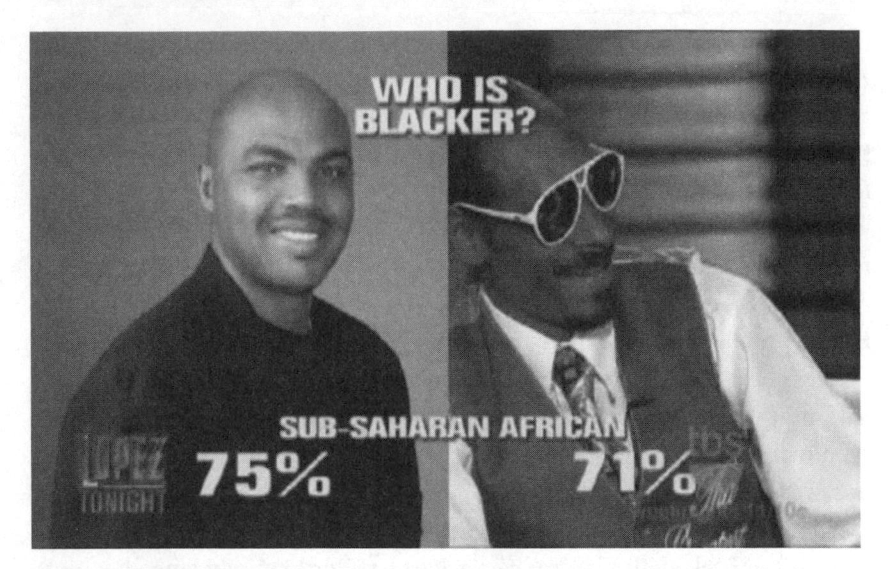

Figure 5.2 "Oh no. Oh no. . . . We got to have a re-vote. This ain't right." —
Snoop Dogg, in response to results of DNA ancestry test
Source: Lopez 2015

At the core of these specific concerns is the conviction that genetic analyses can subvert human autonomy and dignity. Biologist Austin Hughes warns against companies such as 23andMe that report on a range of phenotypes, from heritage to one's propensity toward earwax or breast cancer. Deployment of a technology to hint that "our entire lives are determined by our genetic inheritance" is "a crude but modern scientific form of fatalism that will not enhance but degrade our self-understanding. Fatalistic doctrines can only undermine individual initiative, making people more apathetic, more easily dominated by tyrants or manipulated by technocrats." Hughes urges people to remember that even if we are subject to genetic influence, we are "free beings, responsible for ourselves, our fates ultimately unwritten." Stephen Talbott issues the same warning: we must keep in mind that genetic processes "remain 'on track'" only because the person is an active agent, "coordinating, redirecting, revising, and sustaining the overall form and coherence of countless interactions," including those with the supposed "explanatory program." We can keep genomics in its place only if we pledge to "understand life by studying life, not by picturing a vague mechanism capable of directing its course."[31]

Rejecters are in a difficult, often defensive position in a country largely populated by Tocqueville's optimistic shipbuilders. Rejection can echo the views of the historical figures whom economist Albert Hirschman describes as reactionaries. Hirschman's wonderful book *The Rhetoric of Reaction* examines political actors from the late eighteenth through the twentieth centuries—Edmund Burke, Vilfredo Pareto, Charles Murray—to show "types of argument" that have shaped Western political thought and action. He focuses on three forms of reactionary (i.e., conservative) rhetoric: perversity ("the attempt to push society in a certain direction will result in its moving . . . in the opposite direction"), futility ("the attempt to change is abortive"), and jeopardy ("a new reform . . . would mortally endanger an older, highly prized one").[32] Not all in the Rejection quadrant are reactionaries, or even conservatives. Nonetheless, a descendant of Hirschman's subjects is inclined to argue that the ambitions of genomic Enthusiasts are futile, that efforts to use genomic science in society (and perhaps other social reforms) have perverse effects, and that too much reliance on any technology jeopardizes cherished values such as human autonomy and unity. Rejecters live in a threatening world.

In putting the basic framework of *Genomic Politics* through its paces, I have done the narrative equivalent of what social scientists call selecting on the dependent variable. That is, I created a framework, then explored arguments chosen to illustrate the meaning and import of its dimensions and quadrants as I had previously defined them. My goal was to bring life and relevance into these possibly abstract ideas. But what if the components, and the framework itself, do not accord well with how people actually think and feel with regard to genomic science? If that is the case, the framework may be elegant, but is of limited use.

I hope to have shown so far that the basic framework has an internal coherence that makes sense of disputes over issues such as BiDil, ancestry tests, forensic DNA databases, or prenatal gene editing—but I have not yet answered the question "So what?" Do people engage with genomics in their work and lives in ways that fit comfortably and productively within the framework and its quadrants? To address that question, I turn in the next three chapters to the independent variables—people and their views.

6

Locating Experts in the Basic Framework

> [In the genomics arena,] political undertones do not follow standard lines.
>
> —Interview subject

> Both this issue and I don't fit into standard patterns.
>
> —Interview subject

The 2007 conference that lured me into studying genomic politics for more than a decade addressed BiDil and its possible successors. I was observing from the back of the room, surrounded by experts whose work I had just begun to absorb. Successive scholars rose to explain the poor statistical techniques that had tricked the FDA into agreeing with profit-minded developers that BiDil would benefit Blacks with congestive heart failure, or to deplore the social impact of racially inflected medicine that was based on unproven and unlikely genetic differences. I found their arguments persuasive—until a member of the Association of Black Cardiologists stood and said in effect (and much more politely), "Sit down, you white anthropologists!" He went on to describe his Black patients dying of congestive heart failure; if this drug would help them, he was not concerned about statistical techniques, corporations' profit motive, or the future of American racial dynamics. The Black woman beside me murmured something to the effect of "Yes, my grandfather died of heart failure, and I was a doctor but couldn't help him."

The mostly White academics and mostly Black medical professionals stared at each other, and the meeting eventually dissolved into circumspect expressions of appreciation and thanks. These were, after all, highly educated, liberal experts with the shared goal of identifying the best medical treatment possible to help eliminate racial health disparities. But the gulf between Black and White, between scholar and practitioner, and between those willing to risk the future to gain benefits in the present and those willing to abjure possible present benefits out of concern about future losses was wide. (Those categories overlapped but were far from identical.) Without sufficient knowledge or insight, I was caught in the middle, and caught in fascination.

In my continuing effort to make sense of and evaluate that debate, *Genomic Politics* now moves from exploring *what* the basic framework entails and *how* its components fit together to *who* resides in the four quadrants and *why* they hold a given view. This chapter starts that exploration by examining three sets of opinions held by genomics experts. Those sets consist of a database of almost 2,000 coded articles by social science and legal experts on genomics, responses from over 100 social scientists to open-ended online surveys sent in 2013 and 2018, and interviews with about sixty genomics experts in various disciplines and professional positions. I have already quoted many; this chapter provides context for and a richer portrayal of their views.

Examination of "who" and "why" begins with experts for three reasons. First, they create the science and technology that this book is about. They know more about genomics than the rest of us and are more invested in its virtues and defects (which is not to say that their insights and perspectives are more important in some normative sense than those of the general public). Second, experts shape the discussion among the rest of us about genomics science and technology, whether deliberately through articles and tweets or inadvertently by bringing their science into hospitals, courtrooms, and stock markets. Third, experts are positioned on the front lines of governance by writing laws and regulations, persuading policymakers and judges, and working in advocacy groups or civic organizations. Experts cannot control the public's views any more than they can persuade one another, but they set the terms for what the rest of us know, experience, and debate.

Actually, *Do* Experts Matter?

Despite my paean to the importance of experts, I begin with a bit of brush-clearing in order to move past the newly conventional wisdom of "the death of expertise." As a presidential candidate, Donald Trump arguably spoke for many when he told a Wisconsin audience in 2016 that "with all the talking we all do—all of these experts, 'Oh we need an expert—' The experts are terrible. Look at the mess we're in with all these experts that we have. Look at the mess." Trump was speaking of foreign policy, but his presidential administration regularly rejected a wide range of conventionally defined expertise.[1]

Trump was, of course, expressing the old trope of what leftist scholars describe as anti-intellectualism. Many Americans share his scorn for the "university types [who] shower before work, not afterwards."[2] A third agree that "people with expertise on a subject area are not any better at making good decisions than other people."[3] Reflecting on American politics during the SARS-CoV-2 pandemic, political philosopher Yuval Levin observes that "our polarized political culture has reflexively approached the pandemic as just another culture-war drama. . . . Worse yet, the very communities of experts we rely on to assess provisional knowledge and provide us their best judgment have failed the test of professional restraint in key moments, giving in to political tribalism themselves."[4]

Even in this skeptical era, however, public support for scientific expertise is strong. In three recent rounds of the GSS, 95 percent of respondents agreed that "scientists are helping to solve challenging problems." Strong agreement rose somewhat from 2012 to 2018, as did agreement that "scientific researchers are dedicated people who work for the good of humanity." At least in the less polarized year of 2005, over three-fifths of American voters wanted Supreme Court justices to have expertise and ability, compared with only a tenth who wanted political agreement. And of particular interest to this book, nursing, psychology, and biology—all not only requiring expertise but also within the ambit of genomic science—were the second, third, and fourth most popular college majors in 2019. They ranked far behind business and management, but ahead of engineering or finance and accounting.[5]

Underlying trust in expertise persists in the Covid-19 pandemic. As of May 2020, two-thirds of Americans held a "mostly positive" view of medical research scientists, while only 6 percent were mostly negative. Four-fifths agreed that thinking about the coronavirus outbreak makes them "see developments in science as more important for society" (compared with 4 percent who said "less important"). The pandemic's associated culture wars, of course, shade this overall pattern of agreement: confidence that medical scientists or scientists act in the best interests of the public rose substantially among Democrats from 2019 to 2020 but remained lower and flat among Republicans. By May 2020, twice as many Democrats and Democratic leaners (52 percent) as Republicans and Republican leaners (27 percent) expressed a great deal of confidence in scientists; the gap was similar for "medical scientists."[6] Expertise is certainly contested, but the view that "experts are terrible" does not quite rule the day.

The Evidence from Experts: Coded Articles

Accepting, then, that experts matter and expertise is politically salient, I start with the most global framing of genomics experts' views. Over several years, a small army of research assistants coded highly cited articles in scholarly journals on genomic science and its uses that were written by legal scholars and social scientists in many disciplines. This analysis starts from the assumption that choice of topic, nature of analyses, and interpretation of evidence may reveal important elements of an author's worldview and perhaps of his or her ideological commitments. That assumption seems most persuasive for humanists, legal scholars, and humanities-oriented social scientists, in whose disciplines the link between scholarship and argument about values and goals is often intentional and explicit. The assumption is less compelling for scientifically oriented social scientists, in whose disciplines there is often a strong norm of value-neutral scholarship.[7]

The coding project yielded two databases; the Appendix to Chapter 6 describes them and their creation in more detail. The first involves 115 social science articles appearing in the earliest years of genomics' move out of the laboratory into the broader society. The goal here is to characterize the most influential articles, as determined by peers' citations, that were shaping scholars' initial reactions to this new phenomenon.

The second database focuses on the most highly cited articles within each of thirteen disciplines, in order to ensure that we obtained the voice of small disciplines (cultural studies) as well as huge ones (psychology). It involves almost 2,000 articles written across almost two decades. The goal of this larger project is to see if residence in the four quadrants of *Genomic Politics*'s basic framework is associated with the disciplinary context in which a scholar studies genomics—and if so, to infer something about the kind of expertise that encourages people to be enthusiastic, skeptical, hopeful, or rejecting. After all, since scholars are teachers, and also blog and give interviews to journalists, knowing how, say, economists or anthropologists view societal uses of genomic science will enable insights into what the next generation's adults are learning about this arena.

Most-Cited Articles

Stances toward genomics in the 115 most-cited social science articles written from 2002 through 2011 are easily summarized. Almost 30 percent of those articles are in the field of education, presumably reflecting disciplinary norms about publication and citation as well as scholarly interest in genetic components of intelligence or achievement. Another 20 percent of the articles are in ethics, suggesting the impact of funding from the federal Ethical, Legal, and Social Implications (ELSI) Research Program as well as bioethicists' interest in the moral issues raised by genomics technology.[8] Sociology, law, and history or philosophy of science each account for about 7 percent, with a few remaining articles (in descending order) in each of political science, psychology, economics, anthropology, and cultural studies.

Almost two-thirds of the 115 articles are coded as "neutral, mixed, or no occasion for valence" with regard to technology optimism. That result reflects the caution of our coding decisions (and coders), the newness of social science engagement with genomics, and the norms of value-free scholarship in most of these disciplines. Nonetheless, a pattern emerges: twice as many of the most-cited articles with a codable valence are optimistic as are pessimistic. Technology optimism is especially strong in education (7 optimistic to 0 pessimistic) and political science (4 to 0). In sum, setting aside purportedly value-free research, social scientists' introduction to genomics mostly followed the Enthusiastic cue of President Clinton's description of the Human Genome Project as the most wondrous map ever produced.

Placing Disciplines in the Basic Framework

Not surprisingly, the second, much more extensive article-coding project yields richer and finer-grained results. As explained in the Appendix to Chapter 6, each article is coded positive, neutral, or negative in two ways, one focusing on genetic influence and the other on technology optimism. The coding results for each discipline, separately for each of these two dimensions of the basic framework, are distilled in Table 6.1.

Note first the disparate number of articles across the thirteen fields. Disciplinary expectations for publication of many short or a few long articles explain some of the difference, as does the way disciplines are categorized in the *Social Science Citation Index*, from which we obtained this material. But the range from 19 to 370 articles in a given discipline also signals how much or little genomic science matters in the topics addressed by that arena's experts.

Table 6.1 Disciplinary differences in perceptions of genetic influence and technology optimism, 2002–2018

A. Genetic Influence					*B. Technology Optimism*			
1. Discipline	*N for discipline in column 1*	*"Value in my discipline or my research"*				*"Value to society"*		
		2. Positive	*3. Negative*	*4. Neutral, mixed, NA*		*5. Positive*	*6. Negative*	*7. Neutral, mixed, NA*
Criminology	71	63%	0	37%	Science & technology studies (STS)	48%	7%	45%
Biological anthropology	370	61	1	38	Law	46	20	44
Economics	38	61	0	39	Ethics	34	13	53
Psychology	271	57	4	39	History & philosophy of science	23	8	69
Political science	75	51	7	42	Sociology	21	10	69
Cultural anthropology	246	39	5	56	Criminology	20	3	77
Sociology	72	38	10	42	Psychology	18	3	79
History & philosophy of science	257	28	7	65	Political science	16	4	80
Ethics	359	23	10	67	Economics	13	0	87
Science & technology studies (STS)	27	22	0	78	Biological anthropology	13	1	86
Cultural studies	19	21	0	79	Cultural anthropology	12	8	80
Ethnic & racial studies	70	19	13	68	Cultural studies	11	11	78
Law	54	19	17	64	Ethnic & racial studies	7	31	62

(in percentages, from highest to lowest percent positive in each panel)

Total = 1,929 articles.

Reflecting what was seen in the coding of 115 articles published between 2002 and 2011—and again reflecting our and our coders' interpretive caution—many articles in the larger database are neutral or mixed on genetic influence or technology optimism (columns 4 and 7). The proportions vary, however. Fully four-fifths of the authors in cultural studies and in science and technology studies (STS) express no view on the value of genetics to their own discipline, compared with the smaller proportions of two-fifths in criminology or biological anthropology.

Authors in seven disciplines are even more reticent about the value of genomics to society—with an increase, for example, from 42 to 80 percent in the number of political science articles with no clear stance about genomics when the focus shifts from the discipline to the wider arena. Scholars in law and in STS, however, reverse the pattern; they are more willing to express views about genomics in society than in their own arena of study.

Even with these variations in the number of articles and the proportion expressing a valence, disciplinary differences emerge. On one hand, columns 2–4 show that just a fifth of articles in ethnic and racial studies, cultural studies, law, STS, and ethics describe genomics as valuable for exploring their subject matter; a substantial share in three of those five disciplines portray genomics as unhelpful or perhaps harmful to their research arena. On the other hand, three-fifths of the articles in criminology, biological anthropology, economics, and psychology depict genomics as important and useful in their research arena; almost none deem it irrelevant or harmful. Articles in the remaining four disciplines, in the middle of panel A, are all more positive than negative about genomic influence.

Columns 5–7 of Table 6.1 show disciplinary variation in optimism about using genomic science in society. Only a tenth of the articles in five disciplines—economics, biological or cultural anthropology, cultural studies, and ethnic and racial studies—depict genomics as beneficial; fully four times as many articles by race scholars portray harm as portray benefit. At the other end of the spectrum, close to half of the articles in STS and law express optimism. Again, the middle group of disciplines are more positive than negative, and in only two (as well as in law) do any scholars argue that genomics is harmful or unimportant in society.

Putting the two sets of codings together enables me to fill in the basic framework of *Genomic Politics*. Figure 6.1 shows the pattern.[9]

	Technology optimism	*Technology pessimism*
Genetic influence	Enthusiasm Biological Anthropology Psychology Criminology Economics Political Science STS?	Skepticism Cultural Anthropology Ethics? Sociology? History and Philosophy of Science?
Not genetic influence	Hope	Rejection Ethnic & Racial Studies Cultural Studies Law?

Figure 6.1 Locating social science disciplines in the basic framework

The logic of the patterning is complex. Designation as an Enthusiastic discipline, for example, is based both on authors' positive valence about genetic influence and on the predominance of positive over negative articles among those expressing any valence about genomics' societal use. More precisely, in the first five disciplines of the Enthusiasm quadrant, authors more frequently point to the value of genetic influence than to the value of genomics in society—but when they do venture into that sensitive terrain, they endorse its use.

Conversely, relative to the other disciplines, ethnic and racial studies, cultural studies, and perhaps law belong in the Rejection quadrant. These are the fields whose practitioners are most likely to depict genomics as unimportant in or harmful to society. They are, not coincidentally, the disciplines of many of the people whom I quote frequently in earlier discussions of skepticism or rejection, such as Troy Duster, Dorothy Roberts, Jonathan Kahn, Alondra Nelson.

The remaining five disciplines are harder to place. STS scholars, for example, belong among Enthusiasts if one focuses on the relative proportion of articles with positive and negative valences about genetic influence (22 percent positive to 0 percent negative) and the high percentage of articles expressing positive valence about genomics' societal uses (48 percent). But some STS articles are negative on the second valence, and STS is far down in the rows of panel A of Table 6.1, so that discipline's profile is only marginally Enthusiastic.

Cultural anthropology is more clearly Skeptical than are the other disciplines. Its articles are in the middle of the range regarding the importance of genetic influence, but close to the bottom in the share expressing optimism about genomics' societal uses. Duana Fullwiley is a fine example of anthropologists' concerns about the practice of genomic science. Ethics, sociology, and history or philosophy of science are too internally split to categorize safely, though each discipline shows strains of Skepticism. That is, although most articles in each field are Enthusiastic, a relatively high share see genomics as harmful to society.

Why Are Disciplines Scattered Across Quadrants?

The patterns of coded experts' views about genomics revealed by Table 6.1 and Figure 6.1 show, reassuringly, that the basic framework does sensibly organize different types of experts' varying perspectives on genomics. The patterns also suggest why disciplines fall into one or another quadrant. The most methodologically individualist disciplines—that is, those most clearly organized around individual behaviors—are the most open to the explanatory power of genetic inheritance and the most sanguine about genomics' societal role. Psychological traits or states, aggressive or unlawful actions, the pursuit of interests, partisan adherence, physical characteristics—these types of definable, measurable, and separable phenotypes are the most available for genetic analysis and, if genomicists are correct, are the most clearly linked to genotypes. It is also plausible that methodological individualists concur with the principle attributed to, among others, management consultant Peter Drucker: "If you can't measure it, you can't improve it." Another plausible, though vaguer, explanation is that scholars who measure, count, and compare people as they go through their lives are likely to feel most empathy and mental correspondence with scientists who measure, count, and compare DNA segments as they go through human bodies.

Conversely, the disciplines most oriented toward collective activity, institutions, practices, identities, culture, or norms evince the greatest skepticism about the value of genetics. Racial or ethnic identities, a community's self-image, rules and judicial proceedings, perceptions and beliefs—these types of interpenetrating, socially constructed, intangible, or symbolic features of humankind are the most difficult to link to genotypes. Collective entities often do not lend themselves to defining, measuring, and counting; scholars may in fact argue that efforts to impose that sort of logic on their arenas of study are simply category mistakes, unilluminating and wrongheaded.

That observation points to a second reason that disciplines vary in their placement in *Genomic Politics*'s basic framework—the ideological proclivities of scholars in each field. Overall and with some caveats, disciplines that are more than usually leftist or liberal are also disproportionately peopled by Rejecters or at least Skeptics. The most extensive recent survey of academics was conducted in 2006 with a stratified sample yielding 1,417 full-time faculty respondents in most disciplines and types of institution of higher education. Fifty-eight percent of social scientists and 52 percent of humanists, compared with 45 percent of biological and physical scientists, identified as liberal. Almost no biologists but 38 percent of sociologists and 15 percent of political scientists and historians described themselves as radicals or on the far left. In contrast, scholars of elementary education, economics, and criminology report the highest proportion of Republicans—and in our data they are among the social scientists most likely to see genetics as valuable in their discipline and in society.[10]

One cannot straightforwardly infer individual authors' ideology from discipline-wide patterns of belief. And the match between a discipline's average ideology and its average level of enthusiasm for genetics in our coded analysis is imperfect. Nonetheless, in the aggregate, social science and humanities disciplines that are more liberal, less Republican, or more active in left-wing causes tend also to have proportionally more practitioners who ignore or criticize genomic science compared with disciplines whose members are more conservative, Republican, or uninvolved in left-wing causes. In short, although we do not see partisanship in laws, judicial rulings, NIH budgets, or public officials' positions, we can begin to discern among academics an ideological or normative distinction with regard to the quadrants of the basic framework.

Placing Articles in the Basic Framework

The 2,000 coded articles permit one more analysis, this time of variation among articles (rather than disciplines) across the two dimensions of the framework. That is, I calculated the proportion of articles in each discipline with either (1) two positive valences for genomics' value, both in the discipline and in society (Enthusiasm), (2) two negative valences, both in the discipline and in society (Rejection); (3) a positive valence for disciplinary and a negative valence for societal value (Skepticism); and (4) a negative valence for disciplinary and positive for societal value.[11] Table 6.2 shows the results.

In most disciplines, the number of articles with substantive codes for both valences is very small, since many fewer articles venture a view on societal value than on disciplinary value. So one must read Table 6.2 with caution. Nonetheless, the results mostly reinforce the patterns found in the

Table 6.2 Combined valences in disciplines, regarding genetic influence and technology optimism, 2002–2018 (in percentage among articles with two coded valences)

	Positive + positive: Enthusiasm	*Negative + negative: Rejection*	*Positive + negative: Skepticism*	*Negative + positive*	*N with 2 valences*
Economics	100%	—	—	—	4
Biological anthropology	94	3	—	—	51
Criminology	93	—	7	—	14
Political science	92	—	8	—	12
Psychology	88	12	—	—	49
Science & technology studies	80	—	20	—	5
History & philosophy of science	76	22	2	—	49
Sociology	69	23	—	8	13
Ethics	68	27	4	1	90
Cultural anthropology	65	26	6	3	34
Law	50	33	—	17	18
Ethnic & racial studies	24	47	29	—	17

Note: No articles in cultural studies fit the criteria of having a "positive" or "negative" rating for both coded valences.

discipline-wide analysis shown in Figure 6.1. In five disciplines—economics, biological anthropology, criminology, political science, and psychology (with a subtheme of Skepticism)—almost every article expressing two valences conveys Enthusiasm for genomic science and its uses. Those disciplines are almost the same as those in the discipline-wide Enthusiastic quadrant of Figure 6.1. The results reinforce the slightly shaky placement in that figure of political science among the Enthusiasts, but not the more shaky placement of STS.

Contrastingly, in five other disciplines—STS, history and philosophy of science, sociology, ethics, and cultural anthropology—up to a third of the articles expressing two valences do not convey Enthusiasm, in most cases because they are Rejecters, denying both the value of genomics to their discipline and its societal value. Enthusiasts still predominate in these disciplines, but not without robust challenge. Articles in law and especially racial and ethnic studies retain their position as the most likely to demonstrate Rejection regarding both genomics and genetic technology.

The article-level analysis both strengthens the discipline-level analysis discussed earlier and offers a new insight into why people hold a given position in *Genomic Politics*'s basic framework. Arguments in most articles are internally consistent; either they portray genomics as influential (valence 1) *and* valuable (valence 2), or they portray it as unimportant *and* unhelpful or harmful to society. Across the twelve disciplines with articles coded on both dimensions, on average 75 percent of articles were Enthusiastic and 16 percent Rejecting. Just 6 percent expressed Skepticism—that is, inconsistency across the two valences (and a third of those were in only one discipline). The pull toward consistency—celebrating what seems important (or seeing as important what one celebrates) or rejecting what seems unimportant (or deeming unimportant what one rejects)—may be a deep, if often unrecognized, drive.

The Long Tail

The coding enables a final conclusion: "The long tail of genomics is eugenics," as one person I interviewed put it. Scholars of ethnic and racial studies have on average the strongest motivation to perceive eugenic stirrings and the deepest knowledge of how they might manifest themselves. Furthermore, legal scholars and scholars of race and ethnicity are especially likely to be social constructivists, working within disciplinary norms that emerged from rejecting the formalisms of group classification or judicial doctrinal analysis. (So are scholars of Cultural Studies, although the strict conditions of our coding does not yield any data from them that fit within the stringent requirements of Table 6.2.) For some intellectual traditions, in short, genomics connotes at least irrelevance and scientism, and at most danger.

Talking Directly with Experts

My other strategy for determining whether *Genomic Politics*'s basic framework usefully categorizes modes of thought and discourse among genomics experts, pulls on the opposite end of the string. For this investigation, I sought insight directly from the experts themselves rather than inferring their views by analyzing their publications.

I queried experts in two ways. The Appendix to Chapter 6 fills in details; here I offer just enough information to set the stage. In 2013 I sent an email to 537 authors of prominent, mostly social science and law, articles published from 2002 through 2012. After briefly describing my research project, pleading for assistance, and promising confidentiality, I asked a set of questions intended to evoke respondents' depiction of genetic influence, their technology optimism or pessimism, and their views about good governance. In the end, 42 people responded. Having not learned my lesson, I sent the same questions to a similar set of 555 experts in 2018 (again, see the Appendix to Chapter 6), and received 104 responses.

Comments were engaged and engrossing. On a vacation to London, reading them almost competed successfully with West End theaters and the London Eye. More systematic analyses yielded insights about genomic politics that shape many of my arguments in this book.

I also sought experts' insights through more sustained interviews. With some participation by Maya Sen, I held close to sixty conversations from 2011 through 2018 with genetics and genomics researchers, medical professionals, government officials, legal scholars, social scientists, and members of advocacy, religious, or interest groups. Our intent was for people to take on the role of anthropological informants, not to be spokespeople for particular positions—hence we sought out staffers rather than elected officials, team leaders rather than organization directors, science writers rather than corporate executives, and so on. As we told the interviewees, they were contributors to our research, not research subjects themselves.

After trying more conventional interview techniques, I settled on a strategy that I do not recommend to students: I explained the project and my credentials, then invited the person to "tell me the three most important things I should know in doing this study." Each discussion developed from there. This technique does not, of course, yield parallel answers that can be systematically compared. Instead, it yields something much more valuable: distinctive, incisive observations. These specialists are, after all, unusually qualified to discuss genomic politics since many are themselves directly or indirectly shaping it. Talking with them has been a highlight of my decades as a researcher and interviewer.

The Political Contours of Genomics

Collectively, survey and interview respondents make several points about political disputes around genomics. First, left and right do not always differ. "The vast majority of this [societal use of genomics] is not contentious. It is just a matter of as much transparency of scientific information as possible, confidentiality of patient information, and integrity of the research process." Or "Republican/Democrat—it doesn't matter except where it affects money or regulation." After all, disease cuts through position-taking: you "can magnify Democrat/Republican instincts, but most [members of Congress] have a family member with long-term issues," so "apart from reproductive issues, medical genomics is not very partisan." "Everyone supports figuring out diseases." Interviewees also see bipartisan consensus in the criminal justice arena: "The idea that the government is storing your biology [in federal and state forensic databases] makes left *and* right nervous." As evidence, as several pointed out, the 2008 Genetic Information Nondiscrimination Act passed both houses of Congress almost unanimously.

Nonetheless, there *are* ideological splits—occurring within both left and right rather than across them. One person I interviewed points to a "fracturing of liberal views—progressive technological enthusiasts from Silicon Valley" ranged against leftists who "see genomics as imposition of authority via information." Another remarks on a "split on the left between autonomy-based and social justice–based arguments (Ms. Jane vs. Miss Jane: strong independent women versus victim)." When it comes to conservatives, "the right should support genomic capitalism—but can't publicly because of reproductive applications."

Not surprisingly, given these commonalities and fractures, the politics of strange bedfellows is a third common theme. Opinions about forensic use of genomics are "not left/right—[the disagreement] is libertarian vs. law enforcement." Or "23andMe users are libertarian, but also left, on the grounds that 'I have right to information.'" One research scientists decries the profit motive in drug development from his stance as a good liberal—but still, "I am in favor of blockbuster drugs for everyone." So far, the core intuition underlying *Genomic Politics*'s basic framework—that disputes largely cut across partisan or ideological commitments—holds up well.

Benefits of Genomic Science

Survey recipients were first asked to identify "the greatest actual or potential benefits of genomic science, taken as a whole." The most frequent and detailed responses refer to medical advances. Some offer a long both/and list:

> Redefining of genetic conditions, pathway connections among recognized conditions, etiology for "usual suspects" not previously identified, refinement of population frequencies and influence for mutations in common disease such as BRCA1/2, new variants that contribute to common disease risk, the start of drug development targeted at specific disease variants, redefining how diagnoses are made from variants rather than from symptoms, and so much more (2013).[12]

Others point to a specific success—the "poster child" of phenylketonuria as the best example of "identification of rare variants with high influence, and the potential of using this information to shape the environment/exposure of the individual (if possible)" (2018),[13] or "the role of genomic diversity in response to treatment, e.g. genomically-personalized medicine" (2018).

Improved understanding of gene-environment interactions is also a commonly identified greatest benefit, as in "parsing the relative effects of genetic factors, epigenetic factors, social determinants of health, and behavioral factors, in order to improve patient care and reduce mortality rates" (2018). Being social scientists themselves, a few delight in pointing out that "genomic science has, ironically, placed the social sciences at the center of nearly all theoretical and methodological models linking measured genotype to health and health behavior phenotypes" (2018). As that observation suggests, even among these experts who perceive a great deal of genetic influence, a benefit of genomic science is our need to "ultimately surrender with view to the great complexity and multifactorial nature of most genetic risks; and then increase efforts to work against/focus more on the environmental factors promoting cancer and other 'civilization diseases'" (2018).

Some respondents move outside the health arena, pointing to "advancements in plant/animal breeding to feed the increasing world population. Use of bioengineered organisms (bacteria etc.) in alleviating environmental disasters (e.g., oil spills)" (2018). Genomics offers social benefits; it "demonstrates that all human individuals share the same 'genetic' history since at least the emergence of *Homo* genus" (2018), and it encourages "increasing tolerance. Appreciating the fact that genetic differences exist from one individual to the next will diminish the extent to which people have biologically drive[n] biases across groups" (2013). Most generally, genomics can be the gateway to understanding "human cultures, human behavior, and our environments" (2018) or even "life on this planet" (2018).

Risks and Dangers

Of course, being scholars, respondents have just as much to say about "the greatest risks or harms of genomic science, taken as a whole." One person anticipates "none that I can see"; another offers "giant mutant bugs destroying Manhattan? . . . In other words, don't know." But most have no hesitation in pointing to problems.

The medical arena again dominates responses. "Premature translation into the clinic" is a common fear: "It's always tempting to roll out cool new technology into the patient care setting and lots of theoretical reasons to think it will help clinical medicine in this or that application. However, we've learned over and over in clinical medicine that we have to first generate data that something helps patients before we apply it." We must remember that "the stakes are high in medicine and when we get carried away with enthusiasm our patients end up paying the price" (2013). Conversely, some—occasionally the same respondent—fear "failure to incorporate genomic information into routine medical care when it's appropriate" (2018). There are, after all, "people wanting to avoid scientific truths" (2013).

Beyond over- or underuse, genomic medicine presents other dangers. There are "ethical problems with gaining knowledge by tests that tell people's future. Should they be informed? Or their relatives?" (2013). There are distortions created by economic incentives, such as "misuse of medical genetics for the sake of increased profits by pharmaceutical companies in the absence of significant reductions in morbidity and mortality of serious diseases" (2018). Institutional hazards abound: "unequal opportunities for care, due to differential medical insurance implications in countries without an effective publicly funded medical care system (e.g., USA, India, many African countries)" (2018). So do social risks, especially about "the emergence of a genomic underclass, whereby the clinical applications to be derived from genomic science will be distributed in an unjust and resource-dependent manner" (2013).

Genomics in the criminal justice system raises more concerns. Genomics performs "poorly" there; harms include "invasion of privacy; collection of genetic evidence by (legal) compulsion or by stealth/lack of informed consent; lack of oversight of labs; exploitation of genetic material without express legal authorization" (2013). The risk of genetic discrimination haunts many respondents. No one refers to *GATTACA*, *Elysium*, or designer babies, but some concerns are close. On the mild end, "misuse of genetic information could lead children to be placed in educational tracks that limit their exposure to the most challenging educational programming and thereby limit their life choices" (2018). On the most anxious end, genomic science could lead to "the use of data for . . . institutions/governments to commit *genecide* (discrimination against, jailing, sterilization, or outright killing of people with particular genes deemed undesirable)" (2018). In between, "the greatest risk springs from the highly racialized ways in which genomics asks fundamental questions about humans. This is not a matter of genome sciences being used for racist ends. It is a matter of genomics being a support beam in the structure of racism itself in society" (2018).

Dr. Frankenstein's monster stirs not far below the surface. Biological weapons, biocolonialism, "the creation of dangerous life forms," "planned breeding programs," "environmental disruption, even devastation," and cloning are suggested, along with "that long-standing narrative in the West of civilized vs. savage, rational vs. irrational, fact vs. fiction" (2013).

Weighing Benefits and Risks

Collectively, these experts' views include the ingredients for all four quadrants of the basic framework of *Genomic Politics*—varying perceptions of genetic influence and varying balances between technology optimism and pessimism. How are the ingredients mixed? When asked to weigh the benefits and risks that they have described, a few respondents abdicate: "No worthwhile opinion" (2013), or "This is a difficult question. In my mind harms and benefits are neck and neck and it all hinges on the execution" (2018). If not the execution, then perhaps the arena must be specified: "Risks and benefits should be evaluated on a case-by-case basis because what constitutes a risk or a benefit varies from place to place" (2018).

But most good-naturedly take the bait I dangle before them, jointly revealing from the inside what we saw from the outside in the article-coding exercise: social scientific genomics experts reside in all quadrants of the basic framework. Enthusiasts are readily recognizable: "I think the positive impact on human health from genomic science overwhelmingly outweighs the possible negative risks" (2013) or "The usefulness of genomic analysis will far outweigh any risks or dangers" (2013). Enthusiasm can be poignant: "Absolutely [the benefit outweighs the risk] and I think it will continue to do so. For very personal reasons, I wish genomic science were far more advanced" (2018).

Skepticism is equally easy to see: "The potential harms far outweigh the potential benefits" (2018) or "A better way to ask the question might be (1) whether our current society has the capacity to take advantage of the potential benefits and (2) whether we have the capacity to curb the potential misuses. My answer to #1 is 'yes,' but my answer to #2 is 'no' " (2018).

Some respondents are Hopeful—recognizing some level of genetic influence but arguing that it receives "too much hype": "Investment in genomics should not be at the expense of investment in public health, or investments in other known beneficial services. It's quite right to fund genomics, but that funding must be in balance with other ways to address health problems, and indeed other ways to promote the public good" (2018). Or, simply, "More good could be done in the world immediately by investing in malaria treatment and prevention" (2018).

Rejecters, finally, see genomic science as actively harmful: "Social determinants of health and development are ignored or bypassed in policy [due to] a greater focus on genes" (2018). Attention to genomic science generates not only "distortion of investment" but also, even worse, "increasing inequalities, lack of access, increased health disparities, a significant financial investment that will help the few rather than focus on better and more comprehensive plans to ensure equal access to healthcare and better environments impacting health" (2018).

Why the Balance Between Benefits and Risks

We arrive now at the most important question that the interviews and social scientists' surveys can address: *why* do experts array themselves across the quadrants of the basic framework? The database of coded articles shows that they do, and titles of those articles sometimes suggest answers. But the authors themselves, in the surveys and especially in the interviews, give more direct and eloquent explanations of why they are Enthusiastic, Skeptical, Hopeful, or Rejecting.

One type of explanation concentrates on a single overriding goal or concern, before which other considerations must give way. A Skeptic fears that, regardless of occasional resistance, "we will not be able to move beyond a reductionistic model of gene action, and will suffer the social consequences of reinforcing the essentialistic and deterministic ideologies that already haunt our social interactions" (2018). But an Enthusiast equally focused on group hierarchies sees genomics as the solution, not the problem: "Ultimately genomics reveals the greater degrees of relatedness amongst all people and reveals that the vagaries of history, both negative and positive, are to be found in the past of every region" (2018). More pointedly, "genetics is not making racism worse; it can make it better potentially. Genetics will eventually move clinical providers away from using skin color as a heuristic for clinical practice. [We are not there yet, but we] will get there" (interview). Most provocatively, says an Enthusiast, "race is [a] terrible proxy for health; most genetic variation is within people of African descent. However, in absence of biomarkers, if race is the best proxy, use it. To save a life, who gives a shit about social context?" (interview).

Privacy is a second overriding issue from which Enthusiasts and Skeptics may derive opposite conclusions. One Skeptic speaks for many in fearing "establishment of a database containing every person's genomic information. The potential for abuse is truly frightening. In addition, imagine all the errors that would creep in and the difficulty in correcting them. Imagine hackers stealing this information. It makes identity theft look benign in comparison" (2013). As an interview subject summarized, "We need a new genetic privacy law—with legal protection for privacy in all circumstances." They speak for many. Half of Americans were not confident in 2020 that the federal government could protect personal records from unauthorized users, including hackers; two-fifths felt the same about public health organizations, and one-fifth even had no confidence that their health care providers could protect personal health records.[14]

But Enthusiasts push back, asserting that "privacy risks are inflated and become part of the negative hype" (2013), or "Yes, [there is] potential for discrimination in employment, insurance, personal relationships; loss of privacy. That said, my feeling is the law is far ahead; indeed, arguably GINA offers protections for a problem that as of yet has not manifested" (2013). Several people I interviewed stood the concern about privacy on its head: "People are worried about invasion of privacy by looking at DNA—but for these diseases we don't understand, we should *want* to know more," or "Suppose a person has a highly penetrant mutation that increases arrhythmia; I would prefer that this person *not* be an airplane pilot. An anti-discrimination law that always protects [medical] privacy is a mistake." And one interviewee sees attention to genetic privacy as simply misplaced: "We aren't really a private species. Everybody knows everything [in most societies or communities]. It's only because of cities that we don't know who lives next door to us." And even in cities, "faces are revealed, and they give away an extraordinary amount of information."

Beyond particularly important issues lies a deeper explanation for quadrant location—evaluations of humans' capacity to learn and to act for others' good. Enthusiasts and the Hopeful start with the rebuttable proposition that people can be trusted: "I think the potential benefits outweigh the risks, but only if the manipulation of genomes is carefully monitored by those who are motivated by altruism and improving the environment, humans, animals and plants" (2018). Or "I see very few risks or harms in genomic science as a whole, as long as the research is conducted in an ethical manner and as long as it is informed by recent advances in molecular genetics" (2018). Given this starting point, Enthusiasts are proactive: "Genomic science needs to be pursued irrespective of such benefits/harms analysis, purely as a scientific endeavour. . . . Just as we wouldn't ban aeroplanes because they can be driven into buildings, we should not set up the potential misuse of genomic scientific insights to hamper the pursuit of knowledge" (2018). After all, "all science comes with known, and more importantly, unknown, risks. At the same time, science delivers in sometimes unpredictable ways benefits not envisioned at inception" (2018).

Skeptics and Rejecters, in contrast, mistrust human capacity or intent unless they are shown otherwise: "Risks depend on human judgment a lot.

And it is about this matter that I tremble. I am not completely comfortable in trusting genetic manipulation" (2018). Or "Based on the current behaviour of humans it is likely that genomic science will cause more harm than benefits" (2018). After all, "our capacity as a society for misusing and abusing genomic science is enormous" (2018). "The science is correct. The problem is the scientists" (2018). But scientists are not the only problem. Other grounds for technology pessimism include "private organizations focusing on profit, or undemocratic state" (interview); "hubris, glory-seeking, and greed" (interview); "those who are largely ignorant [who] will . . . use it for bad purposes" (2018); and "how advocates [in the criminal justice system] want to use the science" (2018).

As a result, Skeptics are precautionary: the great danger of genomics is "the risk of things going wrong—the unknown consequences which we can't factor into the science because we do not know yet" (2018). In the end, "it really depends on how things will evolve. A good purpose can easily be turned into a bad one" (2018).

A final frequent theme—the sense of inevitability invoked by the third epigraph of this book—stands at least partly outside the basic framework of *Genomic Politics*. This type of observation is sometimes a tactic to make Enthusiasm seem self-evidently appropriate: "Potential benefits outweigh the risks, and anyway the cat cannot be put back in the bag" (2018). But it often evokes resignation rather than excitement: "The clock cannot be turned back. Develop all the good methods/tools and be vigilant on how information about individuals is stored and used" (2013). Or "If we could put the genie back into the bottle, should we? Probably, but that isn't going to happen. Knowing seems to be a snowball that is not under any particular control. We might be able to speed it up or slow it down or direct it a bit, but that is about all deliberate collective action can do" (2018). As George Church and Ed Regis sum up, "Prohibitions are generally ineffective and counterproductive, and have negative unintended consequences. . . . In general, concerns about a new technology arise mainly during the transition to it."[15]

The Politics of Experts

We now have rich evidence that social scientists do in fact unknowingly reside in one or another quadrant of *Genomic Politics*'s basic framework. Disciplines whose subject matter and analytic methods focus on counting, measuring, and comparing discrete individuals or behaviors tend to find genomics important and useful. Conversely, disciplines whose practitioners are especially likely to study collective actions or nonmaterial concepts tend to find genomics unimportant or even inimical. Neither of those conclusions is necessarily a causal statement since some other factor—such as relative conservatism or liberalism of a discipline's practitioners—may explain subject matter, methodological tastes, tolerance for abstractions such as "culture," and views about genomics alike. Individual idiosyncrasy must never be discounted. Nonetheless, the disciplinary mix of perspectives, epistemologies, and views of genomics is suggestive.

Direct questions to social science experts yield different kinds of answers to the question of why a person holds the views represented by a given quadrant of the framework. Enthusiasts allow themselves to be captured by the excitement and power of genomic science because they start from the premise that people or institutions are usually well-meaning, moral, and capable of learning. Genomic science offers an astounding new vehicle for engagement with and knowledge about one another. Skeptics seek to constrain the power of genomics because they start from the premise that too many people or institutions are ill-intentioned or incapable of solving the problems that the new science creates. This new vehicle is too prone to collision or to veering off the road. People in the quadrants of Hope and Rejection see little or no excitement and power in genomic science; the vehicle is at best overhyped and at worst an obstacle to genuine forward movement.

Social scientists' views, while not partisan, are deeply political. How that matters, however, is variable; as a reader of an earlier draft of this book pointed out, "There is a President's Council of Economic Advisors; there is no President's Council of Cultural Anthropology Advisors." The disciplines most important for public decision-making—economics, psychology, criminology—are also most likely to be in the Enthusiasm quadrant, while disciplines lacking prominent public roles— cultural studies, ethnic and racial studies—tend to be in the Skeptical or Rejecting quadrants. That pattern might matter for how the genomics revolution helps to shape American society.

But even experts, who almost by definition influence new policies for societal use of new technologies, are only one set of relevant actors. In the purportedly democratic United States, residents' views and actions also matter. Public opinion seldom determines policy outcomes in any straightforward way, but majority preferences may win out; as a politician once reminded me when I was urging what I earnestly described as a good proposal, "The good proposal is the one that wins 50 percent plus 1 of the votes." We therefore need to ask the same questions of ordinary Americans as of experts—who resides in each quadrant of the basic framework, and why? Chapter 7 addresses those questions.

7

Locating the Public in the Basic Framework

Justice is worth a swab

—Respondent, GKAP 1

Get your swabs out of my face!

—Respondent, GKAP 1

The experts I surveyed differ on how much the American public should be able to shape societal uses of genomic science. Views on what actors are best suited to making policy decisions in this arena range from "General public. Because they can empathize with the victims while others over-intellectualize" (2013) to "Scientists with extensive experience developing and applying genomic analytic techniques, and governmental regulatory bodies which listen carefully to the advice of these scientists" (2013). Although elected officials might share this range of views, any who want to keep their job must express at least tempered enthusiasm about public involvement in policymaking.

I see two reasons beyond the expression of democratic pieties to examine public views on genomics. First, most policy arenas—say, tax rates or military spending—are dense with committed actors, complex histories, multiple laws and rules, elected officials with public records, and partisan auras. Not for nothing do policy analysts talk about "iron triangles," loosely defined as a close relationship among relevant bureaucratic agencies, members of congressional committees, and interest groups. The public at large has a hard time turning triangular relationships into a square—that is, influencing the policy's development or implementation—so their views may matter relatively little. Genomics is not like that. It has few rules or laws and little public history; advocacy organizations tend to be small and tightly focused; elected officials have almost no record of engagement; partisan valence is missing. In this atypically empty milieu, constituents' views may matter more than usual to elected officials; given their own lack of record, commitments, or expertise, politicians may find in this arena a good opportunity to follow their voters' lead.

Second, genomic science is close to unique in the intimacy as well as extent of its impact on people's lives. Climate change, fracking, and teaching evolutionary theory will of course affect individuals as well as societies, but seldom with the immediate wallop that comes with sequencing a lung tumor, finding a match in the CODIS database for a crime scene sample, or discovering an ancestor from a previously unconnected part of the world. As Jennifer Doudna and Samuel Sternberg put it, "It won't be long before the repercussions of this technology reach your doorstep."[1] If genomics can change anyone's life chances and—thanks in part to Doudna and Charpentier's CRISPR discoveries—eventually those of their descendants, then everyone's views about it matter.

I turn, then, to the issue of whether *Genomic Politics*'s basic framework helps to organize and make sense of what people think, who holds particular views, and why they hold those opinions.[2]

Genomics: Knowledge, Attitudes, and Policies—GKAP 1 and 2

I analyze public opinion about genomics through two new surveys. Maya Sen and I designed an online survey called Genomics: Knowledge, Attitudes, and Policies (GKAP 1) that was fielded in 2011. I developed a second survey, imaginatively titled GKAP 2, in 2017. The Appendix to Chapter 7 explains the details; the crucial points are that the sample sizes are 4,291 and 1,777 respectively, the survey did not use an opt-in set of respondents, the weighted samples are representative of the adult American population, and each survey is stratified by race or ethnicity. Both surveys ask questions about genetics knowledge, attitudes toward the benefits and risks of several genomics technologies, perceptions of the amount of genetic influence on various phenotypes, trust in key public and private actors, and policy preferences. GKAP 2 includes questions about the new technologies of gene therapy and gene editing, and about governance of genomics.

Despite the huge increase in research activity around genomics and its considerably greater reach into society between 2011 and 2017, results in the two surveys are similar. GKAP 2 has the advantage of being more recent; GKAP 1 has the advantage of a larger sample size, which is especially important when we want to compare groups within the sample. GKAP 2 includes new questions, although it lacks some questions asked in GKAP 1. Mainly in order to show how much views (mostly didn't) change over time, where possible I provide the results of both surveys.

Americans' Knowledge of Genomic Science

To no one's surprise, Americans are not well informed about genomic science or its uses. In both 2011 and 2017, seven in ten GKAP respondents had heard or read "not very much" or "nothing" about "issues having to do with genes or genetics." Fewer than a tenth in each year had heard "a great deal" or "quite a lot." Even fewer have any awareness of medical or scientific biobanks (which the survey had previously defined). Only 6 percent reported in 2011 that they or a family member had taken a genetic medical test, and only 2 percent had taken a genetic ancestry test. Repercussions from this technology have not yet reached most Americans' metaphorical medical or scientific doorstep.

Nevertheless, societal uses of genomics are leaving discernible tracks in people's lives. Most importantly, in both years, about half had at least heard of forensic databases (which the survey had defined). Three factors plausibly explain that relatively high level of knowledge. First, respondents or people they know may have had direct encounters with the criminal justice system. In 2011, Blacks were more likely than non-Blacks, people with less than a high school education were more likely than those with more education, and the poor were more likely than all other income groups to report "a lot" of knowledge about forensic biobanks. I find similar results with regard to race, but not education or income, in 2017.

Second, the issue of mass incarceration had moved into the public arena by 2017, so even many of the nonpoor and those with more years of schooling have heard of forensic databases at the time of the fielding of GKAP 2. Further suggesting a political explanation for awareness is the fact that, slightly in 2011 and especially in 2017, strong Democrats and self-identified "extremely liberal" respondents were more likely than others to have heard a lot about forensic DNA databases.

Finally, although we have no evidence on respondents' media use, the decades-long popularity of *CSI: Crime Scene Investigation* and its successors is a third plausible explanation for having heard something about forensic DNA databases. *CSI* was the most watched television series in the world for five years in the early 2000s. One of its many successors, *NCIS*, is approaching twenty seasons; scholars such Judge Donald Shelton and his co-authors, or psychologist N.J. Schweitzer and his colleagues, have conducted research on "the CSI effect," defined as jurors having either "exaggerated faith in. . . the forensic sciences" or "greater expectations about forensic science than can be delivered."[3] (They mostly don't find it.)

Even the medical arena shows hints of recognition among respondents that genetics affects their lives. In both GKAP 1 and 2, about a third of respondents report that they, a family member, or someone whom they take care of has one or more characteristics (usually a disease) that are generally understood to be genetic. (The Appendix explains how we obtained this result.)[4]

In the two decades since the Human Genome Project's debut, genetics has increasingly been taught in high school and college biology courses, and promoted through activities such as NHGRI's annual National DNA Day.[5] So most GKAP respondents know at least something about it. Each survey asks where in the body DNA can be found, and (in random order) how similar genes are between Black and White humans, and between mice and humans. In 2011, 72 percent correctly answered the relatively easy question about the location of DNA ("in every cell in the human body" rather than "only in specific organs and cells").[6] Forty-four percent answered correctly that "more than half" of a White person's genes are identical to those of a Black person (alternative choices were "about half" and "less than half").[7] Only 17 percent gave the right answer (more than half) to the difficult question about human-mouse DNA overlap. Answers in 2017 were slightly more accurate—76, 48, and 20 percent correct, respectively, across the three questions. (In both years, more than half of respondents did not venture a substantive response to the question about mouse DNA.)

If these three questions constituted a test, students' grades would have been widely dispersed. At the bottom end, roughly a quarter in 2011 would have failed completely, with no correct answers. In the middle, a third got one right (usually "where is DNA in the body?"), and roughly another quarter got two right—perhaps a B + on this tough test? At the top end, 15 percent would have received a gold star on their forehead—three out of three correct answers. Proportions were similar in 2017, although fewer would have failed completely.[8]

How should we interpret this broad but shallow knowledge? One line of research argues that when respondents know little about the topics being probed, their responses are best interpreted as nonattitudes—choices that are essentially random, having been invented for the occasion in response to an importunate interviewer and as quickly forgotten. However, that need not be the case. People can develop a plausible and meaningful response by using verbal cues in the question, recalling a comment from a trusted source, drawing an analogy to something they do know about, or taking the opportunity to express a normative or political stance about the more general issue.[9] I conclude that whether survey responses about novel issues are in some sense real and to be taken seriously is an empirical question, to which the answer is likely to vary from case to case.

In this case, the evidence justifies a claim that respondents give answers in GKAP that are meaningful enough to have moral and political import. Three-quarters in both years answer at least one genetics question correctly. And as we shall see, in a series of related questions asking for perceptions or evaluations, a given individual's responses are internally coherent and consistent.[10] Open-ended queries about why they just gave a particular answer elicit intelligible and thoughtful—sometimes very moving—answers. Most important, GKAP respondents sort into the four quadrants of the basic framework in ways that enable us to discern not only what they think but also why. I turn now to that sorting.

What Do Americans Think About Genomics?

Genetic Influence

I begin, as in earlier chapters, with the vertical and horizontal dimensions of *Genomic Politics*'s basic framework. Asked about the overall importance of genetic influence, 17 percent of GKAP 2 respondents focus on environmental or social causes of most diseases, while 31 percent focus on genetic causes (half express no opinion). Respondents similarly disagree on a more focused question, on whether most traits come from "family life, environment, or lifestyle choices" or from genetic inheritance.[11] Nineteen percent choose non-genetic causes, while 39 percent choose genetics (the rest express no opinion). Thus among those brave or obliging enough to answer my questions, respondents are twice as likely to choose genetic as non-genetic explanations of human diseases, traits, or behaviors.

To see if that overall view holds up for more clearly specified traits, GKAP 1 and 2 both ask about genetic inheritance compared with environmental or lifestyle causes of eight phenotypes. Maya Sen and I chose four diseases and four traits or behaviors because of their salience, moral importance, and range along a rough continuum from clearly genetic to clearly not genetic.[12] Respondents were asked (in random order) about sickle cell anemia, cystic fibrosis, the flu, heart disease, a particular eye color, level of intelligence, sexual orientation, and aggression. Figure 7.1 shows the results.

Figure 7.1 reveals that respondents distinguish among phenotypes by collectively creating three categories. Almost none perceive the flu, and few perceive aggression, to be primarily genetic. Conversely, substantial majorities (appropriately) agree that genetics is the main cause of sickle cell anemia, cystic fibrosis, and eye color. The sensitive issues of sexual orientation, intelligence, and heart disease comprise a third cluster, with about a fifth of respondents in each case focusing on genetics. Note that attributions to aggression, intelligence, and—oddly—heart disease are especially cautious, with three-fifths or more of respondents choosing the indeterminate "mixture." These patterns change very little across the six years, except for a slight but uniform decrease in genetic attributions.[13]

The choice of multiple causes, especially for heart disease, aggression, and intelligence, has two plausible interpretations. It may express caution in the face of ignorance or of anxiety about difficult questions, or it may be the closest option available for respondents seeking to eschew the too-simple dichotomy in favor of a response pointing to gene-environment interaction or epigenetics. We cannot track respondents' thinking. But it seems likely that respondents seeking to signal epigenetics or interaction would be the most genetically knowledgeable, whereas those signaling ignorance or anxiety would know the least. I therefore examine the likelihood of choosing the middle ground depending on one's genetics knowledge, based on a scale that uses the three questions already discussed (where DNA appears in the body, and how much DNA is shared by Black and White humans and by mice and humans). (See the Appendix for an explanation of the knowledge index.)

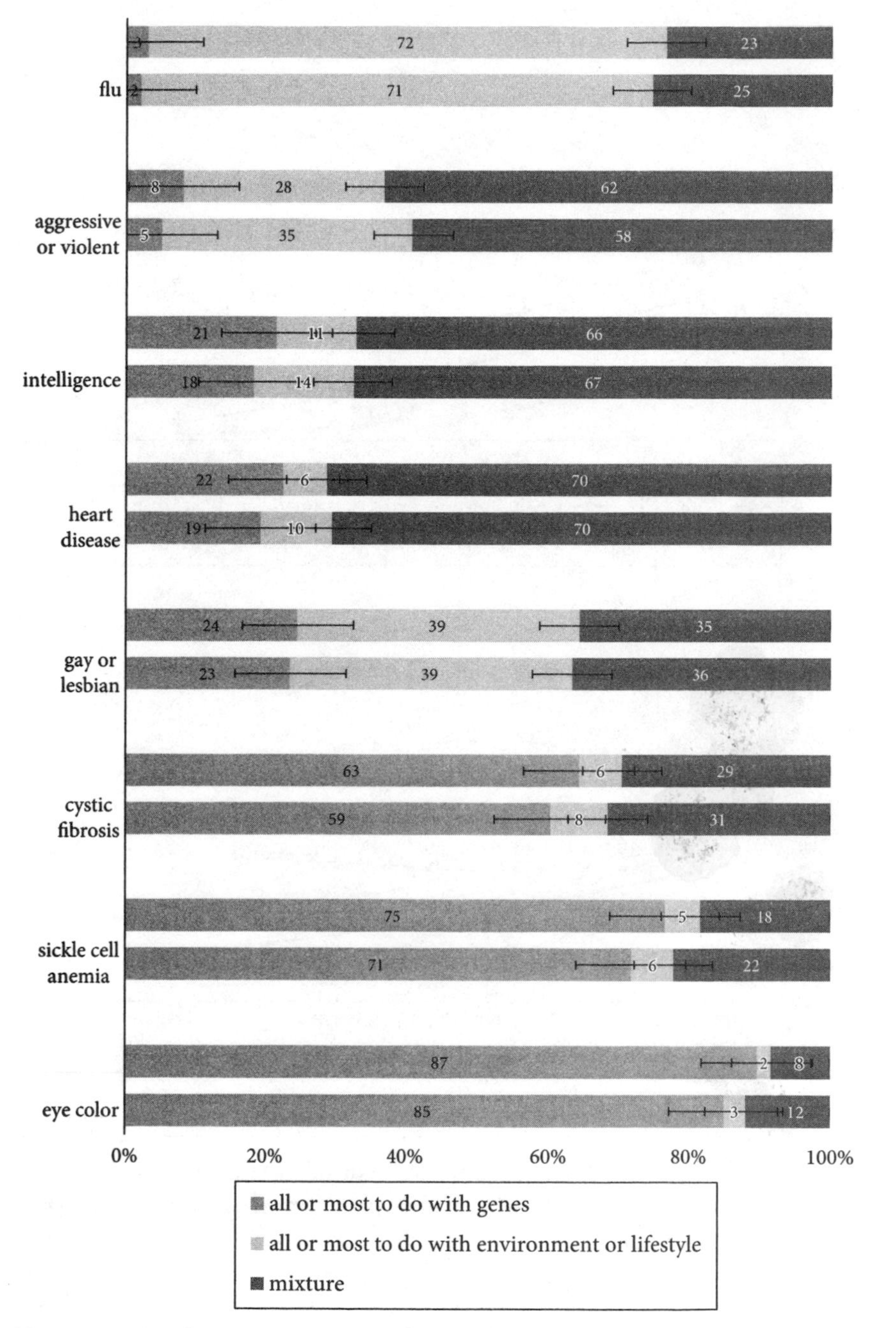

Figure 7.1 Attributions to genetic influence for traits, behaviors, or diseases, GKAP 1 and 2 (top row in each pair is 2011; bottom row is 2017)

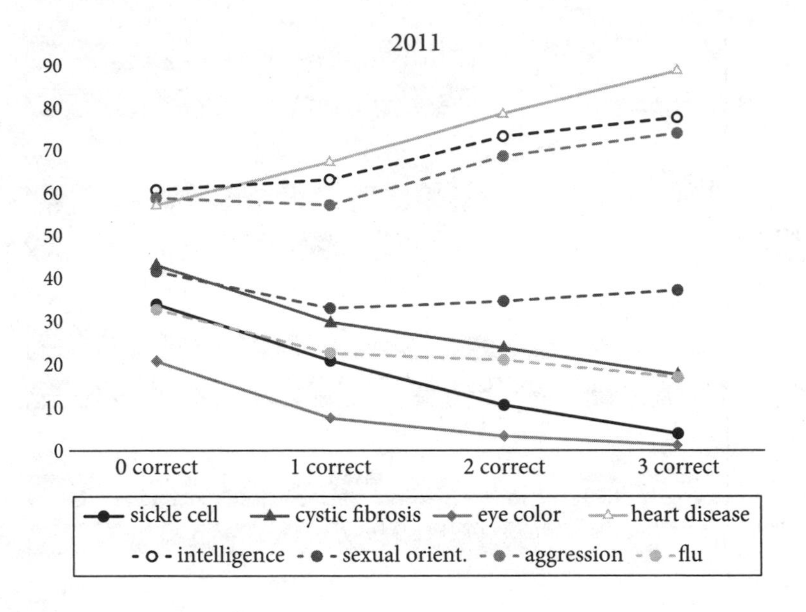

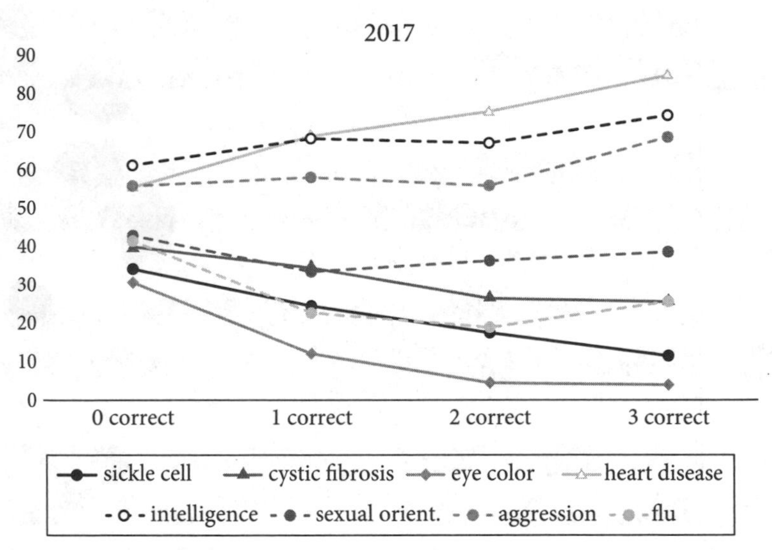

Figure 7.2 Proportion choosing "a mix of genetic inheritance and environmental or lifestyle choices" to explain phenotypes, by levels of genetics knowledge, GKAP 1 and 2

Both caution and interaction seem to be operating in the choice of middle-ground answers, depending on one's level of genetics knowledge. Figure 7.2 shows how I come to that conclusion.

On the one hand, in both GKAP 1 and 2, the explanations of those who correctly answer none of the genetics questions make relatively few distinctions among the eight phenotypes, as the left side of both graphs shows. I interpret that as a signal of uncertainty. On the other hand, with increasing genetics knowledge comes increasing differentiation in explanations for the eight phenotypes, as movement toward the right of the graphs shows. In both years, the most knowledgeable were *more* likely than other respondents to choose a mix of genetic and nongenetic causes for heart disease, intelligence, and aggression, but *less* likely to choose a mixture of causes for cystic fibrosis, sickle cell disease, flu, and eye color. I interpret that as a signal of deciding when interactions are or are not responsible, depending on the phenotype. Only for what is arguably the most fraught characteristic—"being gay or lesbian"—were the most and least knowledgeable respondents equally inclined to choose the middle ground, whether out of caution or some other reason I cannot say.[14]

Overall, we can be reasonably confident that Americans are giving meaningful answers to questions about genetic influence. In the aggregate, views are consistent across time, make substantive sense, and vary as one would expect depending on level of genetics knowledge. Luckily for my narrative, the basic framework's vertical dimension holds up to scrutiny of whether it is a useful tool for analyzing how the American public views societal uses of genomics.

Technology Optimism

The next step of the analysis is to see if regular Americans, like experts and like policies themselves, vary in their proactive and precautionary stances toward use of genomics. Two sets of GKAP evidence show that Americans are on balance, though not always, technology optimists.

The first tranche of evidence focuses on depth at the expense of breadth. After a definition of somatic gene editing, GKAP 2 respondents were asked their level of agreement with a battery of statements (presented in random order). The procedure was repeated for germline gene editing. Table 7.1 presents views of the statements that focus on risks and benefits, in order from greatest to least optimism.[15]

Table 7.1 Do potential gains of developing gene therapies outweigh risks? GKAP 2017 (percent agreeing)

	Gains outweigh risks	**Risks outweigh gains**
Successful gene therapy would demonstrate benefits of science	60%	7%
Should develop gene therapies even without clear predictions for use	40	19
Humans always seek improvement; germline gene therapy is exciting next step	35	21
Society will find solutions to problems of germline gene therapy	27	30
Gene therapy would risk less acceptance of people who are different	27	27
Even if it seems safe, gene therapy will create too many health risks	22	23
Germline gene therapy would deepen unfair inequality because only the rich could afford it	11	50

Although Americans have had little reason to think about gene therapy, at least half are willing to venture an answer to almost all of these questions. Respondents are divided—neither technology optimism nor technology pessimism is routed by its opposite. Instead, particular considerations evoke strongly positive ("benefits of science"), balanced ("will find solutions"), or strongly negative ("only the rich") views.

Luckily for my purposes, genomics technologies other than gene editing have moved a little further into public consciousness, so we need not rely solely on possibly volatile views of gene therapy. A second series of questions addresses a broader array of these better-known uses, albeit at the expense of depth about any one of them. Respondents in both surveys were asked if a given genomics technology would provide more societal benefits, more costs, or equal amounts of benefit and cost. The technologies (presented in random order) are scientific biobanks, forensic DNA databases, research on racially inflected inherited diseases, and research on an individual's likelihood of getting an inherited disease.[16] GKAP 2 adds parallel questions on somatic and germline gene therapy. Figure 7.3 shows the results.

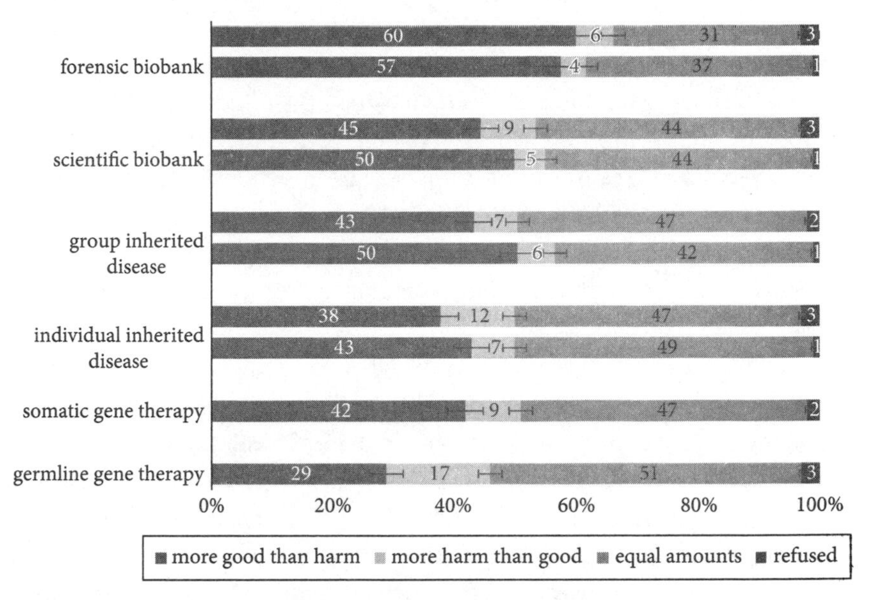

Figure 7.3 Proportions of technology optimists and pessimists, GKAP 1 and 2 (top row is 2011; bottom row is 2017)

As Alexis de Tocqueville's eager boatbuilders would lead us to expect, Americans lean much more toward technology optimism than pessimism. For example, ten times as many people express optimism as pessimism about forensic DNA databases. Even the most underdeveloped and radical technology, germline gene editing, enjoys almost twice as much support as opposition. Results remain largely stable over the six years between GKAP 1 and 2, except that respondents gain a little confidence in the medical technologies and lose a little confidence in the legal arena.

Figure 7.3 also shows that up to half of respondents opt for the anodyne "equal amounts of good and harm" for most genomics technologies. Appropriately, the 2017 question about germline gene therapy elicits, along with the greatest concern, the most ambivalence. As we did for genetic influence, we can explore the meaning of the "equal" category by comparing answers from respondents with greater and lesser amounts of genetics knowledge. Choosing a balance of risk and benefit might signal uncertainty and caution in the face of unfamiliarity with these arcane topics—or it might signal recognition of the complexity of the issues, echoing *The Economist*'s warning of "both good and ill."

I need no figure to convey the results of this analysis. For all of the technologies about which GKAP respondents were asked, the more they know about genetics, the less likely they are to choose the middle ground between technology optimism and pessimism. In both years, 50 to 70 percent of those who correctly answer none of the knowledge items choose "equal amounts," compared with 30 to 50 percent of those who correctly answer all three items. Forensic DNA databases elicit the fewest balancing responses at all levels of knowledge, but the decline in "equal amounts" as knowledge rises is roughly parallel for all six technologies.

Finally, the more genetics knowledge GKAP respondents have, the more likely they are to be technology optimists. Again we need no figure to grasp the point: for each question, there is a 25 to 45 percentage point difference in risk acceptance between those giving no correct answers on the genetics knowledge scale and those giving three correct answers. I discuss the meaning of that pattern later; what matters here is that the more Americans know about genetics, the more they anticipate benefits from it—even if they do not own stock in pharmaceutical companies.

Locating Public Opinion in the Quadrants

We now have the information needed to locate public opinion in the quadrants of the basic framework. I do so by consolidating responses on the vertical and horizontal dimensions into influence and optimism scales, respectively, and then mapping the intersection of the two scales.

Seven of the eight genetic influence items in GKAP 1 and 2—all but the flu—constitute the influence scale. All four items constitute the technology optimism scale in 2011, and all six items constitute the corresponding technology optimism scale for 2017. (As before, see the Appendix to Chapter 7 for details.)

The scales for both dimensions of the basic framework range from 1 to 3. A high value on the vertical axis (influence) shows perception of strong genetic factors, while a high value on the horizontal axis (risk) shows strong technology pessimism about genomics. Combining these two scales shows respondents' locations in the four quadrants, as in Figure 7.4. Each dot indicates a respondent.

As the downward-sloping regression lines in the two panels of Figure 7.4 show, genetic influence and optimism are linked: the more traits or diseases that a respondent sees as genetically caused, the more likely he or she is to be optimistic about genomics' societal benefits. (See Appendix for explanation.) Table 7.2 gives numerical precision to these visual results.[17]

In both years, a healthy majority of respondents score high on the genetic influence scale and low on the pessimism scale—they are Enthusiasts. Tocqueville's boatbuilders do indeed represent the majority of Americans. Their proportion, however, is declining slightly by 2017; whether that is a harbinger of growing caution about genomic sciences remains to be seen. The Hopeful are the second-largest group, with over a quarter of GKAP respondents; their share of the population is rising slightly. This is the most ideologically complicated group. They share a preference for promoting benefits over protecting against harm, but they disagree among themselves about what causes human traits and behaviors, with views ranging from

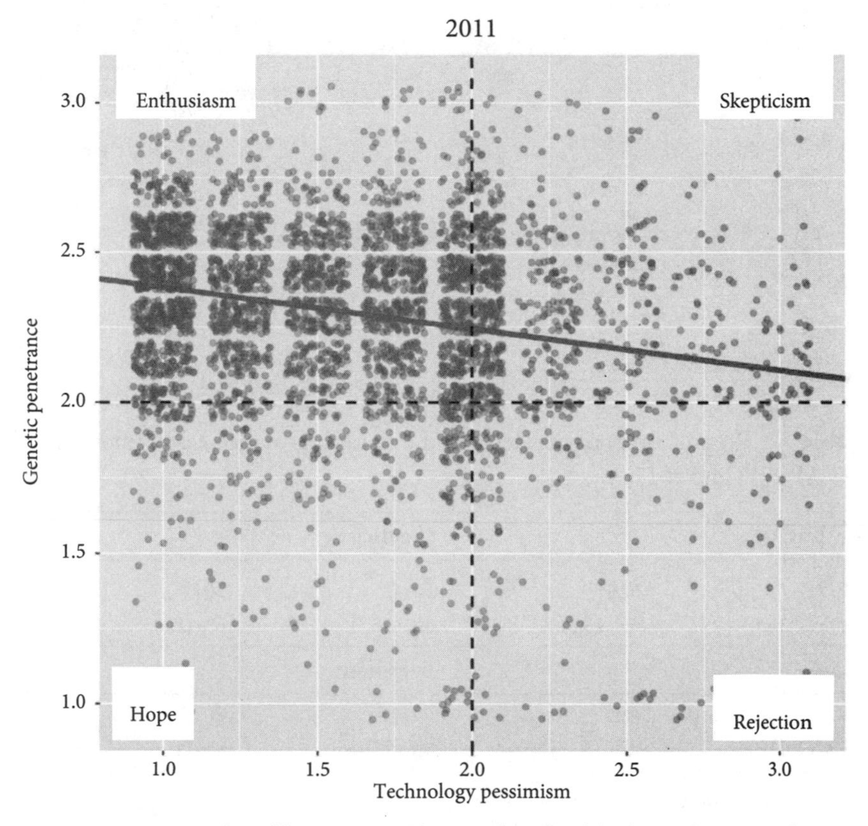

Figure 7.4 Respondents' locations in the quadrants of the basic framework, GKAP 1 and 2

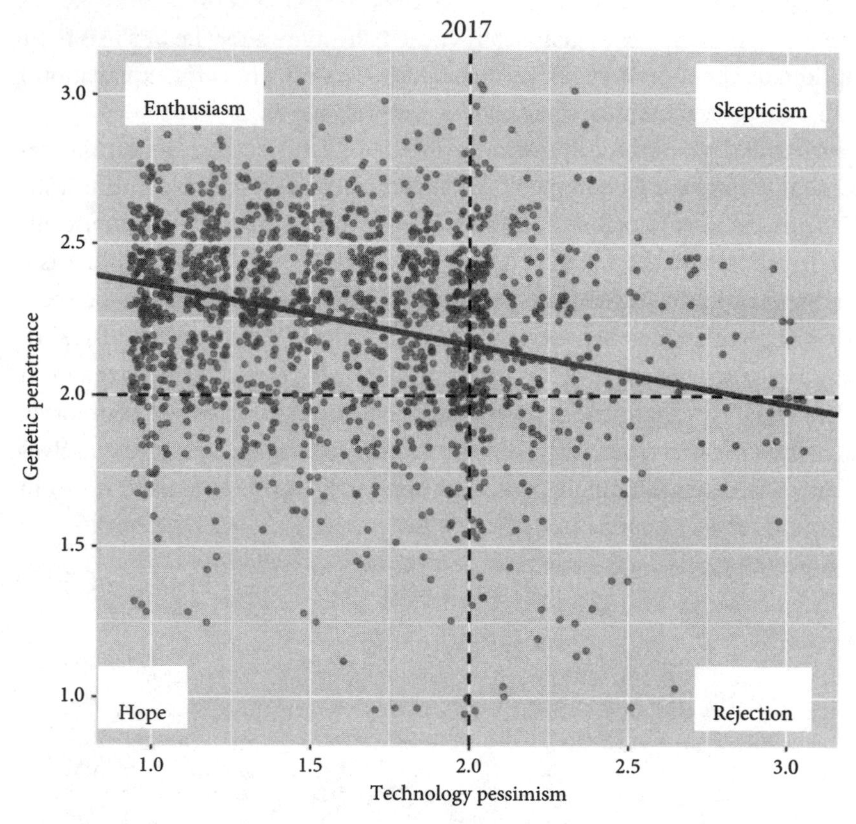

Figure 7.4 Continued

Table 7.2 Proportion of respondents in the quadrants of the basic framework, GKAP 2011 and 2017

Enthusiasm		**Skepticism**	
2011	*2017*	*2011*	*2017*
64% (2,360)	56% (931)	6% (237)	6% (86)
Hope		**Rejection**	
2011	*2017*	*2011*	*2017*
25% (1,012)	30% (515)	5% (220)	8% (106)

Note: 462 respondents in GKAP 1 and 139 respondents in GKAP 2 are not included here, because there are missing values for them on one or both scales.

God's will to environmental stress. Both Skeptics and Rejecters are small tranches of the American population, with fewer than a tenth in each group, although the Rejecter group may be growing. Their small numbers belie their analytic and political importance, as we shall see throughout the rest of *Genomic Politics.*

Why Are Americans Enthusiastic, Skeptical, Hopeful, or Rejecting?

Given that Americans can be statistically and conceptually sorted into the four quadrants of *Genomic Politics*'s basic framework, the next question is why. If quadrant location differs from so many other moral and policy disputes in the United States in *not* being connected with identification as a Democrat or Republican, what does lead some people to agree calmly that "justice is worth a swab" while others furiously insist, "Get your swabs out of my face"?

Characteristics of Respondents in Each Quadrant

The first clue to why lies in the fact that residents of each quadrant show some distinct characteristics; Figure 7.5 portrays their profiles in the 2011 survey. (See the Appendix to Chapter 7 for more detail, and for parallel results in 2017.)

The three-fifths of GKAP 1 respondents who are Enthusiasts are disproportionately White, and a higher share are over age fifty than in the other quadrants. Enthusiasts are especially likely to have, or to be closely connected with someone who has, a genetic disease. Compared with the other two quadrants, Enthusiasm and Skepticism both show higher proportions of college graduates and lower proportions of people with less than a high school degree. The Enthusiasm quadrant contains the smallest share of people who would completely fail the genetics knowledge test, but not the largest share of people who would end up with a gold star.

In the sharpest contrast to Enthusiasm, the Rejection quadrant disproportionately comprises Latina/os and people with very low income. Rejecters are much more likely to have no genetics knowledge than people in the

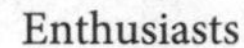

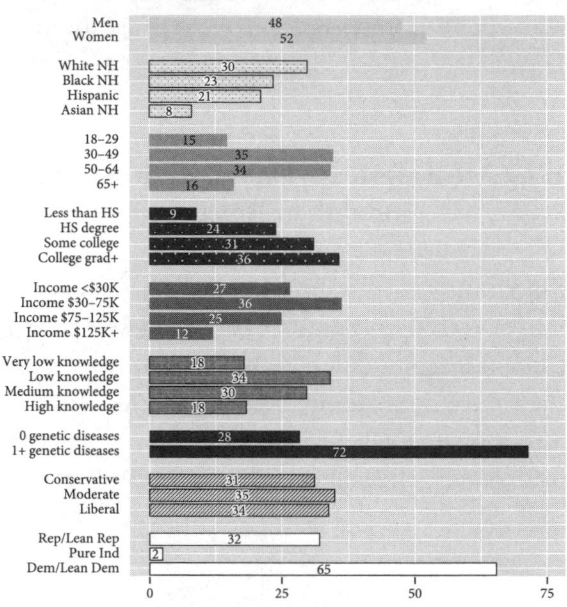

Skeptical

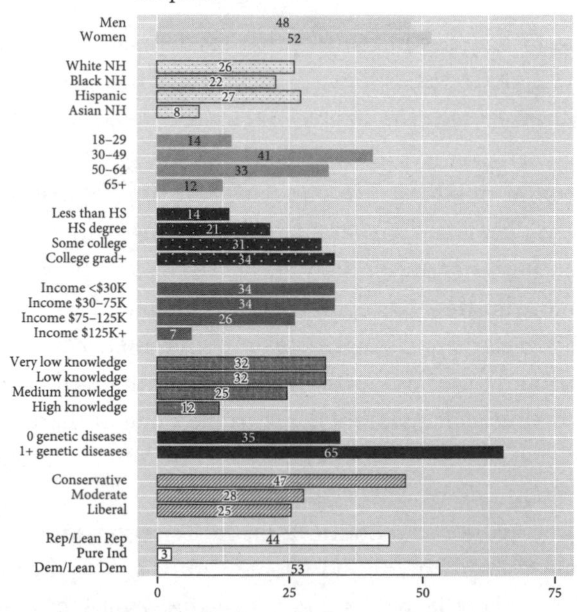

Figure 7.5 Demographic profiles of respondents in the four quadrants of the basic framework, GKAP 1

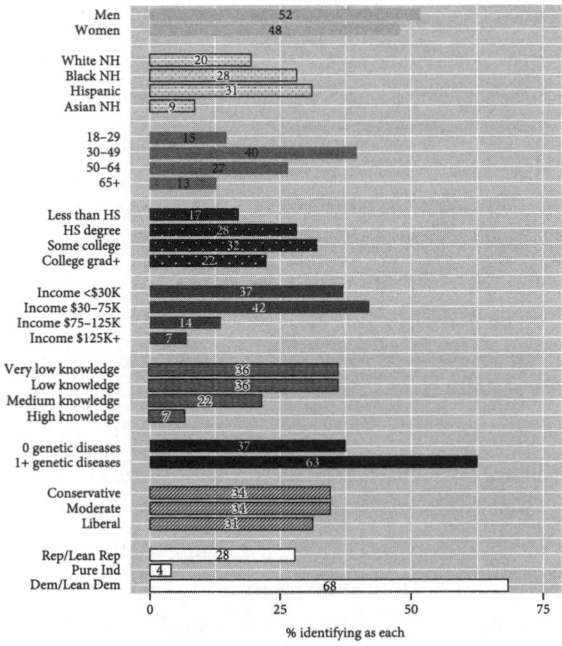

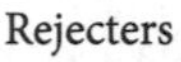

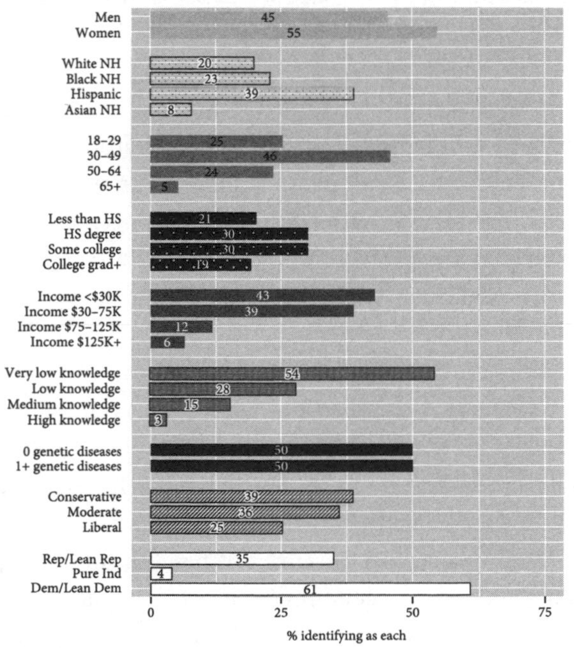

Figure 7.5 Continued

other three groups, and are much less likely to have, or to be connected with someone who has, a genetic disease. Skeptics are more conservative than the other three groups and less likely to be Democrats; Rejecters follow that pattern to a lesser degree. Enthusiasts and the Hopeful do not differ in ideology or partisanship.

In sum, Enthusiasts are most, and Rejectors least, likely to have the profile of someone with high social status. The other two groups are in between in one way or another. But one has to squint to see that pattern; all of this is fairly weak tea, to mix a few metaphors. The message that shines through in Figure 7.5 is that demographic characteristics do *not* sharply differentiate among the quadrants. (That point is even clearer when one examines the error bars in Figure A7.1 in the Appendix; they show indeterminacy in all quadrants except Enthusiasm, even in the large GKAP 1 sample.)

A regression analysis, which reveals the importance of each characteristic compared directly to the importance of all others, clarifies these descriptive patterns. Figure 7.6 presents its most important results in the form of predicted probabilities for one analysis in 2011.[18]

This figure makes three points, so it is worth lingering over. The most important result is by now familiar: comparing across the four quadrants, it is clear that the likelihood of being an Enthusiast is much higher than that of being in any other quadrant, and the likelihood of being Hopeful runs (a distant) second.

The second point is subtler: within the Enthusiasm quadrant, all of the lines slope upward as they move toward the right, from no genetics knowledge (0 on the x-axis at the bottom) to complete knowledge on the GKAP scale (3). That result shows that people with more genetics knowledge are even more likely than people with less to be Enthusiasts.[19] The opposite pattern holds, more weakly, in the Hope quadrant; the more one knows about genetics, the less likely one is to be Hopeful.

The final point shown in Figure 7.6 is actually an important nonresult. Within each quadrant, the darker and lighter lines represent partisanship, ranging from strong Democrat (darkest on top) through Independent to strong Republican (darkest on the bottom). But it is not worth trying to sort out which shade is associated with which partisan stance: all of the lines are close together and run almost parallel.[20] Whatever is nudging people into one or another stance toward genomics, it is not their partisan identity.

Because the implications of genomics for race and ethnicity are so central to how people view the science, it seems worthwhile to examine the role

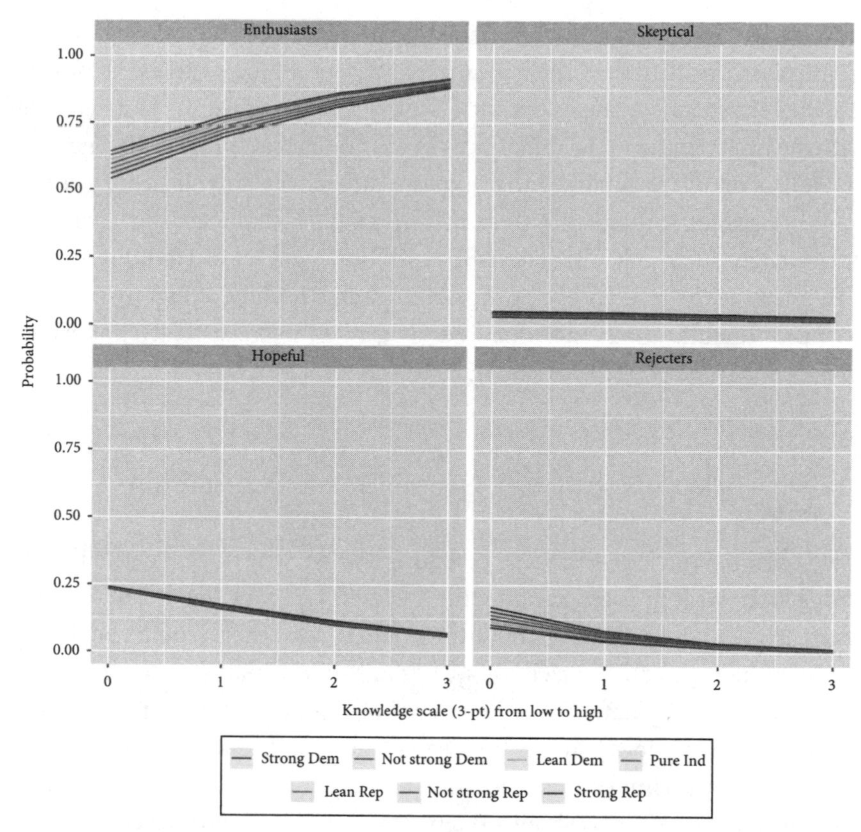

Figure 7.6 Impact of genetics knowledge and partisanship on location in the basic framework, GKAP 1

of self-identified race or ethnicity in determining quadrant location. The predicted probabilities that result from focusing on race and knowledge, rather than partisanship and knowledge, show the same three patterns as we see in Figure 7.6. (See Figure A7.3 in the Appendix.) First, the highest proportion of respondents are Enthusiasts. Second, more genetics knowledge is associated with an even greater likelihood of being Enthusiastic and a lower likelihood of being Hopeful. Third, within a given quadrant, the share of the probability of being in that quadrant that depends on level of knowledge is pretty much the same for Blacks, Whites, and Latina/os.

In sum, race matters about as little as partisan identification does—another surprising feature of genomic politics. It is not quite irrelevant: in 2017 (though not in 2011), Blacks are a bit less likely than Whites and Hispanics to be Enthusiasts and a bit more likely to be Skeptics at every level of genetics knowledge. There may be a trend here—knowledgeable Blacks are slightly less Enthusiastic in 2017 than their counterparts were six years earlier. These disparities are small; we must wait for a future iteration of GKAP to determine whether meaningful racial differences in quadrant location are emerging.

Across all of the analyses that explore differences among people in the same quadrant through the filter of respondents' characteristics, genetics knowledge is the only characteristic worth paying a lot of attention to. No other trait of the type shown in the demographic profiles of Figure 7.5, such as party identification, race, income, or gender, is consistently important across quadrants, years, and statistical tests.[21]

Science Knowledge, Genomics, and Partisanship

Two questions emerge from that conclusion. First, why does genetics knowledge loom large in characterizing Enthusiasts? Just over 50 percent of those with no genetics knowledge are Enthusiastic by my criteria (believing in genetic influence *and* being technology optimists), whereas close to 90 percent of those with full genetics knowledge are Enthusiasts. The more pointed version of "why?" is: does genetics knowledge cause Enthusiasm?

Some say yes; a line of argument sometimes termed scientific literacy holds that the more one knows about a science topic, the more one develops favorable views about it. That is the impetus behind curriculum requirements for public school science classes, PBS shows about the wonders of nature, and the conviction that with sufficient explanation from experts, all Americans will be willing to wear masks to reduce Covid-19 transmission. But others see a reverse causal arrow: enthusiasm about a subject leads one to seek out more information about it. That is the logic of science fairs, hands-on museums, and the board game **Metanon** for young children.[22]

Without detailed evidence that might enable more incisive answers for this particular case, I opt for "both/and"—mutual causation. The social science experts in Chapter 6 illustrate that two-way causal arrow at a high level. Scholars who have chosen to learn a great deal about genetics tend, with important exceptions, to become excited about its impact and use—*and* scholars who are technology optimists about this powerful new tool for changing lives tend to develop expertise about it. The public arguably has the same two-way motivation. On one hand, as we see in the demographic profiles of Figure 7.5, Enthusiasts overall have more years of schooling than do the other three groups, which suggests that higher levels of education in someone's younger years generate positive responses to this new science later in adulthood. On the other hand, Enthusiasts also report more direct, personal experience with genetic diseases than do the other three groups, which suggests that firsthand engagement with medical uses of this new science motivates one to learn more about it. The crucial point is that although knowledge is not necessary in order to be in the Enthusiasm quadrant (half of those who would have scored a 0 on the genetics knowledge test are Enthusiasts), it is close to sufficient (nine-tenths of those who would have aced the test are in that quadrant).

The second remaining question is whether partisan identity, to choose the most salient respondent characteristic in a book on genomic *politics*, generates distinctions among quadrant residents with a given level of knowledge.[23] That is, do knowledgeable Republican Enthusiasts hold different views than equally knowledgeable Democratic Enthusiasts? That is sometimes the case, as elegantly shown in legal scholar Dan Kahan's research on views of climate change. Overall, his survey respondents who score poorly on an eighteen-item measure of "ordinary scientific intelligence" differ little from high scorers in acceptance of evidence that recent global warming is largely due to human activity; one's level of knowledge by itself has little impact on policy views. The picture changes, however, when one focuses only on self-identified liberal Democrats and conservative Republicans. Among those low on scientific intelligence, the opposing partisan groups again do not differ in views of climate change. However—and this is the punch line—the higher their scientific intelligence score, the further apart liberal Democrats and conservative Republicans move in beliefs about climate change's causes. Knowledgeable conservative Republicans are *less* likely to accept human action as an explanation of climate change than are ill-informed conservative Republicans. Conversely, knowledgeable liberal Democrats are *more* likely to accept human action as an explanation of climate change than are ill-informed liberal Democrats.[24] In short, the more one understands what the fight is all about, the more one sees how to align his or her partisan loyalty with the "right" position—which differs for liberal Democrats and conservative Republicans.

Compare those results with the patterns in Figure 7.6 (and Figure A7.2 in the Appendix) on the predictive impact of genetics knowledge and partisanship on location in the basic framework. Although the data and methods differ somewhat, so the comparison is not perfect, the pattern of results is clear: GKAP shows the opposite of what Kahan finds with regard to climate change. In all four quadrants of both survey years, the closely parallel lines show that Republicans and Democrats track one another as respondents move up the genetics knowledge scale. (Substituting political ideology for political party identification shows the same results.) I find no evidence for Kahan-style partisan cultural commitment regarding genomics.[25]

To conclude so far: about three-fifths of Americans would empathize with the researchers in Mary-Claire King's lab who are banging down the doors at 7 a.m. to see how their experiments developed overnight. The more genetics knowledge people have, the more keenness they display. Republicans are very slightly less excited by genomics than Democrats at all levels of knowledge, and by 2017, Blacks were slightly less excited than Whites and Hispanics, again at all levels of knowledge. As of 2017, statistically speaking, the people most likely to resist the Enthusiasm bandwagon are those uninformed about genetics, extreme conservatives, strong Democrats, African Americans, and young adults. That set of characteristics makes for an unlikely, possibly unheard-of coalition; no wonder it is difficult to parse the politics and ideology of genomic science.

In Their Own Words

> Because I no longer trust the government, or politicians, or banks, or just about any institution. Period.
>
> —Respondent in GKAP 1

> If you are an honest law-abiding person, what is the problem [with donating a DNA sample to a forensic biobank]?
>
> —Respondent in GKAP 2

Analysts typically explain respondents' choices as I just did, by identifying statistical associations between other questions in the survey and the outcomes of interest. What correlates with what, and how strongly? Can we tease out a causal argument from correlations? Such an analysis tells us the type of people who make a given choice and how their views are linked, from which we seek to infer why they choose as they do. Or, one can just ask people why they made a particular choice. Although that strategy also has flaws (who among us really knows, or will tell a stranger or a text box why we think a certain way or have taken a certain action?), GKAP followed it for several topics.

To ensure that the forest is seen through the many trees, let me start by pointing to three conclusions from this exercise. First, rather to my surprise and relief, GKAP respondents are willing and able to explain why they would act in a particular way with regard to issues that most of them have never heard of. Their comments are necessarily shallowly rooted, but they are not shallow. Also to my relief, residents of the quadrants present distinct profiles. There is, of course, overlap and noise (qualitative evidence cannot be corralled as quantitative evidence sometimes can be), but clear and explicable differences emerge. Second, as all previous analyses in *Genomic Politics* would lead us to expect, differences across quadrants are not partisan and not even clearly ideological, at least by conventional measures of liberalism and conservatism.

Finally, in ways that *Genomic Politics* has not to this point anticipated, although the vertical and horizontal dimensions of the basic framework are symmetrical in theory, as is shown in Figure 2.2, they are not symmetrical in practice. The confidence of technology optimism or the reservations embedded in technology pessimism divide respondents much more sharply than whether they perceive genetic influence to be a primary, secondary, or nonexistent explanation for human traits and behaviors. Let us examine the evidence behind these conclusions.

Characterizing Opinions on DNA Databases

Both online GKAP surveys include questions about scientific and forensic DNA databases. After scientific biobanks are defined, respondents are asked about their awareness of such entities, support for government funding and regulation of them (separate items), and hypothetical willingness to contribute a DNA sample. They are then presented with a text box and a request to explain why they would (or wouldn't) be willing to contribute. The same sequence followed for forensic DNA databases (see the Appendix for definitions, question wording, and coding details).

I anticipated a smattering of comments that would be useful for illustrative purposes and to liven up the narrative. To my surprise, only 729 and 886 of about 4,300 respondents in GKAP 1 left the text boxes, for medical and forensic biobanks respectively, empty. The comparable numbers for GKAP 2 were 370 left empty for medical and 443 for forensic biobanks out of almost 1,800 respondents. That left more than 11,000 responses in the text boxes. Many completed comments were brief, ambiguous, or otherwise hard to interpret—but torrents of people were generous enough to explain why they would or would not be willing to contribute a DNA sample to a database.

Setting aside comments that focus specifically on biobanks, many point in a wider or more general direction. Some are telegraphic and incisive: "Get your swabs out of my face!" (forensic, 2011), or "science have come a long ways" (medical, 2011). Others are poignant and revealing: "I have had friends who experienced brutal rapes. To stop a serial raper/serial killer having DNA is critical. I can't understand why someone would object to having their DNA on file" (forensic, 2017). Or "My son has a blood disease that produces blood clots in his body at any reason. The doctors at Mayo Clinic don't know what causes it. So he has to take blood thinners for the rest of his life. I would do anything to help figure out why or what reason has caused this disease" (medical, 2011).

To understand how Enthusiasm, Rejection, and the other quadrants feel from the inside, I cluster respondents by their quadrants (determined by their views on genetic influence and technology optimism, as explained earlier in this chapter), then examine each cluster's set of comments. I use a coding scheme developed and revised over six years. Three sets of student research assistants coded, recoded, and reconciled responses into what ended up as ten substantive "buckets." I reviewed all coding decisions (many times!), and a graduate student reviewed a lot of my choices about ambiguous statements; she made the final decisions.[26] (See Appendix to Chapter 7 for more detail on the coding.)

One must handle any numerical analysis, even simple percentages, of coded comments gingerly. Most respondents presumably offer the first consideration that occurs to them, rather than deeply considered views.[27] In addition, there are many plausible coding systems, and coders' judgments differ no matter how careful the system and detailed the instructions; coding is more an art than a science. Nonetheless, simple counts point us toward distinctive patterns of attitudes about genomic technologies across quadrants, as well as toward especially prominent reasons for those attitudes. Beyond that framing, careful reading and interpretation take over.

Table 7.3 shows respondents' respective willingness, across the four quadrants, to at least consider contributing to a biobank. As the table shows, Enthusiasts are a little more willing to participate in medical than in forensic biobanks, while Skeptics are the opposite. But we see relatively little difference within a quadrant in willingness to contribute to one or the other DNA database, despite their distinct purposes and uses. Instead, the gap lies between Enthusiasts and the Hopeful, on one hand, and Skeptics and Rejecters, on the other. The two groups on the left side of Table 7.3 are at least twice and sometimes several times more willing to donate a DNA sample to both types of biobank than are the two groups on the right side.

Given the basic framework's symmetrical structure, it is as plausible *a priori* that Enthusiasts and Skeptics would resemble one another as much as Enthusiasts and the Hopeful would do. That is not the case: Americans are more divided by technology optimism or pessimism than by perceptions of genetic influence. We will see that pattern many times over the rest of this book.

Table 7.4 takes the next step in understanding how people view the quadrants from the inside. It deconstructs the abstract dots of Figure 7.4 by presenting the proportions of each quadrant in specific coding buckets (some more substantively meaningful than others) that show reasons for eagerness or caution about biobanking.

Table 7.3 Proportion of respondents in the quadrants willing to consider contributing to a scientific or forensic DNA database, GKAP 1 and 2

Enthusiasm		**Skepticism**	
2011	*2017*	*2011*	*2017*
Would contribute to scientific biobank			
72% (1659)	72% (754)	27% (81)	27% (34)
Would contribute to forensic DNA database			
65% (1741)	66% (666)	32% (101)	34% (39)
Hope		**Rejection**	
2011	*2017*	*2011*	*2017*
Would contribute to scientific biobank			
63% (647)	56% (209)	20% (66)	32% (24)
Would contribute to forensic DNA database			
62% (410)	59 %(213)	22% (50)	31% (24)

Table 7.4 Proportions of each quadrant in specific coding buckets, GKAP 1 and 2

	Enthusiasts		*Hopeful*		*Skeptics*		*Rejecters*	
	2011	*2017*	*2011*	*2017*	*2011*	*2017*	*2011*	*2017*
			Medical					
Mistrust	8%	8%	8%	10%	14%	44%	5%	19%
Privacy	8	8	10	9	35	9	11	13
Research	43	49	26	33	10	18	10	17
Self-interest	13	11	16	16	12	10	17	15
Value driven	13	10	16	13	8	1	8	13
General negative	8	8	17	17	17	14	44	18
General positive	6	4	4	3	4	5	4	2
Unclear	1	1	3	0	0	0	0	4
			Forensic					
Justice	26%	23	20	20	10	13	6	11
Mistrust	7	12	9	13	21	23	11	15
Privacy	7	8	6	9	23	10	11	5
Research	4	5	4	5	2	2	1	0
Self-interest	10	13	14	11	12	5	13	20
Value driven	10	12	12	13	4	6	5	7
General negative	19	16	20	21	27	25	45	34
General positive	13	9	14	7	14	8	5	8
Unclear	0	1	2	1	0	1	0	3

Medical Biobanks

Enthusiasts stand out for their commitment to research. In 2017, almost half offer it as the reason for their willingness to submit a sample—between two and four times the proportion of Skeptics and Rejecters and roughly one and a half times as many as the Hopeful. (This is the first of many GKAP analyses in which the Hopeful are in the middle of the four groups.)

"Research" encompasses many ideas. At the narrower end of its meaning lies a hope that biobanks can alleviate particular diseases or conditions; as an unusually knowledgeable respondent writes, "This sounds like a valid area of research, especially as it ties to the understanding and treatment of diseases related to genetics such as sickle-cell anemia and Charcot-Marie-Tooth disease. If I can help with a mouth swab, I would" (2011). Another would contribute because a medical biobank "could be used for testing on high blood pressure in African Americans" (2011). A broader formulation of "research" links to other societal uses of genomics: respondents would contribute in order "to advance mankind's understanding of the human body and genome" (2017) or "to be able to add to the body of scientific research. I believe much good will come out of biobanks in the future" (2011). Others point to even broader convictions about scientific knowledge: "Me gusta contribuir con los avances para un futuro mejor. Y ser parte de nuevas curas para generaciones futuras seria un honor" (2017) or "I think genetics are a key to society knowledge needs." (2011). Most simply, "science is exciting" (2017).

In addition to focusing on knowledge, Enthusiasts are not the only group to reach for values to explain their willingness to contribute to a medical biobank, but they do stand out in the particular values that they adduce. In GKAP 1, compared with fewer than 3 percent of Skeptics or Rejecters, fully 11 percent of Enthusiasts use the word "help." "I am in favor of helping the medical field in future success"; "To help protect our children and advance products for country and the world"; "Anything that will help humanity"; "Porque todo lo que nos pueda beneficiar, lo considero importante."

The Hopeful resemble Enthusiasts not only in expressing willingness to contribute to a medical biobank but also in their reasons. More than a quarter invoke research. The range of comments resembles those of Enthusiasts, from cures for a particular disease to promoting knowledge to benefit humanity. However, this presents a puzzle: given that, by definition, the Hopeful perceive less genetic influence than do Enthusiasts, why are so many willing to contribute to a DNA database? The answer lies mainly in their technology optimism: "I think my participation would benefit scientific research and progress the understanding of the human genome"; "The outcome of this study will help many millions of people's life"; "Any research that can help us understand each other better, where we come from and how it affects us all, is important. If all it takes is a swab from me to support that effort then I am willing" (all 2017).

As the coded responses in Table 7.4 show, the Hopeful are marginally more likely than the other three groups to invoke a value to explain their biobanking choice. Over a tenth, second only to Enthusiasts, offer some version of "Anything that helps others helps me." The Hopeful are most distinctive, however, in their curiosity (coded as a subset of self-interest)—which helps to explain why people who perceive little genetic influence nonetheless see value in biobanking. Fifteen percent in 2017, compared with 10 percent or fewer of the other three groups, chiefly want to know what a DNA test would tell them about their own genomic profile, sometimes including ancestry. Others among these technology optimists are simply eager to learn more: "No real reason. Could be interesting"; "It'd be COOOL"; "For the experience and being able to participate" (first two are from 2011; third is from 2017).

Skeptics are distinctively mistrustful, and are most concerned about privacy. Although we sought to distinguish them in coding, these two reasons are often closely linked: "Privacy. Mistrust of government," or "I want to protect my privacy and I don't trust the government to do that." And not only government: "Snoop. Snoop. Snoop. It's bad enough that companies keep track of my shopping information just so that I can save a nickel on a bar of soap" (all 2011). In both GKAP 1 and 2, the combination of privacy and mistrust comprises half or more of Skeptics' reasons for *not* being willing to contribute to a medical biobank—and the comments are usually what some respondents themselves call "spicy."

Privacy is sometimes broader even than mistrust of government or other organizations. Many Skeptics simply wrote "privacy" in the textbox, while others make the same point more expansively: "An invasion of personal privacy (not that we Americans have any left as it is)" (2011) or "NONE OF ANY BODIES BUSINESS" (2017). Furthermore, Skeptics, along with Rejecters, are concerned about genomic privacy in arenas other than biobanks. GKAP 1 includes a vignette that describes a hypothetical Maria/David (same gender as the respondent) who takes a medical test to determine the probability of inheriting a disease. The survey then asks respondents whom they would tell about its results if they were Maria/David. Three-tenths of both Skeptics and Rejecters say "no one," compared with 6 percent of Enthusiasts and 13 percent of the Hopeful. Skeptics vie with Rejecters for being least likely to share their test results with most of the seven remaining categories of possible confidants, such as family and friends, medical professionals, clergy, or co-workers.

Conversely, mistrust of the (usually generic) government is sometimes independent of or moves beyond privacy concerns: "I think that the government has way TOO much liberty as it is. We (the people) are not slaves nor owned by them. It should be illegal for them to have that much control over people" (2011). Or "my DNA is my personal property and no one has the right or need to use it. Furthermore it is not just used for these tests it is used for things such as the police department and FBI and NSA to track you it is wrong and unethical to collect and store any ones DNA for such purposes" (2011). Respondents can make favorable references to the government—"Possibly donate to research but only for an accredited university or scientific organization or a U.S. government agency" (2017)—but those are rare; almost every reference to government is excoriating.

This expression of hatred for government is especially striking in the medical arena, since the scientific biobank questions precede those about forensic DNA databases and refer to "organizations" and "scientists"—not to any sort of public sector actor. Nonetheless, a quarter of Skeptical respondents in 2011 and a fifth in 2017, compared with fewer than 5 percent of Enthusiasts and Hopeful in both years, refer to the government in explaining their unwillingness to contribute to a scientific DNA database. (As always, take these numbers with a grain of salt. I searched open-ended responses for "government," "govt," govmt," and "gobierno," but I may have missed some spellings. So these estimates are likely to be low.)

Skeptics' mistrust is, in the end, protean. The government is voracious ("Would never help a greedy government without compensation. Simply because they are greedy and only do anything that will get them more money, or land" [2017]) or ideologically biased ("I don't trust any government to do anything where there is not a motive bias toward any current PC thinking" [2011]). Elites are corrupt ("I don't understand what this will be used for. Generally speaking, scientists nowadays are doing immoral research, and putting much of the results to immoral purposes") or racist ("African Americans have a long scary history with governmental agencies and labs. Need I say more???" [both 2017]). Or simply "Don't trust anyone to do right" (2017).

Although, as the percentages in Table 7.3 show, Rejecters resemble Skeptics in unwillingness to contribute a DNA sample to a medical biobank, their profile is distinct despite their small numbers. As we see in the two final columns of the coded buckets in Table 7.4, Rejecters evince no strong concern about privacy or mistrust in 2011, and are only moderately concerned in 2017; their distinguishing characteristic is a global, terse, indeterminate refusal to engage. Forty-four percent of Rejecters' comments in 2011 were coded as "general negative," compared with 8 percent among Enthusiasts and 17 percent among the Hopeful and Skeptics. Typical Rejecter comments are "not interested," "I DO NOT CARE," "no," "nothing," or "not going to do it."

If Skeptics are furious at their powerlessness and vulnerability, the modal Rejecter simply opts out—of societal involvement as well as GKAP's invitation to communicate. I have no good explanation to connect this divergence between the two quadrants with shared technology pessimism but different perceptions of genetic influence; my best guess is that Skeptics see a lot of societal harm in the growth of genomic science and are motivated to fight it, whereas Rejecters dismiss the whole enterprise as a waste of time and effort in which they have no stake.

Forensic Biobanks

In broad strokes, each quadrant's profiles in the forensic arena resemble those in the medical arena. There are, nonetheless, some variations consequent to the different policy purpose and political context.

Not surprisingly, Enthusiasts are most willing to contribute to a forensic DNA database, although not quite as much in the medical arena. As the coded responses in the second section of Table 7.4 show, Enthusiasts' distinctive emphasis is "justice," which takes several forms. Some express faith in genomics' capacity to prove their own innocence: "In case I get in trouble they'll have my DNA to prove if I'm innocent or guilty," or "Because if I didn't do a crime but someone says I did and DNA could prove I didn't, then it's useful" (both 2011). Many offer a broader remit, arguing for the general value of genomic science in determining guilt or innocence: "Because I believe DNA is more evident than things like eyewitness testimony in proving guilt or innocence," or "If it shows the differences in DNA to rightfully convict someone I would be willing to donate my DNA" (both 2011).

Some focus on exoneration or the need for it: "There are incidents/facts that have been identified as innocence of prisoners with the DNA test results after many years of sentencing," or "If someone is wrongfully convicted and my DNA can help exonerate them then that is a good thing" (both 2011). And a few invoke justice to explain *un*willingness to donate to a forensic biobank: "IM NOT TRYING TO SEND NO ONE TO JAIL" (2011).

Finally, some Enthusiasts in the "justice" bucket place their emphasis more on optimism about genomics' broad societal benefits than on a particular desire to improve the criminal justice system: "To build a better future and if my DNA helps in any way to the criminal justice, I would be happily willing to provide my DNA," or "If it could help solve mysteries, I would be willing to help" (both 2017). In sum, "The truth shall set you free!!" (2011). The same faith in genomics technology may be expressed with the opposite trope—a figurative shrug and a literal "Why not?" in the text box. (Enthusiasts are ten times more likely than Skeptics in 2011 to endorse forensic DNA databases with that exact phrase.) "Why not?" is sometimes preceded by "I have nothing to hide, so . . ." or "I'm not guilty of anything and believe in the science of DNA, so . . ." And contra the Skeptics' fear of government overreach, the general insouciance implied by the phrase can be explicit: "Why not? If they mess up . . . oh well, for me" (all 2011).

In the forensic as in the medical arena, the Hopeful follow the same patterns as Enthusiasts, but less so. Table 7.3 shows them to be almost as willing to contribute a DNA sample. The details of the second section of Table 7.4 show them to be the second-most-likely group to invoke justice and its variants in explanation. The Hopeful are also the second-most-likely group to dismiss pessimism with a metaphorical shrug and a literal "Why not?" As in the medical arena, the Hopeful want to help ("Because I would like to help that person why not nothing is going to happen to me I wont get sick or anything") and to satisfy their curiosity ("Para enterarme de una experiencia nueva" [both 2011]) or, simply and powerfully, "To see what it look like" (2017).

For Skeptics too, their profile in the forensic arena resembles that of the medical arena. They are the most likely to express mistrust and concerns about privacy, corruption, and governmental overreach. They may be conservatives who prize law and order, but not at the cost of risking misuse of their DNA. And in any case, the government is incompetent if not malevolent:

- I do not trust the Federal Government, and those that speak out often are framed or disappear entirely. (2011)
- I don't trust that the information gathered would be kept private, confidential and not used against me in the future. For example, because business (corporations) own the government, they would find a way to obtain the information and use it for profit or an invasion of my privacy. (2011)
- I just don't trust the government and law enforcement in these manners. I, paranoidically admit, see a Hitler and medical experimentation and future control of who lives or dies being based on this sort of thing. (2011)

- Because as an African-American, I know the history of the United States exploiting the use of cells from African-Americans. (2017)

Rejecters, as in the scientific arena, specialize in vague, terse, or summary statements about their unwillingness to contribute to a forensic DNA database. Almost half of their answers in 2011—compared with a quarter of Skeptics' and a fifth of Enthusiasts' and Hopefuls' answers—were again some version of "no," "not my job," "not interested," "no time," or "I've told you I have better things to do." Once again, Skeptics tend to be deeply and unhappily engaged with the issue of societal use of genomic science, while Rejecters tend to have no investment. Rejecters may be rejecting society—and governance, as we will see in Chapter 8—as well as technology and genomics.

Three conclusions rise above the evocative details: First, even Americans who care little and know less about societal uses of genomic science separate into groups that are genuine, multifaceted, and intelligible. The quadrants cohere statistically, and comments are meaningful and sometimes moving; they remind us forcefully that there are real people behind survey data. Respondents who know more about genetics differ from those who know less, but deeper disagreements have to do with levels of optimism and beliefs about causation.

Second, once again, these deeper divisions map poorly onto conventional political categories. Except for the possible emergence of shared views among extreme conservatives and strong Democrats, regression analyses show only marginal impacts of partisan identification or conventional ideology on location in the quadrants. And in 12,000 opportunities to present views of biobanking in Americans' own words, the terms "Democrat," "Democratic," "Republican," and "conservative" are completely absent.[28] "Liberal" appears once, in a comment about gun control. Instead, residence in a quadrant is associated mainly with words or phrases such as "to help," "why not?," "privacy," "mistrust," and "research."

Nor do GKAP respondents echo advocates' and experts' concerns about the invidious impact of genomics on particular racial groups or sexes. Across the 12,000 opportunities for comment, "Black" appears six times, "African" ten times, and any variant on "Latin," "Asian," or "Mexican" once each. A few refer to their race to justify refusal to contribute, but most of these (few) references point in the opposite direction—to possible benefits of DNA research for people of different races, or as a counter to racial discrimination in the criminal justice arena. Variants of "race" or "racism" appear about thirty times in the 11,000 responses, mainly for the same two optimistic reasons. A few would condition their contribution on an assurance that a forensic DNA database would not be used in any racially biased way, a few praise the use of DNA for exoneration and hint at a racial inflection, and a few condemn racism outright. One seems to suggest that DNA testing will show that blacks commit more crimes. Several people use "White," but always to describe themselves as possibly useful controls for data analyses in the medical arena. About twenty people seek to help "the human race."

"Sex" appears once, in a comment that DNA evidence can help to solve sex crimes; "gender" appears once and "female" not at all. "Male" appears three times—once in reference to male pattern baldness, and twice as a reason for a Black man not to contribute to a forensic DNA database. There is one reference to religion.

This mismatch between the public's lack of focus on group identity (at least as expressed in a survey) and the apprehensions or commitments of experts and advocates does not, of course, mean that the apprehensions are wrong. Almost by definition, experts and advocates have spent much more time than the public at large pondering the risks as well as the benefits of DNA databases. The mismatch mainly underscores my point that genomics, although deeply contentious, is not (yet?) politicized in the United States along predictable lines, unlike so many scientific arenas. Whether the Covid-19 pandemic—with its disproportionate harm to people of color and its entanglement with mistrust of vaccines—combines genetics, race, government, and fear in a new, polarized, public discourse remains to be seen.

Returning from speculation to evidence, my final conclusion here points back to the basic framework. Both dimensions—genetic influence or its absence in explaining phenotypes, and technology optimism or pessimism—contribute about equally to the statistical distinctions among GKAP respondents shown in Figure 7.4. But the coded open-ended comments reveal that levels of technology optimism do much more to distinguish among views of societal uses of genomics than does the extent to which people believe in genetic influence.

These three conclusions—intelligibility, unusual clustering, and the prominence of risk calculations—and the ways that they echo among experts, advocates, and public officials as well as in the general public, are the terms on which managing and governing genomic innovations will be built. That is the subject to which we now turn.

8

Who Should Govern?

> Can we use genetics/DNA to create new capacities and functions? . . . We are looking at catastrophe if the political arena doesn't catch up to the last ten years of science. People *care* about bodies and science.
>
> —Interview subject

> All of these issues are better framed in mixed terms—whether the development and utilization of new techniques is worth the benefits or not. The question then becomes whose insight should be trusted and . . . who should ultimately decide about how the new technologies are used.
>
> —Heather Douglas, 2015

> It is inherently a distributed system.
>
> —Respondent to social science survey, 2018.

These speakers, whether they know it or not, are inheritors of the twentieth century's experience with the new science of physics. As *The Economist* observed in the passage quoted as an epigraph to *Genomic Politics*, explorations of the physical world by Albert Einstein, Niels Bohr, Werner Heisenberg, Marie Curie, and others generated "a feeling of advancing into the unknown" a century ago. The epigraph's invocation of "both good and ill" outcomes turned out to be exactly right for physics; that new discipline gave us the horrors of nuclear warfare as well as the wonders of space flight. If *The Economist*'s analogy between twentieth-century physics and twenty-first-century biology is warranted, we may indeed be both "creat[ing] new capacities and functions" *and* "looking at catastrophe," as one of my interview subjects put it.

So, "who should ultimately decide about how the new technologies are used?" philosopher Heather Douglas asks. Here is where physics and genomics diverge. Control of the most important societal uses of physics began and mostly continues to rest in national governments, often surrounded by secrecy and seldom with any influence by outsiders. Genomic science has not taken that course; in the United States and many other nations, it remains largely the domain of bench scientists, private corporations, state or local criminal justice systems, genealogical hobbyists, and garage-based hackers. But as Americans increasingly wrestle with less reversible genomics choices—genetically modified food, gene drives, prenatal genetic editing, synthetic biology—they will need to collectively decide "whose insight should be trusted." *Genomic Politics* cannot answer that question; it is too big for anyone to predict or adjudicate. But this chapter provides ways to organize, empirically and normatively, possible answers, while Chapter 9 tiptoes into my own.

To preview my findings about Americans' views on who should govern: the most accurate answer at this point is "no one"—or possibly "everyone" or doctors. That is, social science experts, interviewees in the policy arena, and the general public collectively offer nothing remotely resembling a consensus in response to Douglas's query about who should ultimately decide. Political partisanship and race are associated with some differences in views, but residence in the four quadrants continues to show much stronger variation. The most deeply rooted societal uses of genomics, in particular forensic DNA databases, enjoy the strongest support, while the newest possibilities—somatic and especially prenatal gene editing—engender widely varying reactions. But I see no emerging central driving principle; governance may turn out to be as difficult as it is important.

Experts' Views on Genomics Governance

After asking social scientists and legal scholars about the risks and benefits of genomic science, the online surveys described in Chapter 6 ask what actor(s) should, and should not, govern societal uses of genomics, and why.[1] The surveys also inquire about relevant potential decision-makers who are left out. Interviews cover the same topics, more conversationally and idiosyncratically.

Responses range all over the place.[2] A few respondents resist public oversight, at least in their own research arena: "I don't think the science is anywhere near the point that policy decisions should be made from any of the findings within my area of expertise" (2018). A scholar who works "mainly in pure research" does not "want politicians or any of these other groups to get involved as there's very little that I do that has any likely way of harming anybody. With genomics in general, it's much too early to tell who should be involved" (2018). Nonetheless, most concurred that "genomic science is evolving so quickly and will start to influence our daily life soon. So it is time that national and international policy decisions are being made" (2018).

Concurrence ends, however, with agreement that some sort of governance is needed. Survey respondents collectively suggest a daunting array of appropriate governance or policy actors. They also collectively suggest an equally daunting array of *in*appropriate governance or policy actors—often the same ones. Consider first the possibility of scientists or other experts as managers of the societal uses of genomics (emphases added to all quotations, unless otherwise noted):

Yes	No
Scientists [are best suited]. In my experience, they are relatively cautious, and they tend to pursue projects due to genuine curiosity about processes of human behavior, not because they are pursuing particular agendas. (2013)	"Discovery" *scientists*—empirical evidence demonstrates their tendency to overvalue benefits of genomic tests and technologies. Too close. (2013)
The NIH priority-setting mechanisms. *Biomedical scientists* are primarily important. (2018)	The worst thing in the world would be if *genetics experts* were overrepresented among those making policy decisions. (2018)
Deliberative groups composed of experts from varied fields of expertise from genomics to ethics acting with public transparency and feedback are going to be a crucial part of putting together sound public policy. (2018)	

Yes	No
Those who have the time and the desire to learn about the issues and consider the range of perspectives. It is difficult to think who that would be—*transdisciplinary bodies* whose deliberations are open would be best. (2018)	
Continental science organisations like the African Academy for Science and the African CDC; continental policy organisations like the AU and more pertinently NEPAD. (2018)	

Along with or instead of scientists, how about social scientists or other scholarly experts?

Yes	No
Among the many bad alternatives, I would choose *bioethicists* as the least bad. (2018)	*Bioethicists*: they are paternalistic parasites. (2013)
Sociologists, because they are the actors in the best position to understand how the social and policy environment impacts gene expression. Geneticists typically do not understand the complexity of the social environment and measure it poorly. (2018)	*Certain wings of the social sciences and humanities*, social anthropology in particular, so depend on biology looking threatening for the perpetuation of their careers that it is very difficult to work with them in any serious way. They are so concerned with looking woker than thou that they don't seem to take their research subjects (scientists) seriously with the regularity that they should. (2018)
	Academics. No skin in the game and they are not accountable to the broader public. (2018)

Perhaps final responsibility should be left in the hands of people elected to govern and their appointed officials. Or not:

Yes	No
Elected officials who are closely advised by scientist experts. (2018)	*Politicians*, because most don't have even the slightest inkling of the science of genomics or why genomic approaches might not be advantageous in certain circumstances. (2018)
[Follow the] example of preimplantation genetic diagnosis [PGD] in the Netherlands. Lots of discussions in *parliament* 5 years ago. Due to media and public opinion, law has changed and PGD is allowed now for hereditary cancer. (2013)	*Parliamentarians*. Do not put too much store in the role of law; it is not sufficiently adaptive to the pace of science, cannot assist in the difficult [choices]. (2013)
Governmental *regulatory bodies* which listen carefully to the advice of these scientists. (2013)	
Regulatory bodies that include representation across racial/ethnic/other minorities' perspectives, as well as lay public representatives. (2018)\	

Should the United States bypass even representative governance structures in favor of direct democracy?

Yes	No
General public—engagement important to understand the parameters within which technologies should be assessed (may offer a broader perspective on utility than simple clinical effectiveness). (2013)	*General public*, because it is subjected to too much fake news, misunderstandings, pseudoscience, and religious orientation bias. (2018)

Yes	No
The *general public* [is best suited to make policy decisions]. Because they can empathize with the victims while others overintellectualize. (2013).	

Other societal actors also have a legitimate stake in managing societal uses of genomic science—or perhaps they do not:

Yes	No
Corporations have a profit motive that I find problematic, but I wouldn't exclude their perspective. Every vantage point has contributions that can help inform just policy formation. (2013)	*Corporations* that are only concerned about making profit. Genomic science can potentially change the health system. If it is managed to show profit it would continue to leave a significant part of the population without access to state-of-the-art healthcare when needed. (2013)
	People with financial interests, like scientists with ties to corporations which develop genetic therapies. (2018)
Nonprofits should have some advisory role as well, including *religious groups*, in articulating socially held values. (2018)	I distrust *nonprofit* agencies more than I distrust corporations. I distrust the "green people nut cases." (2013)
	Religious figures and organizations. They operate on superstition and not science, logic, or rationality. I would rather trust a random group of people with a high school degree. (2018)

Yes	No
Judges might need to deal with genomic evidence and assess how it is relevantly similar to or different from other kinds of evidence. (2018)	
I believe it crucial that *advocacy groups* put pressure on elected officials, regulatory bodies, and scientists to ensure that the ramifications are understood. (2018)	
	I think *anyone who is ignorant of or actively hostile towards science* should not be making important policy decisions that affect the scientific enterprise. (2013)
	POLICE! (2018)

A perhaps fragile agreement is found with regard to enabling medical professionals and their patients to act as de facto policymakers:

Yes	No
Physicians and other front-line *health providers*—they represent the gatekeepers to tests and technologies, for better or worse. (2013)	
Patient advocacy groups are the only responsible bodies I work with. (2013)	
Genetic counselors who are truly in the trenches and see how this information is being used. (2013)	
Medical anthropologists, public health PhDs, and *MDs.* (2013)	

But even this agreement is subject to strong challenge, as offered by my interview subjects in comments discussed later. So, although I cannot foresee how genomics governance will eventually develop, I am confident in one prediction: this level of disagreement over the roles of close to two dozen plausible public actors presages complication. (Sadly, no one suggested political scientists!)

Some survey respondents offer pathways through the thicket of competing opinions. One set of suggestions comprises variants on "all of the above." This may take the form of a both/and list, such as "clinicians, health system leaders, health care payers, genomic scientists, population health scientists, regulatory bodies, deliberative groups" (2018). The contents of the list may emerge from experience, or be derived from social science theorizing: "One of the primary results of research in . . . philosophy of science is precisely the claim that it takes *all* such parties, in dialogue, in order to successfully execute democratic science policy. We must foster ways in which any and all such actors can be convinced to come to the table" (2018).

Moving a step beyond lists that call for "collaboration among all the stakeholders" are proposals involving temporal sequencing of decision-makers—"medical doctors and scientists who understand the field. *After that,* judges" (2013, emphasis added)—or different approaches for distinct genomics arenas: "*It depends on the issue.* Some, such as criminal law or government benefits, are exclusively in the domain of government. Other areas, such as the proper role of genetics in evaluating the families, depend on a mix of laws and values" (2018, emphasis added).

The social scientists' most sophisticated governance proposals differentiate among policymaking functions or specify the role appropriate to particular actors. Thus "elected officials are the ultimate source of all decision boundaries; regulatory bodies do some of the work. Judges intervene in some ways. Corporations determine what will be available. Publics make decisions among the choices as they are offered to them. It is inherently a distributed system. Elected officials should and must set boundaries. Individuals should and must choose within those boundaries" (2018). Similarly:

> Physicians and other front-line health providers—they represent the gatekeepers to tests and technologies, for better or worse. Health technology assessment agencies—well placed to provide unbiased assessments of utility, potential impact, and resource implications of emerging genomic technologies. General public—engagement important to understand the parameters within which technologies should be assessed (may offer a broader perspective on utility than simple clinical effectiveness).

> Regulatory bodies—depending on the genomic technology—public safeguarding role among other aspects. (2013)

"Distributed system" comments can themselves be further differentiated. Some argue for enabling scientists to set the rules, with public officials joining in to the degree that they can function as scientists: "Scientists, and elected/governmental officials WITH SUFFICIENT SCIENTIFIC BACKGROUND [should make policy decisions about genomics]. You need to understand the material policy is being crafted for" (2013). In this view, the logic of scientific literacy determines the primacy of experts; they hold authoritative knowledge that needs to be conveyed to a possibly ill-informed or irrational public:

> The meaning of risk varies between scientists and the public in general, so it is important for scientists, and communicators of science, as well as the regulatory agencies and governments, to engage in meaningful conversations with the public about the benefits and potential risks. Within the field of GMOs for food, often the debate is driven by fear, and heightened risks, despite there being little or no evidence of any risk to humans. Psychological factors such as ideologies, identities, and other pre-existing attitude roots are likely to influence risk and benefit perception (as well as attitudes) more so than [actual] risk and benefit (as defined by scientists). (2018)

Some are more blunt: "The general public has a right to be informed, but in a society where the teaching of evolution in schools is a controversial issue among some, where some people continue to avoid vaccinating children, and where many regard genetically modified plants as 'Frankenfood,' I am really concerned about letting uninformed individuals make decisions" (2013).

In something of a contrast to the scientific literacy logic, other social science experts would cede priority to public officials, so long as they are primed by scientists and lay actors: "This science, like nearly all science, is best regulated by government entities (elected officials and appointed professionals within regulatory bodies) acting with the advice of scientists and clinicians and with input from industry, advocacy groups, and the general public" (2018). Some move even further from the scientific literacy logic, with calls for deliberative democratic forums or citizen panels to make policy decisions. The call for democratic forums is usually associated with a commitment to promoting equality or social justice (although there is no necessary link between them): "I see a need for communities who participate in genetic science via their genetic material to have a voice in how policies should be constructed and implemented" (2018).

We cannot expect social scientists or law professors who study genomic science to resolve thorny questions of governance, especially since few are political scientists. Nonetheless, they matter, both because of their prominence and intellectual quality, and because many are in policy-relevant disciplines or have connections with public officials. Collectively, they offer three lessons. First, we should not expect a consensus to emerge on its own regarding either appropriate or inappropriate actors to manage genomics' societal use. We can instead expect many criteria and much disagreement, possibly sharp. Second and relatedly, we should not expect consensus on whether genomics is a broad society-changing innovation, like movable type or the steam engine, or a focused research tool with many separate applications and implications for public and private sectors. So meta-questions may themselves emerge, about how authority should be allocated across substantive arenas, how the arenas are to be determined, and how winners and losers are to be identified in the inevitable conflicts.

Third and more reassuringly, the United States already has well-developed channels for developing science policy and managing some of its social uses; these can plausibly be expanded, replicated, or revised to govern genomics. Genomics is in this regard different from nuclear power, in which the institutions for creating and managing the technology emerging from the science had to be invented at the same time that the power was discovered and deployed. Genomics is new, but expert respondents pointed to relevant precedents in national and international agencies and organizations, scientific associations, and established policy routines. If genomics, or particular genomic technologies, can be slotted into ongoing structures and processes, or if new structures can use predecessors' patterns, the call for the political arena to "catch up to the last 10 years of science" in genomics may not be hopeless.[3]

Actors in the public arena show even more disagreement, or simple fragmentation, about existing and future governance structures than do social scientists. Members of advocacy organizations perceive regulations to be too lenient for genetically modified organisms or too strict for new drugs; people with disabilities argue furiously over whether regulatory agencies should promote some forms of gene editing or bar it; DNA ancestry companies have no idea how to protect clients' privacy while cooperating with criminal justice agencies seeking to catch horrible serial killers. The underlying point is that disruptive technologies disrupt institutions and processes.[4] Genomics experts closer to or embedded in those institutions and processes, to whom we now turn, have a lot to say on the subject, albeit with little certainty or agreement.

Interviews

Like a few social science experts, a few interview subjects wistfully hope or confidently assert that since their work is independent of politics, ideology, or other societal distortions, they and their colleagues need no additional governance. In this view, "the vast majority" of genomics uses "are not contentious or subjects of public awareness." Little genomics information "is controversial—[it is] just enabling technology." Soon "most genomics uses will become mundane and routine." Most interviewees, however, especially those with public sector jobs, disagree, pointing to a shared "recognition by everyone that issues need to be dealt with. Where to do it is where differences lie."

As I discussed in Chapter 6, interviewees do not perceive differences about governance to lie in partisanship or ideology. One official must be reminded of the party affiliation of members of Congress before whom he is testifying or with whom he is meeting. Another prepares in a different nonpartisan way for going up to Capitol Hill: "I had to know the disease obsessions of every member of the Appropriations Committee." Since "politicians of all political persuasions get sick, have family members who are sick," the NIH is a "bastion of bipartisan support" (budget evidence for which we saw in Chapter 3). Conversely, public funding for DNA testing with an eye toward exoneration is sparse not because of partisan calculations about law and order, but because "the elite world doesn't quite believe 'I could be caught up in the criminal justice system.'" In short, despite the fact that politicians will pursue partisan advantage "every chance they get," even they perceive that "the technology is nonpartisan."

As we saw in the discussions of BiDil and the other cases in Chapter 1, different types of controversy quickly fill the vacuum left by the absence of partisan antagonism. On the left, prenatal testing with its implied possibility of pregnancy termination is fraught: "Pro-choice groups see any effort to control the choice to abort as ultimately about controlling abortion. Any distinction feels like a loss. But the wave of sex selection is nervous-making. [This is a] hard issue for feminists—sex selection is troubling, but government restrictions on abortion are also troubling." Complicating matters further, "professional societies don't want [legislation] to close the door on gender selection because they want to leave the door open for [freedom of choice regarding] X-linked diseases." As we saw earlier, another interviewee sees a different "split on the left, between autonomy-based and social-justice-based arguments—strong independent women versus victim." Most generally, conflict on the left is shaped by enthusiasm about genomics' "identity politics [through ancestry testing], the promise of genetic diagnosis, the right to prenatal testing and decision about the fetus"—in the face of, always looming, "the link with eugenics." How to thread governance through that tangle is not clear.

Conservatives have their own internal controversies around genomics' governance and policy. Disputes about forensic DNA databases "are not left/right. They are libertarian versus law enforcement." Outside the forensic arena, "the right should support genomic capitalism—but can't do so publicly because of genomics' reproductive applications, which imply [the legitimacy of] abortion." Some conservatives see genomics as less a moral issue than a "business venture, which can be important commercially. [Consider] Tommy Thompson, [then Republican governor of Wisconsin], who embraced stem cell research because it was good business. He did not get into trouble with [his conservative constituency]."

Moving beyond intra-partisan disputes, as I pointed out in Chapter 6, interview subjects point frequently to strange-bedfellows alliances. But most don't rely on them when they ponder questions of more permanent governance. For one thing, alliances can be fleeting and unstable: even if they bond over shared hatred for vaccine mandates, for example, "the anti-science left and libertarian right won't generate an alliance—they dislike each other so much." More importantly, however, the governance issues run deeper, in at least two ways.

First, argues one of the most thoughtful interview subjects, genomics raises a deep philosophical conundrum: "You can't leave out free will from the political arena. . . . In the long run, 'partisan/nonpartisan' is less important than the free will issue—what do people want to hang on to, what are they willing to risk [in engaging with genomic technology]? GMOs matter in France because they challenge a way of life, not just types of cheese." Our usual modes of governance are ill-suited, at best, to sorting out that sort of question about moral and cultural sanctity.

Second, genomics is making even the debate over free will unstable, with its terms shifting over time along with the technology. As another points out, "technology innovation is first framed by the public at large as a moral statement. Some things are unacceptable. But once . . . the technology works, then it's FDA-approved and the morality flips 180 degrees. People have trouble envisioning that things will actually work out. . . . [For example,] IVF—once it works, now it's [perceived to be] unethical to deprive parents in need of that service." He does not move to a conclusion, but I have difficulty in envisioning governance structures nimble enough to deal with seeing "the morality flip 180 degrees."

Not surprisingly, then, interview subjects are frequently at a loss about who should govern. In contrast to the social science experts, everyone whom I interviewed—except physicians themselves—argues that doctors should *not* be responsible for managing genomics. "The assumption is that doctors know how to use information—wrong." "Most primary care MDs are overwhelmed with what they have to deal with, so they are not interested in genetics till they can use it." Doctors lack time, incentives, training, and experience with genetic information; "don't go to a 60-year-old doctor."

Interviewees are usually sympathetic or at least polite regarding physicians' genetic incompetence. Not so for elected officials—not even the political actors or staff members whom I interviewed have anything good to say about government leaders' management of genomics. They are "clueless"; Congress shows "total dysfunction" in this arena; "When they hear genes they think cloning" (this, from a member of the House of Representatives); "It's all about the money, honey"; "Policymakers—that's a tough one, legislators work hard but only on what's in front of them." Asked how politicians deal with genomics, one health staffer laughs, then continues, "Read the House Energy and Commerce hearing on 'synthetic cells' with [Craig] Venter. Member of Congress [X] couldn't think of any better question than 'how did you do it?' despite lots of staff background and questions provided. Venter tried to explain synthetic cells, but did not get very far." The staffer goes on to spell out the obvious, as he sees it: "There is very limited understanding on how these folks think about these things. Many members of Congress were born before the discovery of the double helix, and were educated before there was any knowledge of genomics" (this interview subject is fairly young). One person does point out that "everything they see about it [genetics] excites them," but that is the best that anyone could muster.

Since neither street-level implementers nor high-level policymakers are suited to govern genomics technologies, interview subjects search for another decision-maker. Some call rather vaguely for collective decision-making:

> We are opening a brave new world. If we sequence a whole genome, we may make incidental findings. They are unanticipated, but have medical significance and may be actionable, so that raises questions regarding what information do you report back, under what conditions, what are protections regarding insurance and employment? It's got to be at the level of society that these policies should be made—not left to doctors—especially regarding employability or discoverability.
>
> Regarding who has a right to predictive information, we as a society need to decide. The doctor-patient relationship is important, confidentiality is an important pillar—but as a society we need to think about those issues of privacy and responsibility.

Another subject offers a more fully developed argument: "We need conversation nationally, while we have the time, while there are still people of good will willing to do this. With things with these ethical implications, we [too often] have answers before questions. We should move into dialogue, scenario planning: 'If we could predict X with accuracy, what to do?'" After starting these conversations, we next need "training. We have explored [the implications of genomics] so little, most people would be talking out of ignorance." Following training, "we could have better conversations, how to maximize value and minimize harm. There is so much unknown potential for good and harm, we need the best minds in science, education, religion, and politics to sort this out before we start using it, which will be within ten years. [The current debate over] health reform is a challenge—but nothing compared with the technology that's going to come out."

Broadly speaking, many of the experts directly participating in the management of genomics concur with many of the social scientists studying it: genomic science is, in slightly mystical ways, different from science as usual. Careful governance will be essential; the most obvious candidates for governing are flawed; some form of public involvement is important. The next question, then, is what the public itself thinks about genomics governance.

Public Opinion

Although few argue that people's unfiltered opinions ought to control societal uses of genomic science, every politician will publicly affirm the importance of voters' views. Americans usually agree, at least in principle, that all individuals' views matter, and communities, organizations, and schools seek to bring nonexperts into the decision-making about genomics.[5] So it behooves us to examine what Americans actually say about who should govern.

Forensic DNA Databases

Forensic databases are already enmeshed in an elaborate network of federal, state, local, and even international laws, judicial decisions, and regulatory procedures. They are consistently funded, frequently used, and visible; half of GKAP respondents have heard or read at least "some" about them. Forensic DNA databases, as we have seen, are also popular. About three-fifths of GKAP 1 and 2 respondents agree that they do more good than harm, while barely 5 percent say the opposite. Beyond that global assessment, close to three-quarters in both years trust police officers, and (in a separate question) judges and juries, to use this information for the public good. More than 80 percent in both years endorse both public funding and (in a separate question) regulation.[6] Three-fifths in each year even claim willingness to contribute a DNA sample to help with a criminal investigation—a fairly stringent test of support. This is not an ambiguous profile or an ambivalent public.

Nor, as I have pointed out, is there much partisan variation. Republicans and Democrats hold similar views in both GKAP 1 and 2 on funding, regulation, willingness to contribute, and trust in judges and juries. They differ a little—three-fifths of Democrats and two-thirds of Republicans see more good than harm in forensic DNA databases, and roughly eight in ten Republicans compared with seven in ten Democrats trust the police. Even with these small differences, given that Democrats support forensic biobanks and their attendant officials, and Republicans support them even a little more, it is no wonder that politicians never speak ill of forensic biobanking.

Racial differences are greater, especially since the Black Lives Matter social movement emerged in 2013. In 2017, just over 70 percent of Blacks, compared with just over 80 percent of Whites, supported regulation as well as more public funding for forensic DNA databases. Three-fifths of Whites would contribute a DNA sample and are technology optimists, while just short of half of Blacks would contribute and only two-fifths see more good than harm. (Nonetheless, only 9 percent of African Americans compared with 4 percent of Whites see more harm than good in forensic DNA databases.) About three-fourths of Whites trust police, and judges and juries, in GKAP 1 and 2, but both sets of actors lost 10 percentage points of trust from Blacks over the six years between the two surveys, ending in 2017 with only 43 percent trusting the police and 54 percent trusting judges and juries in their use of these databases. Overall, a consistent majority of Whites support forensic biobanks and their users, while a majority of Blacks waver between support and caution, with perhaps a downward trend.

These differences are politically and morally important—but they pale in comparison to the patterns presented by the basic framework's four quadrants. Figure 8.1.shows the patterns for both years.
The numbers on the vertical *y*-axis show the percent of respondents agreeing with particular claims, which are shown on the horizontal *x*-axis. The I-shaped lines within each bar show the confidence interval for each result—that is, whether one bar is statistically significantly different from the other bars that represent responses to a particular survey question. For example, Enthusiasts and the Hopeful are significantly different from each other with regard to whether forensic databases produce more good than harm, but they do not differ significantly with regard to whether they would contribute a DNA sample to a forensic database.

The patterns of views about forensic databases in Figure 8.1 are almost eerily easy to interpret. Comparing the two panels, we see virtually no change over time despite the dramatic increase in the use of DNA forensic databases and the slow growth of familial searching over this period. On all questions in both surveys, Enthusiasts show the strongest support for forensic DNA databases and the most commitment to active public governance of them. Hopefuls follow close behind; then, after a considerable gap, come Skeptics and finally Rejecters. Differences between the greatest and least support sometimes approach 60 percentage points.

Although it is not shown in Figure 8.1, note that for the first time we see active opposition, not merely ambivalence or uncertainty, as is usually the case for those who are not technology optimists. Thirty-five percent of Rejecters in GKAP 2 expect more harm than good from forensic DNA databases. The same proportion do not trust the police at all to use this technology, and slightly more do not trust judges and juries. (By comparison, mistrustful proportions of Enthusiasts are 2 and 3 percent, respectively.) Although Skeptics are not quite as antagonistic as are Rejecters, they are not far behind.[7] These two quadrants jointly account for only 14 percent of the American population, but as political actors, their voices are distinct—and in the highly volatile political climate in which I write, they are important.

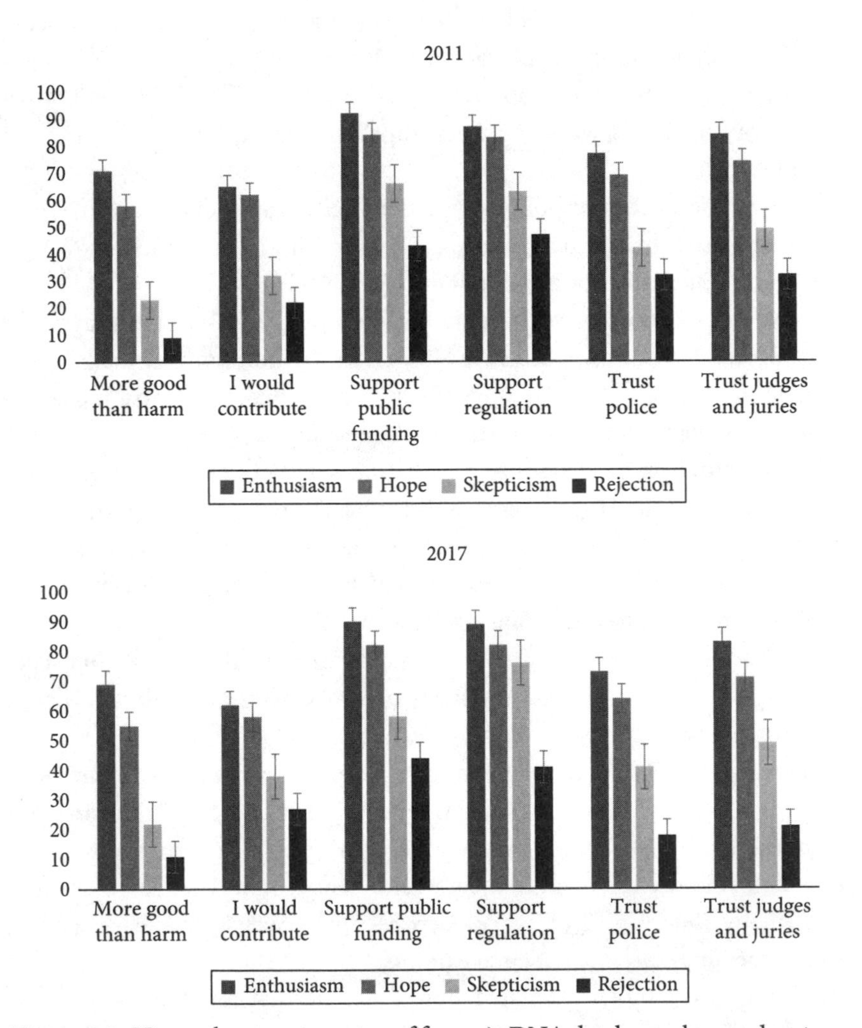

Figure 8.1 Views about governance of forensic DNA databases, by quadrant, GKAP 1 and 2

Medical and Scientific Research and Use

Although scientific and medical research in and clinical use of genomic science are not intrinsically public, they still involve the public sector, at least for regulation and funding. Here too, public opinion does not track conventional partisan or demographic divides.

GKAP 1 and 2 included questions on medical biobanks, research on genetic diseases that might be associated with a race or ancestry group, and tests for individual genetic inheritance of a disorder. As we saw in Chapter 7, between half and three-fifths of Americans are technology optimists in each arena, with a slight increase from GKAP 1 to GKAP 2; fewer than a tenth anticipate more harms than benefits. Three-fifths endorse public funding for biobanks in 2011, with a small increase in 2017; up to three-quarters endorse funding in 2011 for work on genetic diseases.[8] As in the legal arena, majorities endorse regulation as well as public funding. In 2011, 70 percent trust scientists or doctors in all three medical arenas, half trust government officials, and just over two-fifths trust private companies. Trust in scientists rose slightly in 2017. In short, despite some uncertainty, Americans are more excited than fearful about medical uses of genomic science.

Nonetheless, we see more partisan variation around these averages in the medical than in the forensic arena. Sixty to 80 percent of Democrats support both public funding and regulation for medical genomics; smaller majorities (50 to 60 percent) of Republicans concur, depending on the arena. (Presumably that difference partly reflects Republicans' general preference for keeping government out of any aspect of the private sector, but I cannot say how much.) Democrats are a little more likely than Republicans to trust scientists or doctors in 2017, and are more likely to trust government officials in 2011; Republicans are a little more likely to trust private companies. Most generally, in five out of six opportunities, a slightly higher proportion of Democrats agree that the technology in question does more good than harm. (Republicans are slightly more likely than Democrats to predict harm, but only once over the two surveys do as many as 15 percent do so.)

Let us look more closely at partisan and racial differences in technology optimism. The first panel of Figure 8.2 shows Republicans' and Democrats' views on whether research in a given arena will do "more good than harm."[9] Democrats are about evenly split between support for and caution about scientific or medical uses of genomics. Republicans are also split, although they lean a bit more negative. Political scientist Heather Silber Mohamed shows a

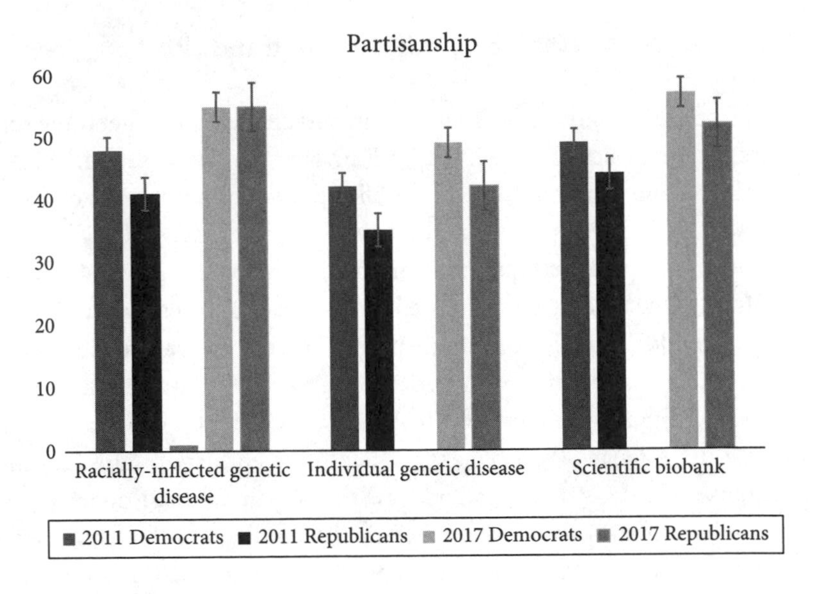

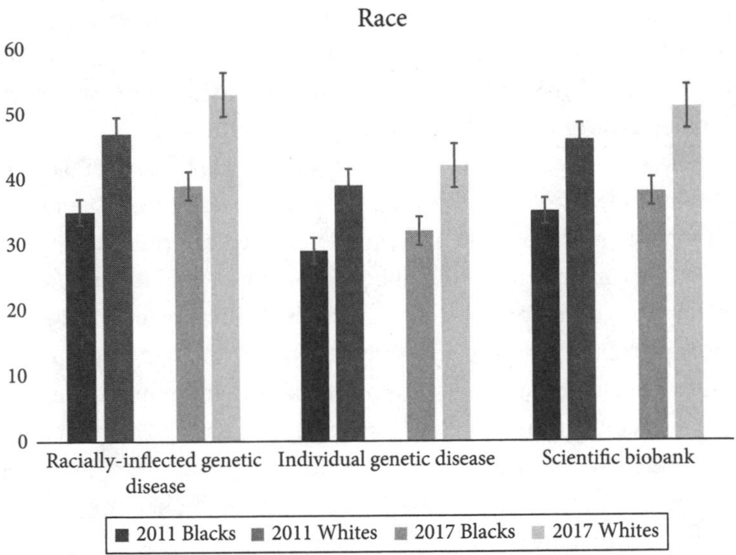

Figure 8.2 Technology optimism about medical genomics technologies, by partisanship and race, GKAP 1 and 2 (percent agreeing "more good than harm")

broadly similar pattern for reproductive technologies; Democrats are slightly more likely than Republicans and independents to accept the morality of in-vitro fertilization and embryonic stem cell research, "yet the substantive effects of these variables are weak."[10]

We see more differentiation by race, as shown in the bottom panel of Figure 8.2. African Americans are consistently less optimistic than are Whites about the three medical technologies. But not deeply: in fact (results are not shown), in 2011 a slightly higher proportion of Blacks than Whites support public funding and regulation, and they are slightly more likely to trust government officials. No more than a tenth of either race expect genomics' risks in the medical arena to outweigh benefits.

It is possible that the two races are edging in opposite directions. Whites' support for funding and regulation, and trust in scientists and government officials, all increase a little from 2011 to 2017, while Black support and trust decrease a little.

A substantial racial divide, if it develops, is for the future. At present, in the medical as in the forensic arena, partisan and racial differences pale beside the differences across the basic framework's quadrants. Figure 8.3 shows the patterns for medical biobanks, which is the only technology in this cluster of three societal uses of genomics that we asked about in both GKAP 1 and 2.

Even more in 2017 than in 2011, the results shown in Figure 8.3 are close to a duplicate of the stair-step pattern we saw in Figure 8.1. In both years, Enthusiasts show the strongest support for medical biobanks and the most commitment to their public governance. Enthusiasts trust scientists more than the other quadrants do, and are tied with the Hopeful in their (lower levels of) trust in government officials and private companies. Where the two quadrants of technology optimists are not the same, endorsement by the Hopeful follows closely on that of Enthusiasts. Skeptics and (especially) Rejecters show much less support—sometimes dramatically less—on all questions. As with forensic DNA databases, differences between the highest and lowest levels of support sometimes approach 60 percentage points.

Also like the case of forensic DNA databases, active opposition, not merely ambivalence or uncertainty, appears in the two technologically pessimistic quadrants. In results not shown in figures, two-fifths of GKAP 2 Rejecters expect more harm than good from medical biobanks (down from 68 percent in 2011); 45 percent do not trust scientists at all to use this technology, and fully three-fifths do not trust government officials. (Only 2 and 12 percent of Enthusiasts and the Hopeful mistrust scientists and officials, respectively.) Skeptics are not quite as antagonistic as are Rejecters, but they are close behind.[11]

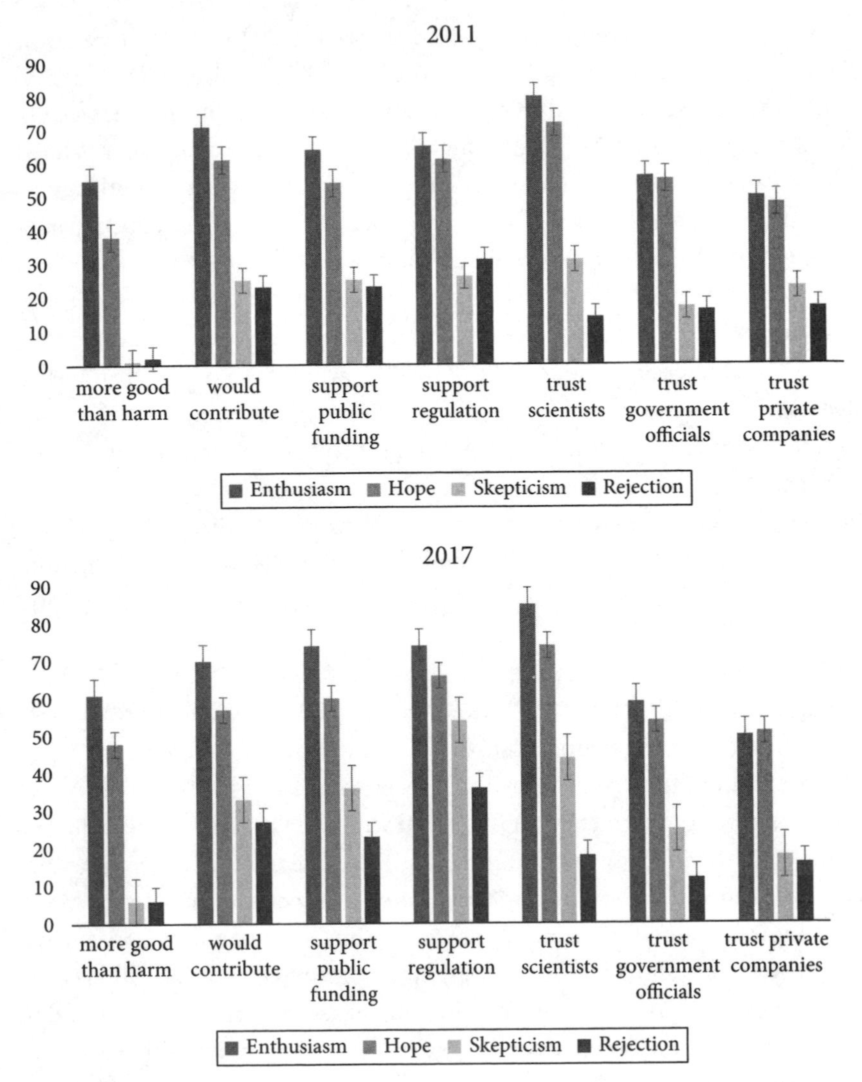

Figure 8.3 Governance of scientific biobanks, by quadrant, GKAP 1 and 2 (percent agreeing)

Somatic and Germline Gene Therapy

Somatic gene editing is in the very early stages of clinical practice; as of August 2020, the NIH states that "gene therapy is currently available primarily in a research setting." The FDA permits only a half-dozen gene therapy products to be used in the United States, and so far it has approved few of the hundreds of research studies or clinical trials that are testing gene therapies.[12]

But somatic gene therapy is becoming more feasible and effective; thousands of laboratories, from well-equipped high schools to multinational pharmaceutical companies, are using CRISPR-Cas9 and its ever more refined successors to edit genes in plant and animal models and in nonviable human embryos. Life-forms ranging from bacteria to rice to cows are being edited; the FDA has approved gene therapies for a few human disorders, such as treatment-resistant mantle cell lymphoma and spinal muscular atrophy in babies.

National laws around the world prohibit, and moral commitments mostly condemn, the much more intrusive technology of germline gene editing in viable embryos. Bioethicist Carolyn Brokowski found very few statements among the sixty-one that she examined from governments, organizations, industry groups, and professional societies that even cracked open the door to research on eventual germline gene modification. She concludes, "Various categories of risk outweigh any potential benefits for now. Overall, much of the international community seems reluctant to proceed with heritable germline editing."[13]

Nonetheless, it will be difficult to halt a technology that is exciting, potentially beneficial, and potentially lucrative. So the operative term might be "for the foreseeable future"; germline gene editing is likely to occur eventually, perhaps in rogue countries or through "ethics dumping".[14] (Ethics dumping is the practice of exporting morally dubious or unacceptable practices from a wealthy country or research operation to a poorer one that may have fewer restrictions or less capacity to control scientists' bad behavior.) Understanding public opinion now, before these innovations are widely known and used, gives us a baseline for tracing and explaining how moral views and political action around them develop.

It is none too soon to set a baseline. In 2018, He Jiankui, a scientist in the People's Republic of China, used CRISPR to alter viable human embryos with a gene variant intended to confer resistance to HIV. Twin girls were born in 2018, and another live birth followed in 2019. An international uproar followed his announcement; scientists described the procedure as—to use only printable terms—"irresponsible," "premature," and a violation of international norms. Prominent scientists and scientific associations have called for, and governments have mandated, a slowdown or even moratorium on research in heritable human gene editing.[15] In 2019, the Chinese government fined He and sentenced him and his team to several years of imprisonment; he has not been seen in public since then.[16]

Although we have seen some of these items in earlier chapters of *Genomic Politics*, Table 8.1 provides the full battery of GKAP 2 questions about somatic, and then germline, gene editing. To facilitate later comparison with the quadrants, it includes responses for the full GKAP 2 sample, as well as for Republicans and Democrats, and for Blacks and Whites:[17]

Table 8.1 Views on developing gene therapies, GKAP 2 (% giving response favorable to gene therapy)

	Total	*Democrats*	*Republicans*	*Blacks*	*Whites*
		Somatic gene therapy			
Should have more research, to treat diseases	61%	67%	65%	48%	65%
Will show benefits of science	60	68	61	39	63
Violates laws of God or nature	45	52	43	27	52
Should develop even though unclear predictions for use	40	49	40	31	42
Disease caused more by environment than genetics	31	36	32	23	34
Risks less acceptance of difference	27	28	31	20	32
Will create too many health risks	22	27	21	16	25
Don't use on those who cannot consent	16	21	13	14	18
		Germline gene therapy			
Do all possible to prevent diseases from being passed down	58	69	56	46	61
Waste of time because traits are not genetic	39	44	40	25	45

Table 8.1 *Continued*

	Total	*Democrats*	*Republicans*	*Blacks*	*Whites*
Exciting next step in seeking human improvement	35	40	32	29	34
Trust society to find solutions to problems	27	31	28	20	25
Violate autonomy because unborn have no choice	20	26	20	11	24
Will deepen inequality because available only to rich	11	12	15	10	11
Somatic gene therapy will cause more good than harm	42	52	42	27	45
(. . . more harm than good)	(9)	(6)	(9)	(15)	(8)
Germline gene therapy will cause more good than harm	29	37	27	20	29
(. . . more harm than good)	(17)	(10)	(23)	(20)	(18)

Overall, optimism about somatic gene therapy is greater than for germline therapy; concern about risks of the latter is substantial, and only one question receives majority support. That accords with the United States' prohibition on federal funding for research on editing embryos. But these results leave open the more immediate issue of how and how much to control somatic gene therapy. More generally, from a risk-averse politician's or policymaker's perspective, this is an unpalatable set of findings.

Between one-fifth and three-fifths of GKAP 2 respondents endorse somatic gene therapy, depending on what considerations are brought to their attention. The more general the question about somatic therapy, the more support it receives. But in the face of particular concerns (for example, about informed consent or impact on people with disabilities), optimism plummets.

These results are not anomalous; four other recent national surveys similarly found more endorsement of somatic than germline editing, a great deal of uncertainty, greater support for curing disease than for preventing disorders or enhancing qualities, and concern about creating genetic inequality.[18]

In sum, in the absence of knowledge, sustained media attention, or consistent cues from trusted elites, public support varies depending on what facets of the issue are given prominence in the moment. Even setting aside germline gene therapy for the next generation to focus on, the politics of this new, dangerous, and exciting technology are wide open and highly volatile. The nature of the first framing to capture wide attention might make all the difference.[19]

Results in the second through fifth columns of Table 8.1 hint at partisan or racial channels into which governance disputes might eventually flow. Even when presented with worrisome considerations, Democrats are slightly but consistently more favorable or less opposed to both forms of gene editing than are Republicans. In the broadest question about relative good and harm, Democrats are more optimistic. Perhaps paradoxically given that they are disproportionately Democratic, African Americans are more likely to see dangers from gene editing than are Whites, and are less likely to agree that good overrides harm in somatic gene therapy. Once again, differences (small so far) between African Americans and Democrats intimate moral and political cross-currents that could, if gene editing becomes a salient societal debate, complicate the Democratic Party's decades-long coalition (Schickler 2016). To put the point a different way, we see a hint of an unusual agreement around managing genomic technologies between African Americans and Republicans.

One more dog that isn't (yet?) barking: To my surprise, religion or religiosity plays almost no role in GKAP 1 or 2. Maya Sen and I find only small and intermittent statistical associations between respondents' religious affiliation and their genomics views, or between the strength of their religious practice and their genomics views. Direct inquiries in GKAP 1 invite respondents to tell us how much their religious faith affects their opinions about various genomic technologies; almost no one responds "a lot" and no more than a tenth say "some." Almost no one refers to religion in their explanation of (un)willingness to contribute to a medical biobank or forensic DNA database.[20]

Nonetheless, the impact of religion bears close watching; a 2016 Pew survey finds a strong association between religion or religiosity and views on gene editing.[21] Shared religious views might help to explain that hint of unusual agreement between African Americans and Republicans about genomics technologies. It is possible that this faith-based link could deepen if gene editing becomes a major societal issue; whether the Republican Party chooses to or can build on it remains to be seen.

As I noted earlier, the absence of optimism can imply either pessimism or a middle position of uncertainty and balance. Up to this point, except from those in the Rejection quadrant, GKAP results show much more uncertainty or balance than active opposition. But germline gene therapy is different. Seventeen percent of GKAP 2 respondents expect it to generate more harm than good, as the last line of Table 8.1 shows. Blacks and Whites do not differ, but more than twice as many Republicans (23 percent) as Democrats (10 percent) anticipate harm.

Once again, however, these suggestive partisan and race-related patterns pale beside differences arrayed in terms of the basic framework's quadrants. We see the evidence in Figures 8.4 and 8.5.

Figure 8.4 shows that a plurality (occasionally majority) of Enthusiasts and the Hopeful are optimistic about gene editing, even of the germline, and very few are pessimistic about it. Skeptics and Rejecters present the mirror image: vanishingly few are optimistic about either form of gene therapy, and pluralities or majorities are opposed. Since Enthusiasts account for 56 percent of the population in 2017 and the Hopeful are another 30 percent, Americans

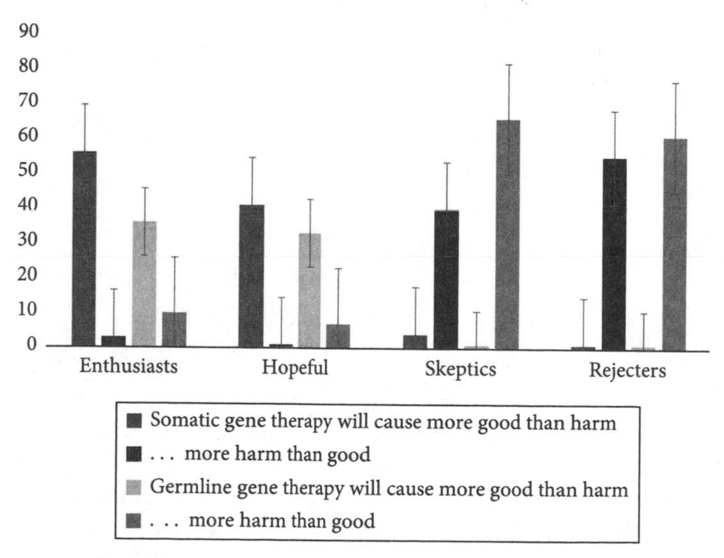

Figure 8.4 Technology optimism and pessimism about gene therapy, by quadrant, GKAP 2

as a whole are more technologically optimistic than pessimistic (although much less so for germline editing), as we saw in Table 8.1. But the patterns in Figure 8.4 give the clearest indication so far of how misleading that average can be, and where coalitions for endorsement and opposition can be built.

More specific questions about gene editing shown in Figure 8.5, however, reveal the fragility of a plausible coalition of support. Where the quadrants differ in levels of optimism—on the first four items regarding somatic gene

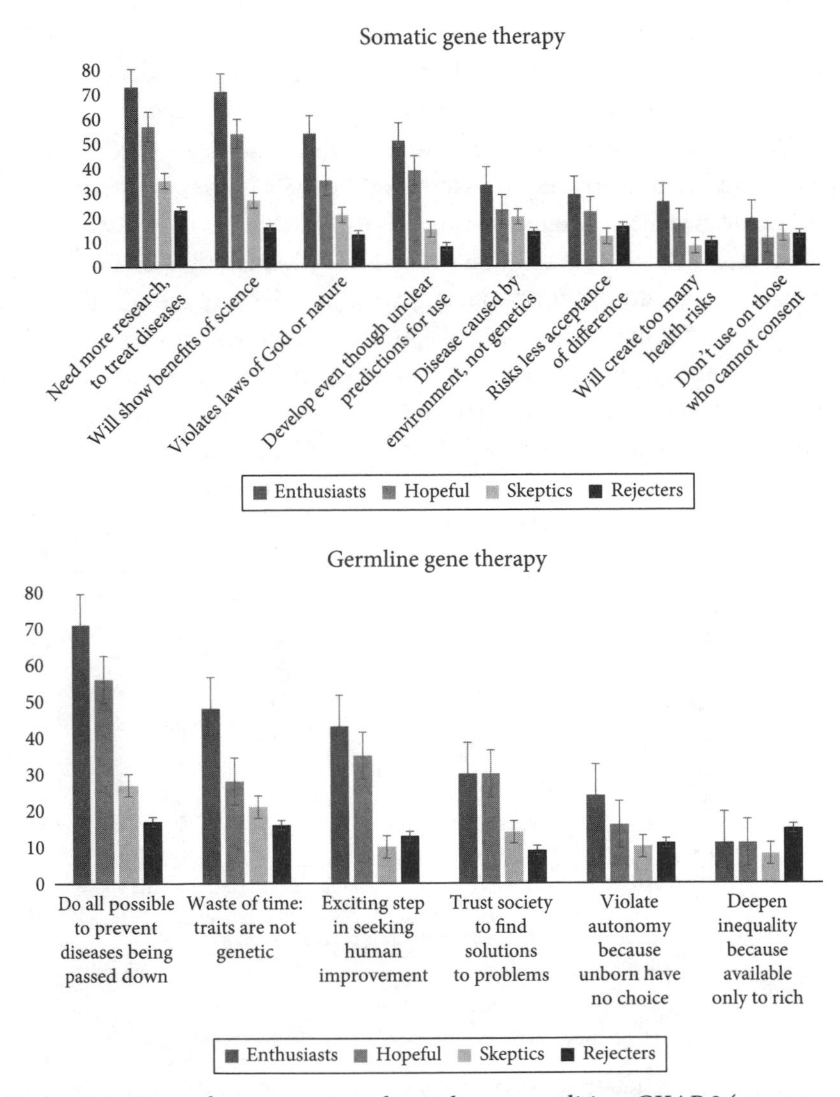

Figure 8.5 Views about somatic and germline gene editing, GKAP 2 (percent favorable)

editing and the first two on germline editing—we see the now completely familiar downward stair-steps from Enthusiasm to Hope to Skepticism to Rejection. But differences in endorsement across the quadrants get smaller and eventually disappear as even Enthusiasts come to focus on issues of environmental causation, disability, health risk, consent, and economic inequality.

Once again, whether the absence of support implies uncertainty, ambivalence, or outright antagonism gives us a window into the moral and political valence of gene therapy. It does elicit opposition, as Table 8.1 shows in its presentation of optimism about gene editing. In particular, up to two-fifths of Skeptics and Rejecters "strongly" oppose gene therapy, especially in the germline. Enthusiastic and Hopeful respondents seldom oppose specific aspects of either type of gene editing—except that roughly a fifth quail when faced with the impossibility of consent, and with the risk that germline gene editing will benefit only the rich.

Genomic scientists express similar concerns as the general public, so the declining levels of support on the right side of figure 8.5 do not reveal a tension between popular and expert Enthusiasts. Instead, they reveal a tension within individuals—between what most see as the promise of gene editing and what they also see as its hazards when they consider it more closely. The Pew Research Center finds the same pattern. Although the share of its 2016 survey respondents who are technologically optimistic is larger than the share who are pessimistic (36 percent expect "more benefits for society than downsides" for DNA editing of embryos; 28 percent expect the opposite; 33 percent abdicate with "about equal"), majorities also agree that several negative outcomes of germline gene editing are likely and that proffered positive outcomes are less likely.[22]

That tension sets the final piece of the agenda for political disputation and difficult policymaking with regard to genomic science in the foreseeable future. Across many aspects of genomics' societal use that we have considered in this chapter, how can we control the hazards and fulfill the promises? Who gets to define hazards and promises, and who will bear the inevitable costs? Perhaps most complicated, how and by whom is the appropriate balance of benefits and harms to be decided?

Who Should Govern?

If those questions have even a faint hope of being answered, it will be through engagement and negotiation among trusted people and organizations—some variant of the lists and considerations produced by my experts and interview subjects. One way to start sorting through those lists is to determine who, if anyone, the public has confidence in to take at least initial steps of balancing promises and dangers. Separately for somatic and germline gene therapy, GKAP 2 asked respondents how much they trust (in random order) each of a series of actors "to help make decisions on policy issues having to do with gene therapy [or germline gene therapy]." Figure 8.6 provides the results.

The short answer is "none of the above." At most, barely a seventh of respondents fully trust even medical professionals, or patients and their families, to make policy decisions about somatic gene therapy. Full trust declines from there across all other actors and both types of therapy, to the vanishing point of 1 percent for public officials and community forums. Over half of respondents have no faith in policymakers regarding germline gene therapy; two-fifths say the same regarding the more imminent somatic therapy. So much for democratic governance; GKAP respondents share interview subjects' disparagement of elected officials and are equally disparaging of their fellow citizens.

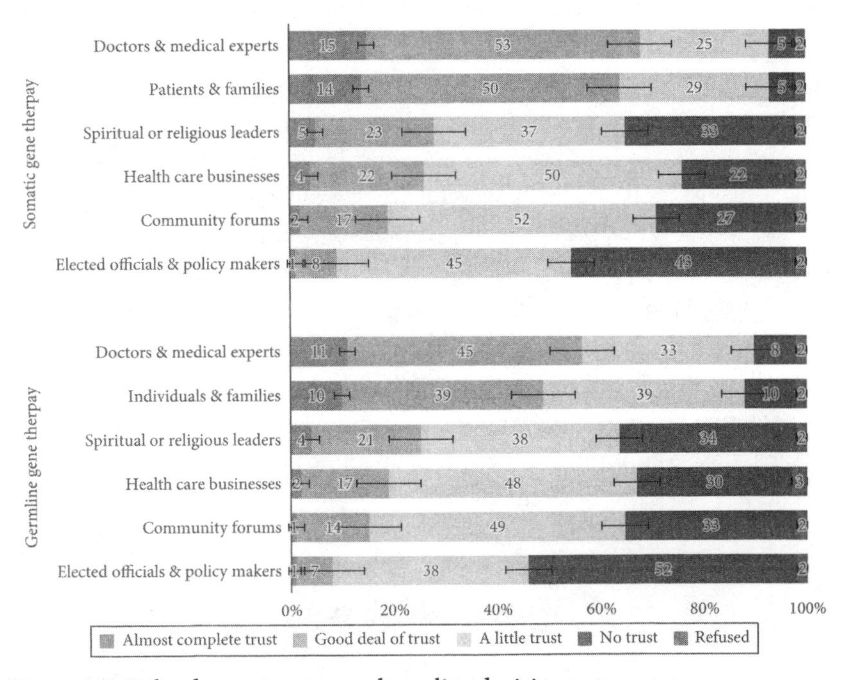

Figure 8.6 Who do you trust to make policy decisions on gene therapy? GKAP 2

Given the dismal reputations of pharmaceutical companies in 2017 (that is, before Covid-19 vaccines),[23] we might not be surprised that at least three-quarters express little or no confidence in health sector businesses or entrepreneurs—but even spiritual or religious leaders and (especially) fellow members of the community receive almost the same rebuff. Democracy, capitalism, and faith—none receives anything close to a vote of confidence.

Perhaps expressed mistrust is really a general statement of opposition to genomics innovation rather than a focused rejection of the possible leaders about whom respondents are asked.[24] That plausible explanation, however, does not receive support, since GKAP respondents are more likely to be optimistic than pessimistic about all four genomics technologies in 2011, and all six in 2017. Even for the most alarming innovation, germline gene therapy, close to twice as many perceive more good than harm than perceive more harm than good.

Alternatively, perhaps the lack of confidence in possible decision-makers is an amalgamation of two distinct patterns, in which Republicans or conservatives trust private sector and religious actors and mistrust others, while Democrats or liberals trust the community and public officials but mistrust businesses and religious figures. That too is not the case, as Figure 8.7 shows.

Republicans probably do trust religious leaders with regard to gene therapy a little more than do Democrats; Democrats perhaps express a little more confidence in community forums. Independents are on balance a little less trusting than either set of partisans. But the fact that the confidence intervals (the vertical lines with endcaps in each bar) overlap for each question shows that these are unreliable as well as small variations in a dominant overall structure. (It is statistically possible, for example, that Democrats trust doctors more than Republicans do.) The main lesson of the results in Figure 8.7 is that there is a much greater difference between Americans' trust in doctors and their trust in elected officials than between Republicans and Democrats with regard to trusting either doctors or public officials.[25]

Do the quadrants provide their usual stair-step structure, in this case for understanding levels of trust in possible policymakers for gene therapy? Not much; we have finally reached the outer bounds of the insights provided by *Genomic Politics*'s basic framework. Figure 8.8 shows the pattern (and lack thereof).

Where there are differences across quadrants, we again see a stepwise decline in support from Enthusiasm to Hope to Skepticism to Rejection.

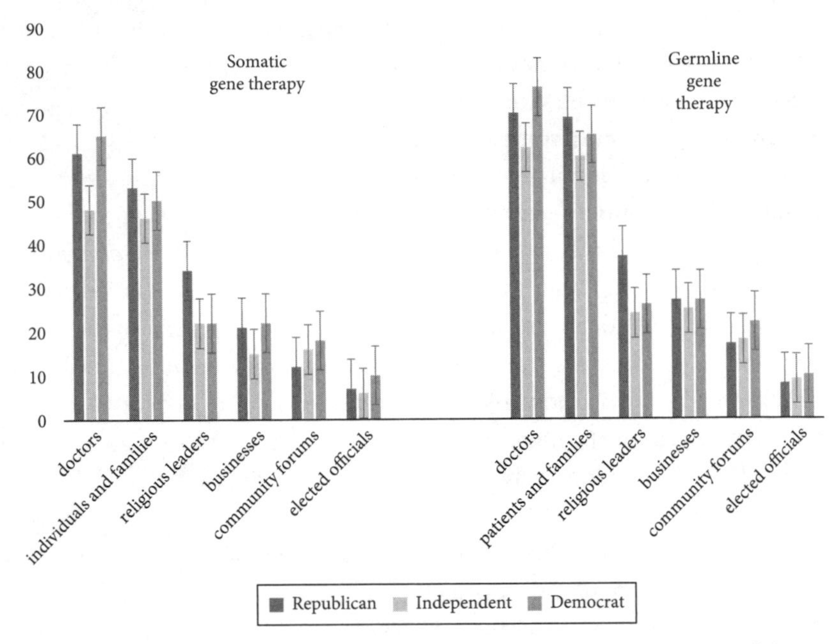

Figure 8.7 Who do you trust to make policy decisions on gene therapy "almost complete" and "a good deal," by partisanship? GKAP 2

But that configuration holds only for medical professionals and perhaps for patients (or individuals) and their families. Beyond that, the pattern disintegrates; neither Enthusiasts nor the Hopeful express much support for any of the other four proposed policymakers for gene editing. In rebuffing all plausible governing entities, they have finally joined Skeptics and Rejecters.

The most established societal use of genomics, forensic DNA databases, receives the deepest support, widest trust, and smallest opposition. The least established uses, somatic and especially germline gene editing, receive the most tenuous support, enjoy only partial trust, and face broad opposition. Medical biobanks, DNA ancestry testing, prenatal genetic testing, and pharmacogenomics are in the middle, in terms of both how well they are established and how much Americans support and trust them.

I do not infer from this pattern any clear causal link between endorsement and timing of becoming institutionalized—or rather, I see three equally plausible causal links. First, as one interview subject suggests, support might emerge out of growing familiarity with a new technology, particularly if

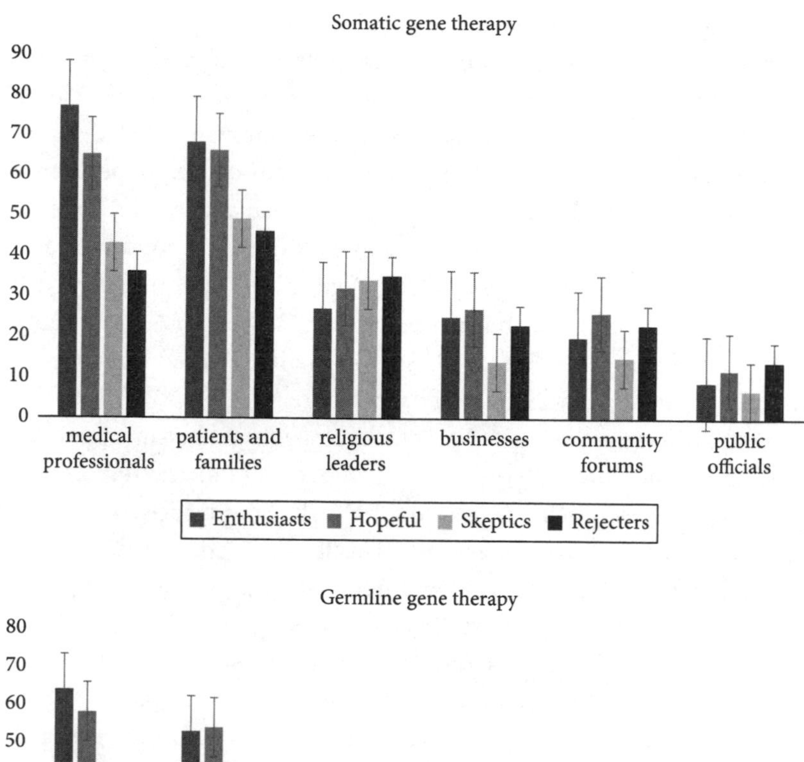

Figure 8.8 Who do you trust to make policy decisions on gene therapy by quadrant? GKAP 2

people discover gains and do not see their worst fears being realized. This is an old story. Initially, Middle Eastern and European monarchs feared and outlawed coffee for centuries; a quarter of the U.S. states banned or licensed margarine in the late nineteenth century; the U.S. Senate removed its dial telephones in order to return to switchboard operators during the 1930s; poet Dorothea Lawrence Mann wrote in 1921 that readers are rejecting "such a blatant advertising scheme as the [newly colorful and decorative] book-jacket, they throw it away with the greatest celerity and never, never read a book until its jacket has been safely disposed of and forgotten."[26] Closer to home, biomedical innovations such as IVF, recombinant DNA drugs, and direct-to-consumer medical tests initially faced considerable opposition. But just as people got used to and then came to cherish coffee, telephones, and paperback books, so they might come to endorse mitochondrial transfer, prenatal gene editing, George Church's Digid8, and mosquito gene drives after they have become established practices.[27] If these historical examples are illustrative, forensic DNA databases are paving the road for many other genomics technologies to attain wide acceptance.[28]

Second, reversing the logic, a new technology might become institutionalized only if and after it has developed public and official support. Given Americans' concern about crime in the 1990s, it is not surprising that DNA's first public policy use was in the criminal justice system during that decade. If this causal logic holds, some genomics technologies (personalized genomic therapies) will become widely established, others (prenatal gene testing and therapy) will remain objects of contention along lines of race, religion, or degrees of risk acceptance, and still others (BiDil) will disappear.

Third, another cause may generate both institutionalization *and* public support. Again consider forensic DNA databases: only after scientists developed the technical capacity to reliably and quickly identify unique individuals through saliva samples, and to create flexible databases including millions of cases, could both implementation through CODIS and public support simultaneously follow. If this is the case, development of a new technique such as the use of CRISPR-Cas9 "causes" both institutionalization and public approval, perhaps at the same time. That, after all, is the history of movable type and steam engines.

Luckily, I do not need to determine exact timelines of and causal explanations for the use and support of various aspects of genomics; my central point is that this is a fluid, indeterminate arena with a lot of simultaneous movements. Nonetheless, several insights from my varied sources of evidence can help us to think through the question of who should govern. I offer the central point here, and develop it in the final chapter.

The choice of good decision-makers—and of those who should be left out of the room where it happens—is a matter of politics, not rationality. The social science experts' long lists of appropriate and inappropriate decision-makers make it clear that we can expect no consensus around a small set of "the correct" governors. Many actors have claims; people have varied evaluations of the legitimacy and priority of those claims; there is no authority or consensual process for evaluating the actors or their claims. That is why political activity is essential—and entirely warranted in a democratic polity. Decision-making about societal uses of genomics will and mostly should include negotiation, emotion, rational discourse, scientific evidence, procedures and precedents, trade-offs and pay-offs, and power grabs. It may never be finally resolved. After all, almost six centuries after Johannes Gutenberg developed the movable-type printing press, we still disagree about whether information wants to be free, and whether it should be; why should something as profoundly influential as genomics be different?

The people I interviewed add to my conviction that the question of who should decide is intrinsically political. Many interviewees challenge the capacity to govern genomics of the two obvious claimants to different types of expertise—medical professionals and elected officials. The general public adds to the tangle by choosing "none of the above," at least for gene editing.

"None of the above" is an intelligible answer to the question "Who should govern?" But it is not very useful. I cannot predict how we will find our way out of this cul de sac, but I can offer a few plausible routes. I consider them in *Genomics Politics'* final chapter.

9
Governing Genomics

> Today a dawn of technological optimism is breaking. . . . If governments rise to the challenge, then faster growth and higher living standards will be within their reach, allowing them to defy the pessimists.
>
> —*The Economist*, January 16, 2021

> Doing social policy is really hard.
>
> —Interview subject, reflecting on decades of work in the genomics policy arena

Francis Collins and his luminary co-authors Eric Lander, Jennifer Doudna, and Charles Rotimi conclude their 2021 panegyric to developments in genomic science with a paragraph more sober than those preceding it: "Many uncertainties remain, however, and not all the big questions can be answered by science alone."[1] Although I am nowhere near such a luminary, I am taking a lesson in assertiveness from these writers—and therefore respond that *none* of the big questions can be answered by science alone. They rightly point to issues of equity, inclusion, the morality of germline gene editing, privacy, cultural norms, and scientific openness; in my view, all of those can be subsumed into the overall questions of who governs, and to what ends? If we are to have any control over how the revolution in genomic science is shaping American society and for whose benefit, it must be through the collective engagement of many types of people.

To bring that proclamation down to earth, I conclude with three observations that emerge from analyses in the previous chapters. These observations do not sum to anything like a prediction or prescription for governance, but they do point to possibilities and likely difficulties for managing the social impact of the revolution in genomic science. They also push me, hesitantly, into explaining my own location in *Genomic Politics*'s basic framework, and into final ruminations about the politics and policy of technology innovation in a democratic society.

No Most-Compelling Approach

In my surveys, more than 100 social science experts propose collectively about two dozen entities that ought to participate in governing societal uses of genomic science—and the same experts propose an almost equal number of entities that ought *not* to participate. Many entities appear on both lists. In my interviews, genomic scientists and public officials outline an array of governance ideas, often with considerable detail and often in contradiction to an equally thoughtful idea from a different interview subject. The American public trusts some public actors some of the time but none all of the time—and many barely at all. The public's attitudes vary a little by partisanship and race (and that might be growing), and a lot by quadrants of the basic framework, a set of distinctions that are not declining. And of course, advocacy organizations and writers of articles, books, blog posts, and tweets fervently endorse and oppose the same array of actors for reasons ranging from direct experience to research-based evidence to material incentives to ideological and normative convictions.

Cacophony is not unique to this policy arena. With the likely exceptions of ensuring military protection of American citizens and reliably disbursing social security checks, most government functions are contested if people are asked to consider them closely. Obvious cases of governance disputes range from the Civil War and the January 6, 2021, attack on the Capitol to lawsuits—as I write—over whether any state government has the authority to set gas mileage requirements for new cars or to mandate vaccination during a raging pandemic.

Nonetheless, possibilities for genomics governance seem even more wide open and inchoate than the possibilities in most other policy arenas. There are as yet few signs of the usual systems of organization—iron triangles of interest groups, regulators, and legislators; dense subgovernments; interest group capture of regulatory agencies; feedback loops between the voting public and elected officials; governance systems metaphorically depicted as elastic nets (purportedly settled issues bounce back into the decision arena), garbage cans (issues stew in their own juices in disorganized clumps), and marble cakes (federal, state, and local authority are entwined in complex ways).[2] Most strikingly, there remains almost no public voice from politicians about most genomics technologies, regardless of their importance for health, food production, abortion, disability, national defense, or the earth's environmental future. The only genomics technology to enjoy sustained public communication from American elected officials— forensic DNA databases— receives almost universal acclaim. This combination of silence and unanimity is, if not unique, highly unusual.[3]

My only confident prediction in this context is that as genomics moves further into shaping American society, contestation will become more visible, politically volatile, and morally urgent. I see four plausible pathways for this contestation to follow.

The first is the rational, or rationalist, approach. Scientists, professional associations, convenings of international experts, and practitioners in particular genomics arenas will take the lead. The United States' National Academy of Sciences and Great Britain's Nuffield Council on Bioethics exemplify this approach, with their considered volumes and judicious recommendations on gene editing or mitochondrial therapy. A quintessential example is the National Academy report, *Human Genome Editing.* It explains the science in almost a layperson's terms, outlines its history, distinguishes among types of gene editing and offers reasons to endorse or oppose each, promotes greater public engagement, and concludes with a "summary of principles and recommendations." It offers most of what one could ask for in a primer; I admire and use it.[4] By the rationalist logic, museums, science educators, university committees, state departments of education, medical and agricultural schools, and other institutions will build on these and other reports to deepen students' genomics training and engage them with complex ethical issues.[5] Civic activists will promote community "DNA days" and programs for citizen science.[6] Scholars will publish books on the politics and ideology of genomic science, and aspire to wide readership.

The logic here is the traditional understanding of scientific literacy. Experts and knowledgeable laypeople will foster genomics knowledge, ethical deliberation, and appropriate societal use. They will not necessarily aim for consensus or expect decision-makers to accept all of their views. But they will expect rational deliberation and attention to facts to be at the core of broad policymaking and particular regulations; expertise will take precedence over perspectives, interests, or passion. The canonical 1975 Asilomar Conference on Recombinant DNA, with its (almost) consensual agreement on a temporary moratorium on experiments and successful call for research guidelines, is, for most people, the model for a mix of expert self-governance and public-private regulatory cooperation.[7]

A second plausible pathway to genomics governance eschews this elegant sequencing from experts to citizenry to a system of rules in favor of disaggregation, with multiple sites of engagement driven by a few people's commitment to particular concerns. Patients' advocacy groups promoting research on cures for rare Mendelian diseases is a quintessential example, as is the Innocence Project. But the set of actors can range from a research lab working on Alzheimer's disease to multinational corporations such as Monsanto or vaccine developers such as Moderna. Organizations committed to a particular process more than substance—promoting citizen science, empowering marginalized groups, or deliberating democratically—also fit into this model. Actors in this route to governance may be just as convinced of the priority of facts in policy-making as are the rationalists, but their deepest commitment is to their issue or to implementing the correct policy as they see it—not to evidence or expertise per se. Strategies for action and forms of governance may vary depending on the nature of the issue in question; actions that benefit the disability rights movement are irrelevant to international monitoring of genetically modified food, and vice versa. Common to all arenas, however, will be the importance of policy entrepreneurs who devote great energy and skill to seeking or creating windows of opportunity to attain their goal for genomics technology.[8]

A third plausible pathway is neither top-down rationalism nor sideways advocacy but rather a bottom-up incident. The model here is a political whirlwind that suddenly and unforeseeably swirls around a person or event in a way that thrusts into view a value-laden, highly symbolic conflict—what economist Robert Shiller calls, to mix metaphors, a narrative contagion that goes viral.[9] Typically the dispute will be understood or structured to be zero-sum and broadly consequential even if the precipitating factor is serendipitous or idiosyncratic. An epitomizing case is that of Terri Schiavo. In 1990 she was diagnosed as being in a persistent vegetative state after cardiac arrest; eight years later, her husband petitioned to have her feeding tube removed. Ms. Schiavo's parents objected, and eventually state and federal courts, Florida governor Jeb Bush, Congress, and President George W. Bush engaged in a long series of lawmaking and judicial hearings and rulings. The images were powerfully motivating (Figure 9.1).

Pro-life groups, right-to-die groups, religious communities, and disability rights groups of both the left and right worked with politicians in the

Figure 9.1 Praying and protesting for Terri Schiavo
Source: Tampa Bay Times

seven-year controversy ensuing after her husband first petitioned for removal of the feeding tube; the passions aroused by Ms. Schiavo's fate have contributed to shaping ethical reflection and political action in all of these arenas.[10]

The point for our purposes of this tragic story is that an unpredictable event may enable political entrepreneurs to draw on cultural and moral tropes to bring a complex and potentially abstract issue into vivid, poignant relief. They are both responding to and galvanizing activism from organized groups, as well as creating a narrative frame that will rouse large numbers of previously uninvolved people into political action. The logic here is to clarify and mobilize—to turn an arcane issue over which experts are debating or advocacy groups agitating into a clear moral battle with bright lines, readily identified allies and enemies, and wide public resonance. Politics inevitably follows.

Genomics is ripe for such an episode—in, for example, a case of prenatal gene testing, a felony conviction, or a direct-to-consumer predictive test, or in a community affected by gene drive or genetically modified food. RNA-based vaccines for Covid-19 looked initially as though they would be a polarizing eruptive case. Epidemiologists have been calling for years for a more flexible vaccine base that can be easily modified to combat new strains of influenza or new coronaviruses. Bench scientists agree, and research was progressing in what looked from the outside to be a slow and steady march. Then came Covid-19, the worldwide shutdown in March 2020, and the consequences, of which readers are all too aware. Chinese scientists posted the SARS-CoV-2 genome in January 2020; building on those years of research, highly effective RNA-based vaccines were developed at "warp speed" before the end of the year.

Here is where a Schiavo-type whirlwind could have occurred. A few opponents warned "all of those considering taking, or being bribed into taking this vaccine . . . [:] You do recognise that this permanently and irreversibly changes your DNA? . . . No one knows what the side effects will be after a few years." Further Facebook posts claimed that the earliest pharmaceutical companies to produce successful vaccines, Moderna and Pfizer, "have made their plan very clear, they want to alter your DNA." After all, "look at their name 'MOD e RNA' they specialise in modifying DNA by using RNA!" These posters insist that they do not oppose vaccination or science; instead, they reject "programming the minds of people in society to 'choose a side' which can then be pitted against each other and via the divide and rule technique ultimately control us all." Protesters gathered, as shown in Figure 9.2. In May 2021, 26 percent of sample of adult Americans agreed that it is definitely or probably true that "the COVID-19 vaccine can alter a recipient's DNA."[11]

Despite having all of the ingredients for a confrontation that would pit people against each other, with the broad social, political, moral, and health-related impacts that the posters predicted, this particular trope did not catch on widely as the Schiavo issue did. As I write, opposition to Covid-19 vaccination persists; the issue is indeed politicized and emotionally fraught, but the battle lines do not revolve around the mysteries of genomics.

A final plausible pathway to governance is something like "none of the above." Governance already exists, of course, in the layering of forensic DNA databases from local to international, NIH and FDA rulings on research proposals and results of clinical trials, European Union regulations on genetically modified food, and GINA. But genomics remains relatively ungoverned in the United States. A very light, even inconsistent governance touch is what many scientists and individuals prefer, as we have seen—and that situation may persist. IVF provides a good example. After great initial controversy, several million babies have been born in the United States as a result of IVF, and the procedure is increasing in popularity, or at least acceptance. Unlike in many European nations, IVF is regulated by no American federal laws. State laws vary widely. After all, " 'It's your body, they're your eggs,' [says one

Figure 9.2 Do mRNA-derived vaccines for Covid-19 change a person's DNA? Although that is not possible, some fear this outcome.
Source: Irfan Khan / *Los Angeles Times* via Getty Images.

successful mother]. In more restrictive countries, Terrie Callahan probably wouldn't have become a mother." As a doctor points out, "Here in the US, everything moves faster than even ethics. Because the patients are paying for this and not the government, we have leeway."[12]

In sum, governance may proceed from the top down, or come in from the side, or erupt from the bottom, or not happen much at all. Or, perhaps, all of the above. That is, there may be multiple forms of governance corresponding to multiple arenas of genomics technology. This takes us back to a question explored in the second chapter of *Genomic Politics*: is genomics a thing, or simply a protean set of tools?

Redux: Is Genomics a Thing?

I raised this issue when asking how far *Genomic Politics*'s basic framework can take us in understanding the politics and ideology of disputes within genomic science around such issues as BiDil, forensic DNA databases, or ancestry or prenatal testing. As I put it in Chapter 2, positing a single framework implicitly asserts that genomics is or will become a transformative technology, linking many particular uses in ways that will change our world and ourselves. Like *The Economist*, and like Jennifer Doudna and Samuel Sternberg, quoted in the book's epigraphs, some people whom I interviewed do indeed see such unity: "The last revolutions were the personal computer and World Wide Web. Same thing could happen with genetics. Most social revolutions take centuries. Now social revolutions are occurring in less than a year." Fifty years ago, an originator of the Human Genome Project, Robert Sinsheimer, stressed "cosmic" unity: "For the first time in all time a living creature understands its origin and can undertake to design its future. . . . This possibility . . . is potentially one of the most important concepts to arise in the history of mankind. I can think of none with greater long-range implications for the future of our species. Indeed this concept marks a turning point in the whole evolution of life."[13]

But other interview subjects see sharp divisions: genomics has "three universes: the legal arena, using genomics to identify people for forensic use in criminal justice; the identity arena, regarding paternity, ancestry, relationship testing; and the medical, health related, and 'recreational' arena, with attention to physical traits." The three arenas have "very different potentials for politicization." Others concurred with her: "You need to break out specific types of genomic uses—advocacy or interest groups are outcome-determinative, not principled [in a way that views move across policy or substantive arenas]." Most simply, "there is no such thing as the general public—publics are constituted around issues."

I said in Chapter 2 that I do not propose to decide in the abstract whether genomics is a thing or a tool. The next six chapters were to be a test of the claim that genomics is a coherent entity: if I found that the basic framework illuminates some core issues but distorts or is irrelevant to others, I would need to conclude that we cannot usefully analyze the politics and ideology of genomic science as a single entity.

Six chapters later, we have now examined a body of evidence that permits at least an interim decision. My answer, perhaps unfortunately, is that the evidence points to a paradox, a tug-of-war between the most and least knowledgeable Americans over whether to understand genomics as a whole with many parts or as many issues with a few unifying threads.

On one hand, particular disputes are firmly attached to personal and community histories, emotional flashpoints, moral commitments, material incentives, political loyalties, group identities, scientific findings and puzzles, and organizational ties. Consider once more the four issues that *Genomic Politics* has been tracking since Chapter 1. Whether the FDA should license BiDil and other drugs for people with a specific racial identity evokes attention to American racial injustice, medical discrimination, the history of eugenics, and the social construction of race and illness—as well as the desire for self-determination, the soundness of statistical techniques, the validity of colorblind policies, the policy impact of corporate incentives, and the promise of personalized medicine. Disputes about DNA ancestry testing evoke a mostly different set of particular forces: the Middle Passage, pogroms, tribal independence, eugenics (again), the history of human migration, group identity, and corporate incentives. A dispute about prenatal genetic testing is "a proxy fight over something else, it is all about abortion or scientized discrimination against the vulnerable. Add in the disability rights movement. There is a powerful emotional impact of supporting the legal and moral right not to have a child who will be disabled—what does that say about me?" Finally, a legislature's decision about funding forensic DNA databases risks pitting crime victim against a person driven by circumstances to crime, poor young Black men against everyone else, police against family members, the falsely convicted against the falsely accused, statisticians against jurors' intuitions and lawyers' skills. Issues of privacy and government overreach come to the fore. And so on; these disputes point to the importance of sideways, advocacy-driven, entrepreneurial politics mobilized around specific genomics technologies.

On the other hand, however, attitudes in the broader public mostly do not track these disputes' specificity. Differences in public opinion across races and partisanship are matters of degree rather than kind regardless of the issue, even in a country riven by political polarization and racial antagonism. Enthusiasts and the Hopeful, who together account for over four-fifths of the American public, differ in the weight they place on genetic influence but substantially agree on most political views and policy support. People willing to contribute a DNA sample to a forensic biobank are usually willing to contribute to a medical one. With the clear exception of germline gene editing and mixed views on somatic gene editing, large majorities of Americans who express a view are technology optimists regarding all of the genetics technologies about which they are asked. They endorse funding as well as regulation, and trust scientists and doctors to a considerable degree, regardless of the specific topic. In short, most Americans express views that enable an inference that they see societal uses of genomics as a reasonably integrated phenomenon.

The paradox, then, is that the relatively ignorant public may sense a broader unity of transformative innovation than do informed advocates, policymakers, or private sector actors. A "revolution in what it means to be human," or development of "a defining, if not *the* defining technology of the century," in the words of one synthetic biologist, calls for general moral and policymaking principles beyond those of particular genomics purposes.[14] Whether such principles can emerge from public discourse, expert deliberation, legislative action, or something else, they imply an imperative that seems intuited by the public even though it cuts against the grain of people with focused outlooks and commitments. In short, I come down on the side of the public - -invoking, for almost the last time, *The Economist*'s "feeling of advancing into the unknown" with the new biological sciences of the twenty-first century.

Positive-Sum and Zero-Sum Conflicts

Assuming for the sake of argument that there will be extensive public discourse about governance moving across specific technologies, where will the deepest fault lines in that discourse lie? That is the central question of *Genomic Politics*, and its basic framework comprises my initial answer. Given the framework's structure as a 2 × 2 table made up of intersections of two independent dimensions, the starting assumption is that each quadrant is equidistant from the adjacent two. Enthusiasm is as different from Skepticism as it is from Hope; Skepticism is as different from Enthusiasm as it is from Rejection, and so on. Chapters 2 through 5, which develop the framework and show how it illuminates genomics disputes, are predicated on that symmetry.

But later chapters call for an amendment to the framework's logic and to my initial answer about the fault lines. The survey results in GKAP 1 and 2 show that the quadrants are not in fact symmetrically distant from one another in Americans' opinions. As we saw in Chapter 8, the Hopeful almost always follow close behind Enthusiasts, while Rejecters almost always follow close behind Skeptics, often with a gap between the two pairs. Except for the final questions about germline gene editing, in which all four groups were equally mistrustful, there was virtually no deviation from that stepwise pattern. That shows that the framework's horizontal dimension of technology optimism or pessimism functions as much more of a dichotomy than does its vertical dimension of genetic influence or its absence. Technology optimists hold similar views about societal uses of genomics even if they disagree on genetic influence—while technology pessimists differ from technology optimists even if they agree on genetic influence. Risk acceptance or aversion trumps genetic influence or its absence.

In addition to the insights it gives into the emotional and cognitive underpinnings of policy views, and by implication policy design, this empirical rebalancing of the framework's analytic symmetry matters for politics. Successful governance typically requires finding unanticipated or previously unaligned allies; if *Genomic Politics*'s findings hold up, Enthusiastic and Hopeful political actors and policymakers will seek to find common ground. This makes substantive sense. For most phenotypes, there is no inherent conflict between genetic and nongenetic causes; that is the old story of the interaction between nature and nurture, or genes and environment, or free will and context—now made more sophisticated by epigenetic science. With a few exceptions of particular faiths, or of severely scarce resources, actors promoting more attention to genetic influence can often work with rather than compete against actors promoting more attention to nongenetic influences. Excessive weight, to choose only one example, may have a genetic element of causation, but that need not get in the way of promoting policies to encourage exercise, self-discipline, bicycle paths, farmers' markets, safe sidewalks, or healthier school lunch offerings.

That in fact is a final message from GKAP respondents. Half of the Hopeful as well as of Enthusiasts agree that research on gene therapy should have equal priority with efforts to change social and environmental causes of disorders. In open-ended explanations, many observe something like "all avenues should be explored" or "anything that will help our society, including gene therapy and environmental research, should be extensively researched." Given that Enthusiasts and the Hopeful comprise most of the GKAP respondents, this is a good starting point for alliances mixing genetic with nongenetic strategies for engaging with societal problems.

In contrast, GKAP results suggest the implausibility of alliances across technology optimism and pessimism. In principle one can balance *The Economist*'s "both good and ill;" even biochemist Erwin Chargaff, whose research was foundational for discovering the double helical structure of DNA, warned that "scientific curiosity is not an unbounded good." But contestation between precautionary and proactive stances is typically deep, broad, and insistent. That choice may also be, or at least appear to be, a zero-sum game. Science studies scholar Sheila Jasanoff spells out the worry: despite its "undeniable value for segments of the human community," contemporary science "may not speak to the fundamentals of the human condition, and they [scientists] may err or produce unintended consequences . . . . [T]he reaches of the human mind extend far beyond . . . synthetic biology." This view seems irreconcilable with the claim by policy analyst Eli Lehrer and his co-author that "permissionless innovation is essential to technological progress. . . . [T]he burden of proof should be on those who claim the new technologies pose risk."[15] The epistemological gap between pessimistic and optimistic standpoints does not bode well for political or policy coalitions, regardless of one's view about the importance of genetic influence.

Mapping the Quadrants onto Governance

Even if the vertical quadrants have more affinity for each other than the horizontal quadrants do, coalitions always have strains. Given their differences, it is likely that residents of each quadrant will have different preferences than the other three about genomics governance. With all of the necessary caveats about the quadrants being ideal types, not delineations of actual people, organizations, or policy choices, *Genomic Politics*'s basic framework enables a few predictions.

Enthusiasts are likely to accept any governance structure that facilitates, or at least does not impede, genomics research and its societal applications. The concept of permissionless innovation will be appealing. Enthusiasts will encourage public funding, with the perhaps grudging recognition that it comes attached to regulation, politicians' electoral imperatives, and engagement with the public as well as with what political scientists call intense policy demanders.[16] Enthusiasts will promote, as much as possible, self-governance by scientists and their research facilities or professional associations. The 1975 Asilomar conference and the 2015 International Summit on Human Gene Editing provide models for scientists' self-discipline united with concern for the public good.[17] The public, advocacy organizations, and policymakers are to be consulted, educated, informed, and perhaps influential—but preferably not included as full decision-making partners. The top-down rationalist model of governance, especially self-governance, works well in this view.

The Hopeful are likely to accept any governance structure that facilitates research on and implementation of a wide mix of strategies to address societal problems. The hallmark here is multiplicity—of policies, activity, stakeholders, and therefore governance mechanisms. Scientists will be among the salient actors, but they might take a backseat to nonexpert identity or environmental groups, public officials, engaged parents, and perhaps religious leaders. Collectively if not individually, the Hopeful perceive so many causes for human disorders, each with different appropriate responses, that mixed governance strategies will emerge despite inevitable conflict over resources and tactics. Hope is most closely aligned with the sideways strategy of advocacy and organizational entrepreneurialism; the rationalist top-down strategy will seem too elitist, confining, and scientistic. Let a hundred flowers bloom.

Skeptics are likely to want the core principle of governance to be "first, do no harm." Scientists, advocates, and policymakers may have the best of intentions and valuable genomics knowledge, but almost by definition they are inclined too much toward proactive rather than precautionary strategies. Good genomics governance, therefore, will be largely a matter of regulation and restraint. Since local communities may perceive risks and moral or cultural flaws in innovations more clearly than do people far removed from direct involvement, community members should have a substantial say and possibly veto power over technologies such as gene drive or genetically modified food. Litigation to halt excessive activity might be an important tool of Skeptical governance, as might bottom-up resistance to centralized control.

Rejecters are likely to start from the premise of "a pox on all your houses." They will have few if any positive goals for or preferred strategies of governance; they will mainly seek to protect privacy, autonomy, rights, and freedom of action. Like Skeptics, Rejecters are likely to turn to local communities, litigation, or libertarian politicians (of both left and right) to give relatively powerless entities the means of preventing too much governance. Some Rejecters are religious, and may seek to ensure that the metaphorical worship of science and progress does not replace the genuine worship of a deity. Or, as many Rejecting GKAP respondents do, they will simply check out.

My Views

Genomic Politics seeks to present protagonists' outlooks as they would want them to be presented; that is one reason for so many citations and—in addition to their eloquence—quotations. The issues that genomics poses are too important, new, and complex to afford the luxury of one-sided or partial viewpoints. And less grandly, my own opinions are changing, mixed, and too often uncertain.

Nonetheless, it is time for me to step out in front of the orchestra's blind-audition screen and state my own views. Despite sharing opinions found in the quadrants of Hope and Skepticism, I am more an Enthusiast than anything else. Only relatively Enthusiastic: in light of my career studying American racial and ethnic hierarchy, the facts that Black GKAP respondents are slightly but persistently less enthusiastic than Whites and that scholars in racial and ethnic studies tend toward Rejection give me pause. Racial disparities in so many genomics-inflected arenas persist—and always there is the long tail of eugenics.

Still, Enthusiasm is my primary residence, for several reasons. First, I have been deeply influenced by Albert Hirschman's analysis of dyspeptic reactionaries (by which he mostly means mainstream conservatives), who see nothing beyond substantive failure, harmful unintended consequences, and threats to cherished values and practices behind almost all societal innovations. That is an unappealing political—as well as normative and emotional—stance, too much shared, in my view, by contemporary leftist, self-contradictory "progressives."[18] Naiveté is dangerous; progress is not inevitable, and what looks like progress might not be. Worse, optimism is perhaps a privilege available in the United States mainly to well-off Whites such as myself. Nonetheless, as Hirschman's life and work make so eloquently clear, societal gains are impossible without hope, eagerness, faith that unintended consequences can be dealt with, cautious confidence in expertise, and belief that humans can choose to learn from past failures.

Second, I have also been deeply affected by genomic scientists and their efforts and achievements. Again, caveats intrude: as biologist Ari Berkowitz points out, "sequencing the human genome failed to produce big breakthroughs in disease" as promised in 2001.[19] And of course, history is full of examples of spectacularly bad science and mistaken scientists. But geneticists' accomplishments, ideals, knowledge, experimental norms, ambition—and, yes, demonstrated societal benefits—are compelling. As one told me, he feels blessed to have contributed to finding a medication for a terrible disease; that is powerful. Furthermore, since most scientists agree that other factors add to or modify genetic influence in human phenotypes, I feel no inconsistency in celebrating what scientists are discovering about genomics and simultaneously insisting that environment, resources, and leadership—that is, politics—matter.

Finally, you can't beat something with nothing. Skeptics' and Rejecters' warnings and critiques must be taken seriously, and some warrant action: government surveillance, racial essentialism, agribusiness control of food production, parents' demands for perfect babies, "playing God" with species, bioterrorism—all range between threatening and pernicious.

However, many concerns of Skeptics and Rejecters tend to be fairly abstract, hypothetical, and inattentive to possible countermeasures. At least at this point in the history of genomics, these concerns do not in my view outweigh the compelling immediacy of using gene therapy to prevent a child's death from leukemia, growing genetically modified rice that will provide vitamins to millions of malnourished people, freeing Black men from wrongful imprisonment that resulted from bad luck and lousy lawyering.[20] Technology optimists can make gauzy promises that match up to pessimists' abstract fears, but so far, the benefits of genomics are disproportionately concrete and its risks disproportionately diffuse. Thus the former outweigh the latter in my judgment.

Yes, But ...

My Enthusiasm is fully considered—I have been working on the subject of *Genomic Politics* for a decade—but also conditional. I do not envision ever becoming a Rejecter or a person in the Hope quadrant who sees any claim of genetic influence to be wrong and dangerous. The difficult question for me—and perhaps also for others if they share my cautious Enthusiasm—is what developments would promote Skepticism. That is, what would transform my technology optimism to technology pessimism?

I have several times referred to a comment by one of my first (and one of my favorite) interview subjects, himself a geneticist: the long tail of genomics is eugenics. Americans are unlikely to return to the eugenic science of the 1920s. But the United States and other countries might devolve into a softer version, in which genomics becomes or is perceived to be a tool for deepening racial or class disparity.

Forensic DNA databases are the clearest manifestation of that risk. A disproportionate share of samples in CODIS come from poor young Black men and Latinos, because they are the population sector most likely to be arrested for misdemeanors or felonies or to be subject to deportation. That does not imply "the new Jim Code"; forensic DNA databases may in fact incorporate less racial bias than other kinds of purported evidence.[21] And reductions in crime and recidivism will benefit vulnerable people of color the most. Nor do more than a tenth of GKAP's Blacks and Latina/os oppose forensic DNA databases.

So in my view, Troy Duster's, Dorothy Roberts's, and others' opposition to forensic DNA databases is not warranted—but someday it could be. It is not hard to envision a widely-shared association between racial disproportionality in CODIS and a crude racially-inflected genetic determinism about the propensity to criminality. That kind of association has been made before in American history.[22] It can happen again.

Racial dynamics could also take a turn for the worse in other genomics arenas. Consider DNA ancestry testing. At present, African Americans tend to be Enthusiasts, since cheek swabs provide a way "symbolically at least, to reverse the Middle Passage," as Henry Louis Gates Jr. puts it. DNA ancestry testing also offers another benefit: in what is only a superficial paradox, the same test that helps people shape their specific group identity also teaches us that "under the skin, we are almost identical genetically. And that is the strongest argument for. . . the unity of the human species."[23] The combination of finding roots and breaking down walls is compelling.

But here too, Enthusiasm should perhaps not be unbridled. Reinvention of a one-drop-of-blood rule through a commitment to intense racial or ethnic identity could offer a competing frame to that of the "unity of the human species"; in the current political climate, it is not hard to see how claims to group purity might get attached to 23andMe's colorful graphs of biogeographic ancestry.

Perhaps of most concern are the discriminatory possibilities attendant on pre- and postnatal genomics testing, therapy, and eventually editing. In-vitro fertilization has proven gratifying, safe—and expensive. Preimplantation genetic diagnosis is gratifying, probably safe—and very expensive. One-time genetic therapies for hemophilia or spinal muscular atrophy are likely to be gratifying, maybe safe—and astronomically expensive. There is a theme here. As sociologists Dalton Conley and Jason Fletcher remind us, "this sort of prismatic effect" from policies "that can generate genetically filtered inequalities" can be expected "whenever opportunity is presented but not 'enforced' —those genetically (or socially) most capable of taking advantage of government-provided (or other) opportunities are the ones who benefit. As some dimensions of stratification are reduced, others are magnified."[24]

Financial support can arguably be found for novel gene therapies; the cost of technologies typically declines with further development and use, and these might be no exception. Nonetheless, the possibility of genetic inequalities that correspond with economic inequalities is daunting, and there is no reason to assume that gene therapy or editing will spread widely enough in the population to offset the headstart granted to the well-off.

In short, I am a cautious Enthusiast, if that is not too much of an oxymoron. Genomic science needs to be governed with both expertise and democratic control, optimism and great care, authority and autonomy, equality of treatment and consideration of specific needs and desires. We are not (yet?) close to finding those balances.

I began this book by pointing to Memorial Sloan Kettering's advertisements for genetically based cancer treatments, promising "More science. Less fear." The aim of *Genomic Politics* has been to make sense of debates over that promise. How genomics is governed over the next few decades will show if Memorial Sloan Kettering got it right.

Appendix

Appendix to Chapter 3

The information in Table 3.1, on pages 70–71, comes from Congress.gov. I searched for "all Congresses," "laws," and "Subject: Policy Area Crime and Law Enforcement," using the keywords "DNA," "DNA database," "CODIS," or "NDIS". After the initial 1994 law, I excluded criminal justice laws with many sections or with a strong focus on a politically charged issue in which only one or several sections addressed forensic biobanks. (In those cases, the genomics-related section always addressed expanding the reach of DNA databases, providing additional financial support, or both. These laws are the Antiterrorism and Death Penalty Act of 1996, the USA Patriot Act of 2001, and the 2006 extension of the Violence Against Women Act.) My reason for exclusion was to keep the table analytically conservative; no one would have voted on these bills because of their provisions about DNA databases. I also excluded reauthorization laws that only addressed DNA by changing dates for a sunset provision or by making an equally technical adjustment to a previous law. All other laws that satisfied the search criteria are in Table 3.1.

Appendix to Chapter 6

Database of Coded Articles in Social Science Journals and Law Reviews

Most-cited articles in first decade of 2000s: Using the *Social Science Citation Index* and *Arts and Humanities Index*, Maya Sen and I identified the 150 articles with the largest number of citations from 2002 through 2011, using keywords "DNA," "genetic(s)," or "genomic(s)." We searched in thirty-three relevant fields, as defined by the two citation indices. We eliminated thirty-five articles as inappropriate because they are not social science as usually understood (e.g., "Genetic Evidence Implicating Multiple Genes in the MET Receptor Tyrosine Kinase Pathway in Autism Spectrum Disorder") or because the coder deemed them irrelevant to genomic science (e.g., the keyword search found a phrase such as "It is in politicians' DNA to promote laws that benefit their constituents"). At that point, research assistants coded 115 articles for their valence about genomic science, using the coding instructions described in the next section.

Most-cited articles in thirteen disciplines, 2002–2018: Although the first search identifies the most-cited relevant social science and humanities articles of the decade 2002–2011, citation and publication norms inadvertently skew the resulting list to favor only a few disciplines. The search also includes no articles from analytically important but small disciplines. In order to see what interlocutors are most likely to learn about genomics from their close colleagues, a second selection procedure therefore controls for discipline.

In this second analysis, Maya Sen and I focused on the most-influential articles in the highest-impact journals within each of twelve social science disciplines and law reviews, identified in Table 6.1. We divide anthropology since its cultural and biological wings are so different substantively and methodologically; in some universities they comprise separate departments.[1] We use the *Social Science Citation Index*'s category "multidisciplinary psychology" as the closest to the general core of that discipline, which is otherwise divided into many specializations. For law, we chose the ten journals that average the highest across three ranking systems—the *Journal Citation Reports*, law review impact factors as determined by Washington and Lee University School of Law (http://lawlib.wlu.edu/LJ/index.aspx), and the top law reviews as measured by Eigen factor (www.concurringopinions.com/archives/2010/10/the-top-law-reviews-eigenfactor.html).[2]

We identified the ten journals in each discipline with the highest five-year impact factors as of 2011, using the *Journal Citation Reports for the Social Sciences*. Using the keywords "genetic(s)," "genomic(s)," and "DNA," we then used the *Social Science Citation Index* or *Science Citation Index* to determine the twenty most-cited papers in each of those ten journals for four periods: 2002–2008, 2009–2012, 2012–2016, and 2017–June 2018. (We searched in separate periods in order to control roughly for time since publication, which is important for identifying the most-cited articles.) We followed an analogous procedure for legal scholarship.

This selection procedure could yield up to 800 articles per discipline (20 articles in each of 10 journals × 4 time periods) for 13 disciplines, or a total of 10,400 articles. Luckily, in most disciplines the search generates only a few articles for many journals and no articles for some. The total number of articles per discipline thus ranges from 19 in cultural studies to 370 in biological anthropology. We collected 1,929 articles through this process.

Over six years of developing and adding to this database, more than a dozen students (trained for this purpose but otherwise not connected with the project) hand-coded the articles.[3] The two key coding instructions are "What is the author(s)' overall valence and intensity with regard to the impact or value of genetics/genomics on the *inside*—that is, within their discipline or epistemological framework?" and "What is the author(s)' overall valence and intensity with regard to the impact or value of genetics/genomics on the *outside*—that is, its actual or likely effect on society, or in medicine, law, racial definition, etc.?" The first question addresses the magnitude of genetic influence; the second addresses the balance between technology optimism and pessimism. Coders used six answer categories: "strongly positive," "positive, "neutral, or no occasion for authorial stance (e.g. irrelevant, technical, purely descriptive, methods only)," "negative," "strongly negative," and "mixed views." The very few articles with only a slight reference to genomics were coded as "no answer" (NA) .

An article is coded as positive or negative only if the author explicitly states a view about genomics; coders did little to infer valence from implicit cues such as the choice of words, location of arguments within the text, or topics given emphasis. That is, the student coders followed my instructions to be conservative in attaching a valence to an article; coders more expert in the discipline and in genomics would certainly have discerned more authorial viewpoint in many—perhaps most, for some disciplines—cases.

Two, or frequently three, individuals coded the articles. Most coding discrepancies involved disagreement over whether a valence is "strongly positive/negative" or "positive/negative." Both for that reason and for ease of analysis and presentation, I combine the two positive coding categories and the two negative ones. Where a deeper discrepancy occurred, a third coder adjudicated. For purposes of exposition, I also collapse "mixed," "neutral," and NA. (The number of articles coded as mixed or NA is well below 10 percent in all disciplines, often approaching zero. And since the important analytic distinctions in *Genomic Politics* are the variation from more to less genetic influence, and from technology optimism to pessimism, the distinction between "mixed" and "neutral" is largely irrelevant.)

Open-Ended Surveys of Experts

In the first open-ended survey, a research assistant and I identified the first five and last authors of relevant peer-reviewed articles published in English from 2002 through 2012 with at least fifteen citations.[4] (The threshold for citations for law review articles was ten rather than fifteen.) The search used the same twelve social science disciplines and law reviews, the same keywords, and the same sources to find the articles as in the coded database; given the selection criteria, many individuals were included in both the database and the survey. We also included a small sample of bench scientists for comparison, deploying the same key words and selection criteria and using the *Science Citation Index*.

We obtained authors' email addresses from the articles or Google searches. With IRB approval, in 2013 I sent a cover letter and brief survey to 537 individuals. Respondents were asked open-ended questions about the benefits and (separately) harms or risks of genomics, the balance between benefits and harms in genomics as a whole and (separately) in their own discipline, and their views of appropriate and inappropriate decision-makers regarding societal uses of genomics in their arena of expertise.

I repeated the survey in 2018. The indefatigable Ryan Zhang and I used the same process to identify the 100 most cited relevant articles in English since 2010 in the same disciplines, and to find the first five and last authors in each article. After receiving IRB approval, we sent the survey to 555 individuals (we could not find usable email addresses for an additional 16). The cover letter, promise of confidentiality, and queries resembled those in the 2013 survey.

Given the small proportion and randomness of responses to both surveys, they cannot be treated as a sample of a population in any systematic sense. I make no attempt at statistical summary or coding of replies; I treat the respondents as interview subjects by focusing on the content of each response independently of the others.

Invitations for the online surveys and both sets of questions are available from me on request.

Interviews with Genomics Experts

After an IRB-approved introductory letter, email, or phone call, I met with the interview subject, almost always in person. The conversations lasted up to an hour; I took detailed notes or made audio tapes that were later transcribed. Although some conversations were on the record, we promised confidentiality to all. I learned as much as I could about each person in advance, with Maya Sen's invaluable help during her involvement with the project.

In early interviews, I used standard techniques for a semistructured interview, with a schedule of questions tailored to the respondent's position and expertise. Responses were typically dutiful rather than deeply engaged or informative. Switching to an invitation to develop arguments from their unique vantage point ("What are the three most important things I need to know?"), after a brief description of the project, galvanized conversations with these extraordinarily talented, knowledgeable, and articulate people.

All quotations are exact, with one small and one considerable exception. I silently correct typographical errors in the written survey responses or uninformative hesitations in a verbal statement. More importantly, I sometimes quote only part of a response or statement, so quotations do not always reflect a respondent's or interviewee's more complex or complete viewpoint. That is intentional; my goal here is to develop and illuminate analytic arguments, not to present the often rich perspectives of a particular (anonymous) person.

Appendix to Chapter 7: Locating the Public in the Basic Framework

Genomics: Knowledge, Attitudes, and Policies (GKAP) 1 and 2

GKAP 1 is a 23-minute online survey, conducted by Knowledge Networks Inc. (now part of the GfK group) in March and April 2011. Respondents were sampled from Knowledge Network's KnowledgePanel, a probability-based web panel designed to be representative of the United States population aged eighteen or older. A total of 4,291 randomly selected U.S. adults provided complete responses (of 7,248 invited, for a completion rate of 59 percent).

The survey is stratified by the racial or ethnic group with which respondents identified when they joined the KnowledgePanel, typically months or years earlier. The choices are White, Black or African American, American Indian or Alaska Native, Asian, Native Hawaiian or Pacific Islander, and two or more races. A separate item asks respondents if they are of Hispanic origin; this analysis treats respondents who identify as Hispanic (regardless of their race) as a group distinct from non-Hispanic Whites, non-Hispanic Blacks, and so on.

GKAP 1 includes 1,143 non-Hispanic Whites, 1,031 non-Hispanic African Americans, 337 non-Hispanic Asians, 1,096 Hispanics (578 of whom took the survey in Spanish), and a small number of members of other groups. The full set of respondents is weighted appropriately to reflect the United States' noninstitutionalized adult population as of the January 2011 Current Population Survey. When relevant, the sample is weighted by particular racial or ethnic groups. I report weighted results unless otherwise noted.

GKAP 1 includes 111 questions ranging from items designed to assess knowledge about basic genetics and genetic influence, to items querying support for government involvement with genomics' applications, views about genomics' benefits and harms, trust in various influential actors, amount of genetic influence for individuals and groups, religious and moral connections with genomics, and preferences about development of the science. Question order and the order of response categories are frequently randomized to avoid response bias. Respondents were told at the outset that the survey would inquire as to their beliefs about genomics and genomic science. That could have had a priming effect regarding the importance, difficulty, or controversial nature of the issue, but I cannot determine how much, if at all. I discern no direct evidence of priming or self-selection.

The second survey, GKAP 2, extends and partially replicates GKAP 1. GfK also conducted it, in September and October 2017. It has 1,777 respondents (completion rate of 63 percent), also sampled from the KnowledgePanel. Median completion time was 19 minutes.

The GKAP 2 sample is also stratified by race and ethnicity so that it includes 725 non-Hispanic Whites, 362 non-Hispanic African Americans, 354 non-Hispanic Asians, and 366 Hispanics (123 of whom took the survey in Spanish). As in GKAP 1, the full set of respondents is weighted to reflect the U.S. population as of the March 2017 Current Population Survey. When relevant, I weight by particular racial or ethnic groups.

Available participants from the 2011 survey were included in the 2017 sample; the remainder of the sample consists of new participants from the post-2011 KnowledgePanel sample. This sampling methodology inadvertently creates a sample of 959 panelists, i.e., respondents who completed both the 2011 and 2017 surveys. The likelihood that they remembered the survey from six years earlier, never mind remembering their answers on it, seems vanishingly small. And after direct comparisons, I find no evidence that the panelists differ in any substantive way from the new sample. Therefore I use the 2011 and 2017 cross-sectional samples. The full instrument for both surveys is available upon request.

GKAP respondents' familiarity with genetic diseases (page 156): Ryan Zhang and I conducted a brief survey in 2019 of 524 individuals on Amazon.com's MTurk platform (that sample size includes only those passing the attention checks). Using the list of forty-four items that GfK sends to each person in its KnowledgePanel respondent pool, along with additional items in GKAP itself, the MTurk survey asks how much each phenotype "has to do with a person's genes compared with the person's environment or lifestyle." Response options are "mostly caused by genes," "equally caused by genes and environment or lifestyle," and "mostly caused by environment or lifestyle." A randomly selected half of the respondents received the additional option of "don't know."

Responses range from over four-fifths choosing genetic inheritance (for eye color and sickle cell anemia) to 5 percent or fewer making the same choice (for obesity, measles, hepatitis C, skin cancer, HIV/AIDS, and the flu). I define as "genetic" the twelve phenotypes that 50 percent or more of our MTurk respondents describe as genetically inherited (excluding "don't know" from the denominator). On the theory that what matters here is which diseases lay Americans understand to be genetic, regardless of what experts say, that is the list used to identify GKAP respondents with direct experience of genetics or genomics in the medical arena.

Genetics knowledge scale (page 156): Political scientist Stephen Ansolabehere and his co-authors give a convincing justification for using a scale of several items, rather than only one, to measure a construct or opinion.[5] I therefore conducted principal components analysis (PCA) and correlations of the three genetics knowledge items. (The PCA does an Eigen value decomposition with varimax rotation of the correlation matrix, and returns loadings for a specified number of components.) In both survey years, the PCA resulted in one factor. The Cronbach's alpha, a measure of scale reliability or internal consistency that ranges from 0 to 1, was 0.59 in 2017 and 0.62 in 2011.

Influence and optimism scales, and respondents' placement in the basic framework (page 164): Seven of the eight genetic influence items in GKAP 1 and 2—all but the flu—comprise the influence scale. Its Cronbach's alpha score is 0.632 in 2011 and an identical 0.633 for 2017. The PCA resulted in two factors in each year; the loading of each item ranged upward from ~0.55 in 2011 and ~0.51 in 2017.

A scale using all eight genetic influence items (i.e., including the flu) had lower Cronbach's alpha scores, of 0.603 in 2011 and 0.607 in 2017. Responses about the flu correlated negatively with responses about several other items in both years (the only item to do so), and as figure 7.1 shows, there was no variation in respondents' views about genetic causes of flu. Thus the genetic influence scale excludes the flu (although in fact, a comparison of results using the seven-item and eight-item scales shows only trivial differences).

All four items comprise the technology optimism scale in 2011, with a Cronbach's alpha score of 0.775. All six items comprise the corresponding technology optimism scale for 2017, with a Cronbach's alpha score of 0.833. The PCA resulted in one factor in each year. The loading of each item ranged upward from ~0.7 in 2011 and from ~0.65 in 2017.

The downward-sloping lines of Figure 7.4 depict in each case a weighted least-squares regression line, where the optimism scale is the outcome variable, regressed onto the genetic influence scale as the explanatory variable. I jitter the observations to make it easier to see the distribution of respondents.

Why Are Americans Enthusiastic, Skeptical, Hopeful, or Rejecting?

Demographic characteristics of GKAP respondents, by quadrant: Figure A7.1 corresponds to Figure 7.5 in Chapter 7, with four additions. It includes the parallel figure for GKAP 2, evidence on employment status and urbanicity, more detailed divisions of the respondent characteristics in the columns, and confidence intervals.

Regression analyses and predicted probabilities (page 167): Multinomial logit regression analyses for GKAP 1 and 2, predicting which quadrant respondents reside in, are available from the author on request. So are additional graphics of predicted probabilities that derive from these regressions.

The predicted probabilities are computed holding all the variables in the regression at their median values. These variables are: race or ethnicity, seven-point ideology scale, genetics knowledge scale, number of genetic diseases experienced by the respondent or a family member, seven-point party identification scale, religiosity, five specified religions, five-point income scale, four-point education scale, gender, and age.

Impact of genetics knowledge and partisanship on location in the basic framework, GKAP 2 (page 170): Figure A7.2 shows the predicted probabilities for location in the four quadrants in 2017, based on the seven-point partisanship scale and the three-item knowledge scale. It parallels the results for 2011, shown in Chapter 7 as Figure 7.6.

A note on causation in these analyses: Links between views about genomics and demographic characteristics cannot be assumed to be causal. In some cases the assumption seems warranted: personal and contextual implications of being Catholic or Protestant are much more likely to lead to attitudes about genomics than the reverse (that is, attitudes about genomics leading to one's religious affiliation). And of course, it makes no sense to say that views on genomics "cause" one to be White or Black, whereas it is plausible that contextual and personal implications of racial identity shape one's genomics attitudes. So, in loose shorthand, the causal arrow from race or religion to views on genomics is one-way.

In contrast, one must be cautious in causal claims about attitudes when both appear in a cross-sectional survey. It depends on the attitudes in question. On the one hand, for most people, it is unlikely that a view of genomics technology shapes partisan identity or political ideology, or even that a third variable explains both political identity and attitudes about genomics. On the other hand, the causal direction between genomics views and genomics knowledge, or between genomics views and direct experience of genetic illness,

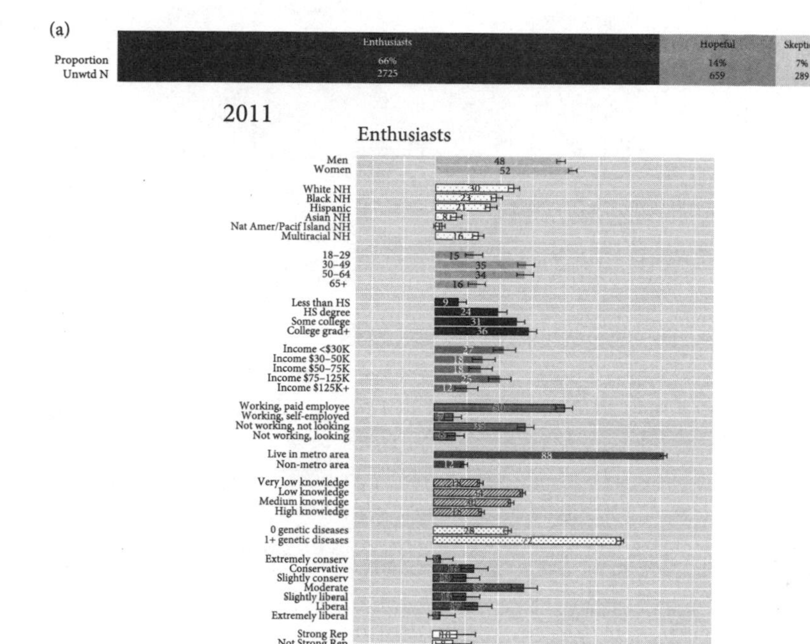

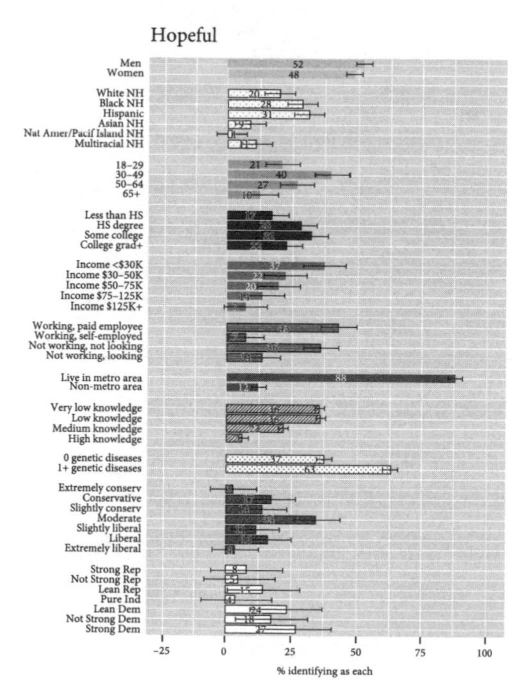

Figure A7.1 Complete demographic profiles of GKAP 1 and 2 respondents in the basic framework's four quadrants

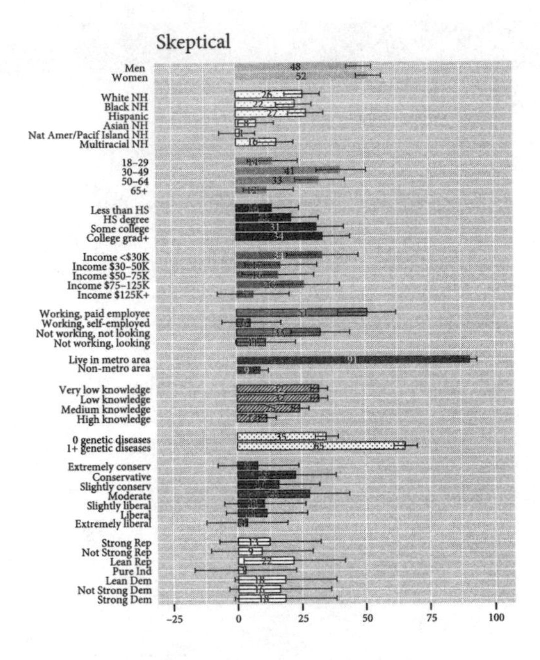

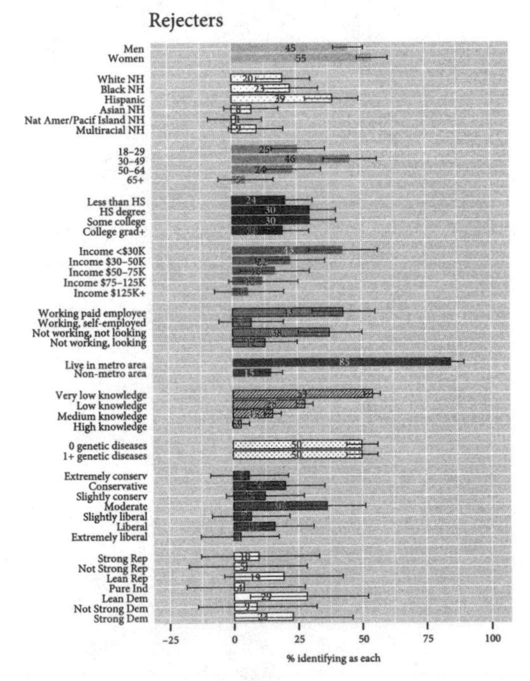

Figure A7.1 Continued

(b)

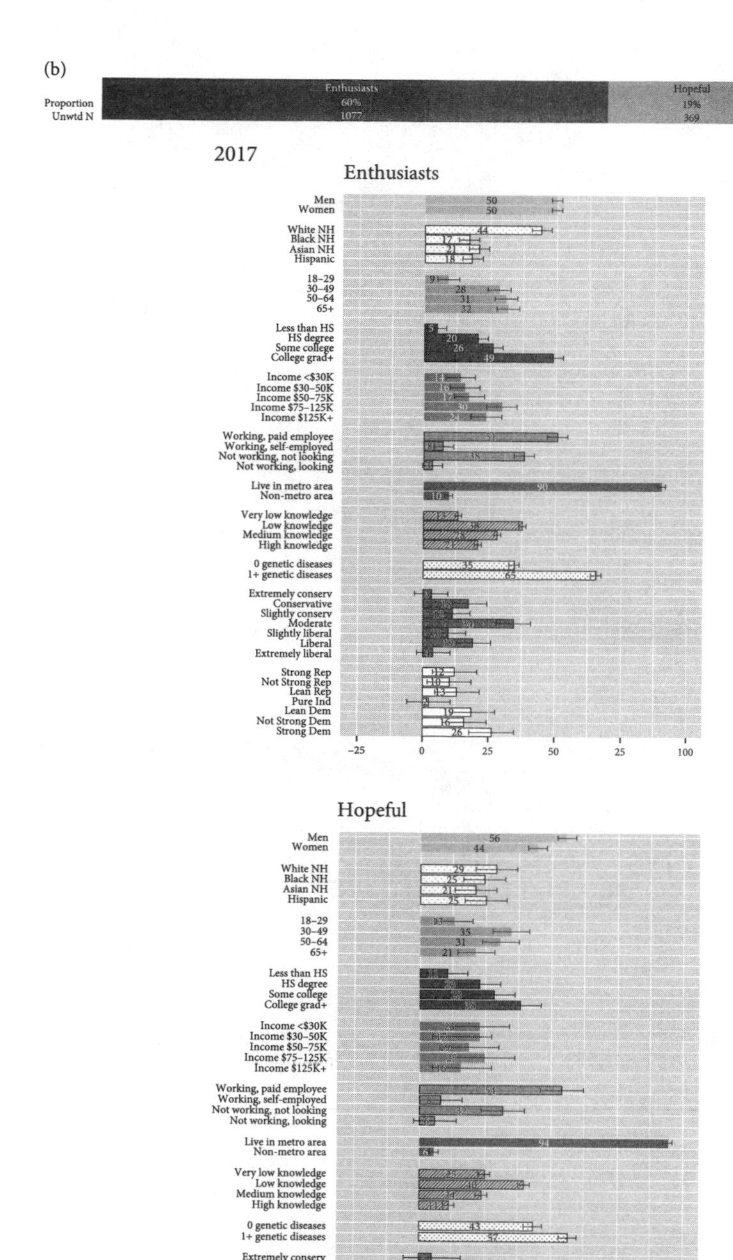

Figure A7.1 Continued

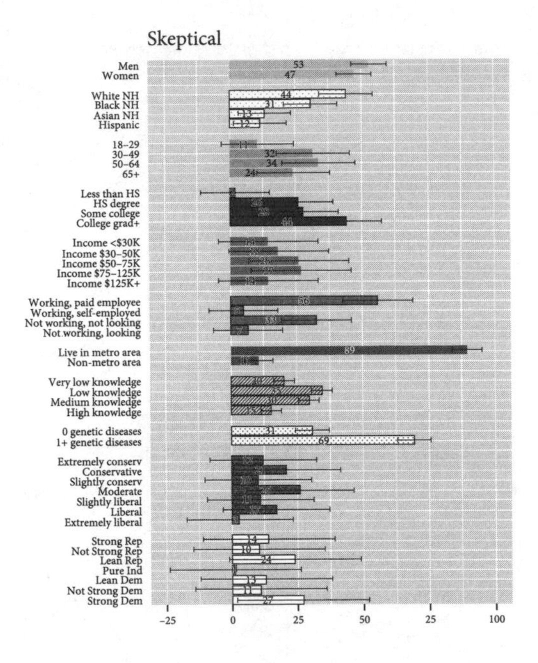

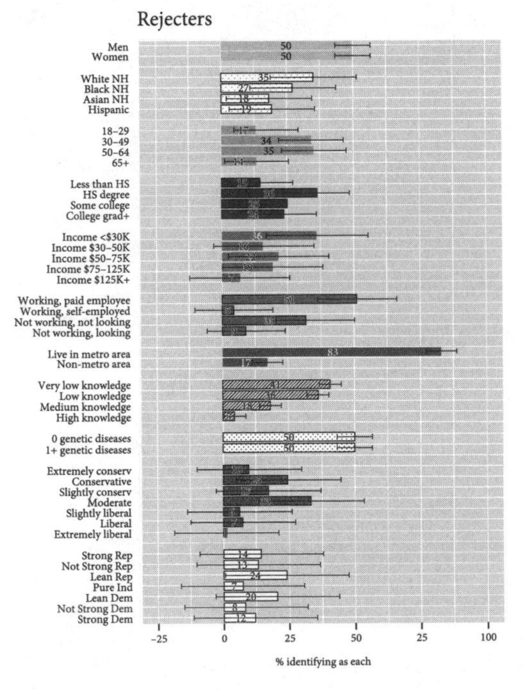

Figure A7.1 Continued

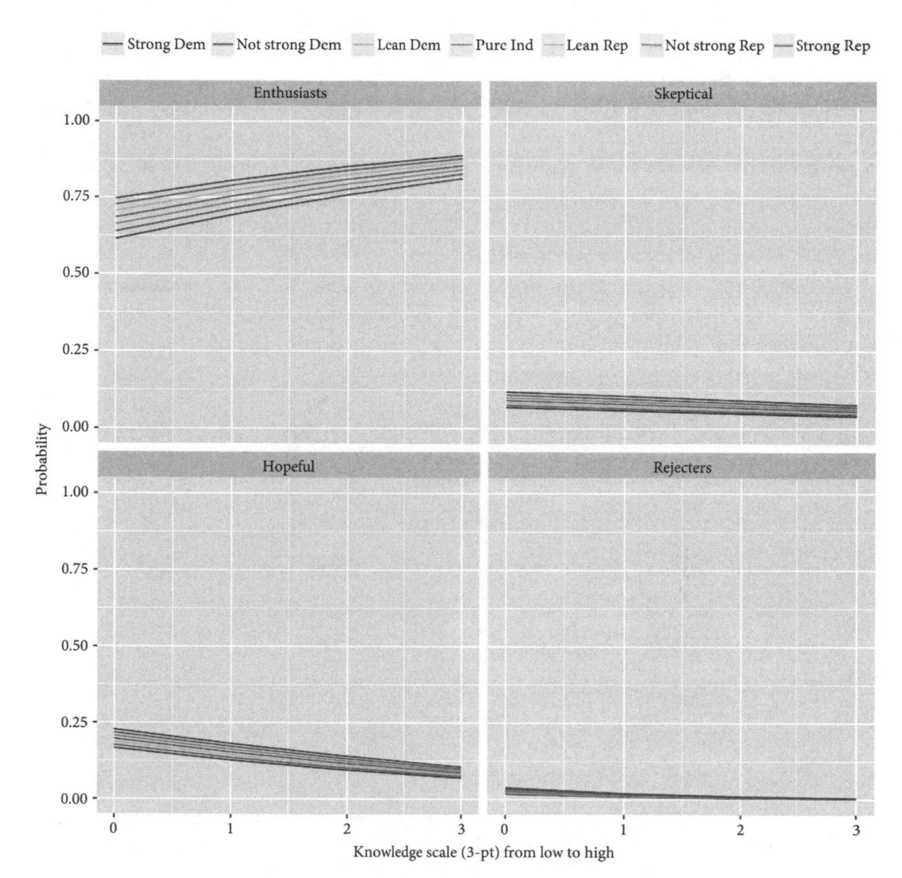

Figure A7.2 Impact of partisanship and genetics knowledge on location in the basic framework, GKAP 2 (predicted probabilities from multinomial logit regressions)

is likely to go both ways. And genomics views, knowledge, and experience may well be mutually intertwined. In that sense, I use the words "predict" or "prediction" as statistical descriptions, not as substantive arguments about how attitudes develop.

It seems worthwhile to substitute race or ethnicity, instead of partisan identification, in examining the predicted probabilities of quadrant location (page 170): Figure A7.3 shows the graphs for the relevant predictive probabilities for location in the quadrants based on the three-item knowledge scale and race or ethnicity. It includes both GKAP 1 and 2.

Genetics knowledge is the only characteristic worth paying a lot of attention to (page 171): Considered as the sole independent variable (other than standard controls for age, gender, education, and income) in regression analyses predicting location in the four quadrants, the genetics knowledge scale is highly statistically significant in both 2011 and 2017, with $p < .01$ for five outcomes, and $p < .05$ for one, with the six outcomes being one computation for each of three quadrants compared with Rejection, in each of two years.

By another measure, the likelihood that the knowledgeable are in a quadrant other than Rejection ranges from a baseline of 3.1 times greater for being Hopeful (rather than Rejecting) among those with no knowledge in 2017, to fully 26 times greater for being Enthusiastic (rather than Rejecting) for those with a lot of knowledge in 2011. Adding other predictor variables to the regression analyses does not reduce the statistical or substantive significance of the knowledge scale. The predictive power of genetics knowledge may decline a little in 2017 compared with 2011, but it remains by far the best predictor of the "risk" of being in a quadrant other than Rejection.

No other trait. . . is consistently important across quadrants, years, and statistical tests (page 171): Consider two examples. Having the maximum number of genetic diseases oneself or in one's family is strongly associated with being in any quadrant other than Rejection in 2011, but the relationship does not appear in 2017. Similarly, after controlling for religious attendance, professing any religious faith (rather than "none") is consistently associated with being in the Rejection category in 2011, but not in 2017.

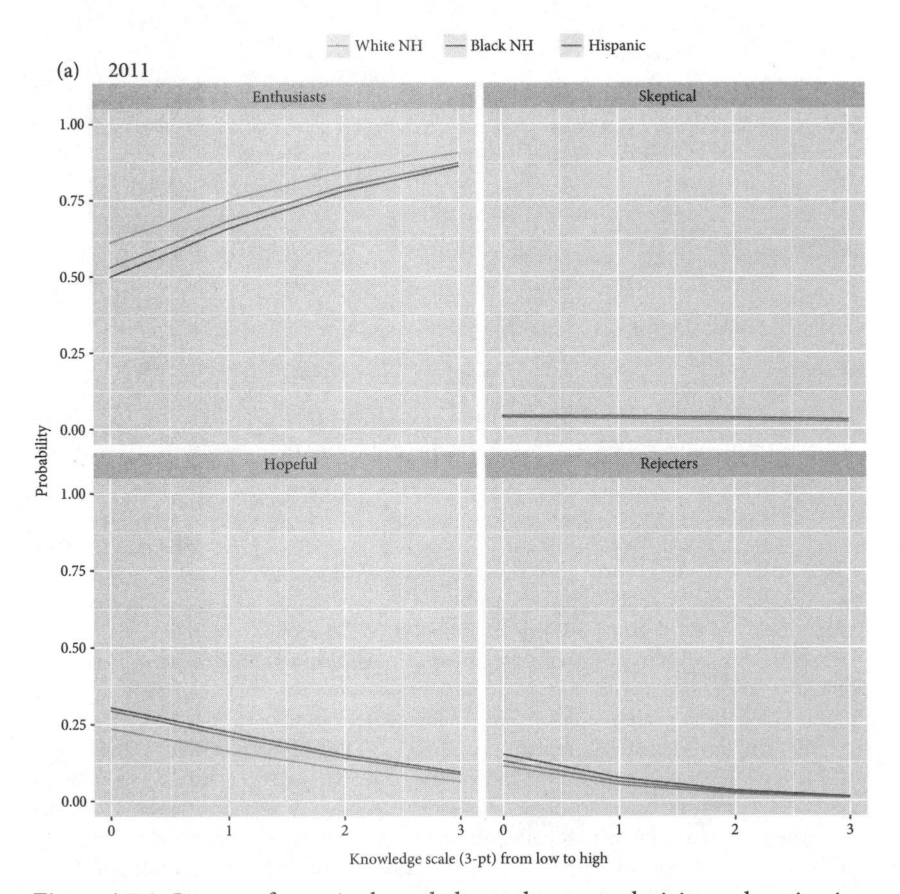

Figure A7.3 Impact of genetics knowledge and race or ethnicity on location in the quadrants, GKAP 2011 and 2017 (predicted probabilities from multinomial logit regressions)

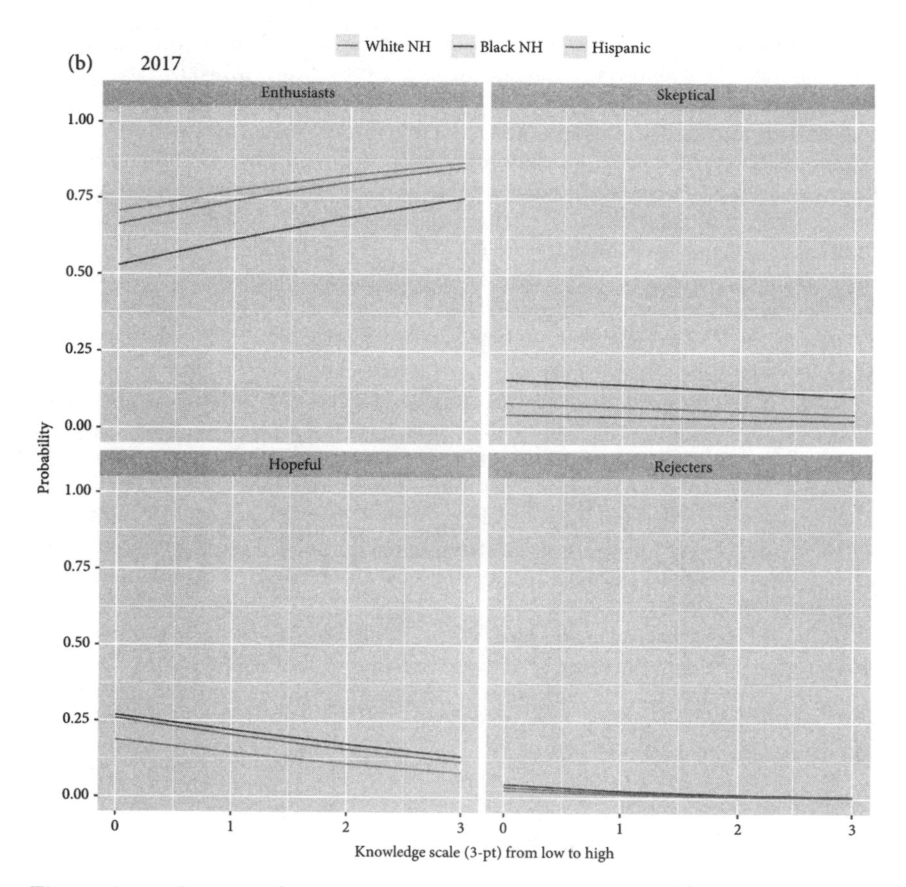

Figure A7.3 Continued

A small refinement: All results reported in the text, and in the Appendix so far, treat the seven-point partisanship scale and the seven-point ideology scale as continuous variables. The same analyses using dummy variables for party identification and ideology show that "extremely conservative" respondents are significantly ($p < .05$) less likely to be Enthusiasts than to be Rejecters, while liberals and strong Democrats are the reverse. The relative risk ratios accord with these tests of statistical significance. These results hold for 2017 only, and do not obtain for other partisans or people with other ideologies.

Forensic and scientific DNA databases: After questions about a different topic, GKAP 1 and 2 each showed a screen saying: "Organizations across the United States and around the world are working together to enable scientists from different laboratories to study thousands of DNA samples collected from patients or the general public. These collections of DNA samples are sometimes called biobanks." Respondents then received questions asking if they had heard of or read about biobanks, and about their willingness to support government funding and regulation (separate questions) of "biobanks for research." They were then asked, "Would you be willing or unwilling to contribute a DNA sample, for example by a swab from your mouth, for use in current or future scientific or medical research?" Answer categories were "willing," "somewhat willing," "somewhat unwilling," and "unwilling." The following screen asked, "Why [would/wouldn't] you be willing to?" with a large text box.

Next, respondents saw a screen with, "The federal government and almost all states require collection of a DNA sample from all people convicted of a serious crime. The samples are stored and may be used in future cases to try to determine a person's guilt or innocence of a particular crime." The same series of questions followed, about awareness, funding, and regulation "of samples in the criminal justice system," then "Would you be willing or unwilling to contribute a DNA sample, for example by a swab from your mouth, for use in current or future investigations to determine a person's guilt or innocence of a particular crime?" The final screen in this sequence again asked "Why [would/wouldn't] you be willing to?" with a text box.

Assistants coded. . . responses into what ended up as ten substantive "buckets" (page 175): The substantive buckets were, in alphabetical order: "general negative or unsure," "general positive," "justice" (forensic biobanks only), "mistrust," "need information," "privacy," "research," "self-interest," "swab etc." (meaningless words or phrases), and "value-laden." The buckets were further subdivided, yielding twenty-seven subcategories in the medical arena and thirty-seven in the forensic arena. I used the subdivisions to develop richer themes within each more general bucket. The subdivisions and coding instructions are available on request.

Quotations are exact except for silent corrections of obvious typographical or grammatical errors. As with the social science experts, I have sometimes quoted only part of a longer statement, since my purpose is to develop an argument rather than portray a given individual's fully developed position. Coding decisions, however, were made for a full statement, not only a segment.

Notes

Preface and Acknowledgments

1. Reardon 2017: 2.

Chapter 1

1. According to Nexis Uni on July 18, 2020, "human genome project" (ignoring capitalization) appeared 125 times before 2000 in the United States newspapers that Nexis Uni was tracking.
2. The game box explains that "DNA keeps making headlines. Kids may not be ready to learn the difficult terminology, but **Metanon: The Biocode Adventure™** prepares them to grasp the basic concepts of DNA. . . . Just playing the game, kids begin to understand how DNA works" (emphasis in original).
3. Reassembling body parts: International Commission on Missing Persons 2012. The "monster inside": see Garde 2017; Kolata 2017; Sheridan 2017a, 2017b. Mika Stump is quoted in Willing 2006: 4A. Deter crime: Doleac 2017: 165.
4. Rosen is quoted in Steinhauer 2010: A14. Shiva is quoted in Specter 2014: 46. Gene drive system: Bull and Malik 2017. Potential neuroweapons: DiEuliis and Giordano 2017. Covid-19 and systemic inequalities: Zakrzewski 2020 (emphasis in original). Experiment gone amok: Heisman 2014.
5. "DNA molecule": Mattei 2012: 503. Synthetic biology: Church and Regis 2014.
6. I include the qualifier "apparently" because there was intense debate over the quality of the statistical analyses showing benefits for Black Americans. That is not my issue, however, so I skirt it here. For the debate, see Taylor et al. 2004; Kaufman et al. 2010.
7. Food and Drug Administration 2005.
8. Center for Drug Evaluation and Research 2005; see also Haga and Ginsburg 2006.
9. Jonathan Kahn: Kahn 2013: ix. Patients are dying: Worcel and Cohn 2012. Race silos: "Illuminating BiDil" 2005. Scientific racism: Bliss 2012: 107, 90, emphasis in original.
10. David Mittelman, quoted in Regalado 2018.
11. "Spit party": Salkin 2008. Reverend Al Sampson: Gibson 2008. Scottish tourist industry: Lei 2007.
12. All quotations without attribution are from my interviews with genomics experts, to whom I promised confidentiality; from listservs; or from emails. See Chapters 6–9 for exploration of the interviews.

13. Middle Passage: Gates 2008: 10. Contemporary Jewish people: Atzmon et al. 2010: 850. Diasporic network: Nelson 2016: 142.
14. Bolnick et al. 2007.
15. An interviewee was equally incensed about direct-to-consumer DNA medical testing: the claim that "'knowing your own genome is empowering' is just silly, [comes from] a political motivation to get money, is wildly oversold, will come back to bite us later. . . . Some tests do harm; returning everything will increase costs [as people demand new tests]. How can we do this responsibly in a context where medical professionals *should* exercise stewardship?" Others express related concerns, such as Ray 2019; Gill et al. 2018.
16. Police chief Charlie Beck, quoted in Sher and Karlinsky 2010.
17. *CSI*: Toom 2010: 390.
18. Obama is quoted in Gerstein 2010. On the UK DNA database, see Levitt 2007.
19. Garrett 2011: 222, 268.
20. Pennsylvania governor Tom Wolf was even more specific in explaining his 2015 moratorium on the death penalty: "Since the reinstatement of the death penalty [by many states and the federal government], 150 people have been exonerated from death row nationwide, including six men in Pennsylvania" (Wolf 2015).
21. Jim Crow's database: Levine et al. 2008. Forensic DNA repositories: Roberts 2011: 264–265, emphasis added. Steinhardt and Annas are quoted in Rosen 2003: 40, 39.
22. De Graff, Buckley, and Skotko 2020.
23. Regional studies: Bowers 2012; Charlotte Lozier Institute 2015; Hill et al. 2017. Comments from survey respondents: Kelly and Farrimond 2012: 73, 76. See Asch and Barlevy 2012 on the ethical issues.
24. Brown 2016.
25. Johnson is quoted in Asch 2000: 235. Rudd is quoted in Garrison 2012, who (jokingly?) asks readers to "go on (assuming you're not already plotting my demise)" when she addresses abortion to prevent the birth of a child with a disabilty.
26. Heighten our differences: Morton 2015, emphasis in original. Your own conscience: Garrison 2012, emphases in original.
27. Peranteau is quoted in Molteni 2019a. Targeted gene correction: Ma et al. 2017.
28. NIH policy is in Collins 2015.
29. Daley, Darnovsky, and King are quoted in Stein 2017.
30. Mills and Tropf 2020: 553. This article summarizes the evidence for genetic influence on traits and behaviors. Roberts and Rollins 2020 disagree, focusing on sociogenomics' implications for racial inequity.

Chapter 2

1. Greely 2015.
2. Quoted in Jaroff 1989.

3. However, some people are coming to perceive that sex chromosomes are not deterministic. Data are imprecise, but a few percent of Americans identify as transgender, and political discourse is newly attentive to sexual indeterminacy or transformation. In the 2018 General Social Survey (the only year with available data), 1398 respondents chose either "woman" or "man"; 3 chose "transgender" or "a gender not listed here" (8 gave no answer).
4. Yengo et al. 2018. See also Locke et al. 2015; Livingstone et al. 2016.
5. Psychologist Robert Plomin (2019) finds extensive genetic determinism; psychologist Steven Heine (2017) finds none; sociologists Dalton Conley and Jason Fletcher (2017) offer nuanced analyses of polygenic risk scores that incorporate both genetic and environmental influences.
6. A too-complicated question in 2016 found that 60 percent of a national adult sample endorsed FDA approval of "gene therapy treatments." There are so many cues in the stem, however, that answers may have been affected in various ways (STAT/Harvard School of Public Health, January 13–24, 2016).

   If not otherwise noted, all survey responses in *Genomics Politics* are in the Roper Center for Public Opinion Research's iPoll, an online database of American public opinion surveys conducted since the 1930s. It is available at ropercenter.cornell.edu/ipoll/.
7. Or more sinisterly, Edmund in *King Lear*: "I should have been that I am, had the maidenliest star in the firmament twinkled on my bastardizing."
8. Sixteen percent of "other" respondents agree that Blacks' relative poverty is due to inborn differences ("other" respondents are disproportionately Hispanic in later GSS years but not earlier ones). Respondents received four questions about possible reasons for Black disadvantage, and they could agree or disagree with each.

   Sociologist Ann Morning and her colleagues (2019) find that with a list experiment, a technique that limits social desirability in survey responses, fully one-fifth of non-Black Americans attribute Black-White income inequality to genetic differences.
9. No less work: Freese 2008: S27. The classic experiment is in Turkheimer et al. 2003: 623; see also Turkheimer 2000.
10. Carpenter and Nevin 2010: 260. Among hundreds of other analyses, see also Krieger 2001; Bearman, ed. 2008; Braveman et al. 2011; Sharkey 2013; Conley and Fletcher 2017.
11. *U.S. News & World Report*, February 1997. See also PSRA/*Newsweek*, August 1999.
12. In Hochschild, Weaver, and Burch 2012, my co-authors and I distinguish between defining a "race" and allocating individuals to a given race.
13. Morning 2011: 18; Heine 2017: 160, emphasis in original. Prejudgments: American Anthropological Association 1998; Kahn 2012: 873–74.
14. Tonry 2009; Cullen and Wilcox 2013; Tonry 2013; Bucerius and Tonry 2014.
15. See also GSS [scitest3], repeated during the 1990s; Pew Hispanic Center National Survey of Latinos, May 24–July 28, 2013; Pew Research Center/Pew Forum Religion & Public Life Survey, August 20–27, 2009.
16. Condit et al. 2009. See, similarly, Jayaratne et al. 2009.

17. This survey warrants replication and revision (which I intend to do, someday). Response options include many societal, family, or environmental stressors, but only one item each to capture individual choice, the law of nature or of a deity, or fate.

    In 2005, the American Public Health Association similarly polled older Americans on the importance of factors that might enable one to "stay healthy as you grow older." Seventy-three percent of respondents agreed that "being born with good genes" is essential or very important. At least three-quarters also agreed on the importance of a long list of salutary behaviors, including among others, good diet, exercise, physical exams, and taking one's medication. As in other surveys, only "having good luck" was received unenthusiastically, with four-tenths agreeing that luck plays an important or essential role in longevity. That is almost certainly mistaken.
18. Cohn et al. 2017.
19. Devaney 2020: S15, S17. Just to reinforce the complexities of the intersection of race and genomics, consider this radically different proposal for engaging with racial diversity in medical research. Jacqueline Stevens, a political theorist, calls on the National Science Foundation and NIH to prohibit their staff or grantees from "publishing in any form—including internal documents and citations to other studies—claims about genetics associated with variables of race, ethnicity, [or] nationality, . . . unless statistically significant disparities between groups exist and description of these will yield clear benefits for public health, as deemed by a standing committee to which these claims must be submitted and authorized prior to their circulation in any form beyond the committee" (Stevens 2008: 322).
20. The book is DeSalle and Yudell 2005. Quote from Mary Jeanne Kreek is on vii; quotes from authors are on 81–83, 90.
21. Reich 2018: xxiv.
22. Chastain et al. 2020. See Martin et al. 2019 for a pre-Covid-19 call for research to offset Eurocentric biases in genomics research.
23. Haunts this field: among others, see Duster 2003; Freese and Shostak 2009; and Nelson 2019. Morning reminds readers in Morning 2011: 15. Contentious subject: DeSalle and Yudell 2005: 92.
24. All quotations are in Summers 2005.
25. Cognitive sex differences: Halpern 2011: xxi. Pinker's evidence: Edge.org 2005. On sex-linked genetic differences in phenotypes or traits, see Voyer et al. 1995; Joseph 2000; Fine et al. 2017; Liu et al. 2020.

    As of 2020, one of Liu's co-authors, Armin Raznahan, is chief of the developmental neurogenomics section at NIH. Even in that position, he expresses trepidation in reporting his results, carefully pointing out that "it could be that there's absolutely no behavioral relevance for what we're finding" (Huckins 2020).
26. Useful starting points are Carey 2012 and Moore 2017.
27. Andreessen: Roose 2014: 62. Oppenheimer: U.S. Atomic Energy Commission 1954: 81.
28. Hjörleifsson et al. (2008: 379) define technology optimism. Hazlett et al. 2011 develop the concept (quotes on 77, emphasis in original).

29. World Economic Forum: Guterl 2013. Genetically modified food: Pew Research Center 2015a.
30. Synthetic biology: Hart Research Associates 2009: 2. "Excited or Worried:" YouGov 2014.
31. Among others, see Fischhoff et al. 1978; Slovic 1987; Taylor and Brown 1988; Quillian and Pager 2010; Kahneman 2013.
32. On experts, see Tetlock 2005, 2016.
33. Mood or attitude: Tiger 1979: 18. Sustaining illusions: Lazarus 1983; Taylor 1991. Delusional optimism: Weinstein 1989; James et al. 1983. "Conscienceless jerk": Comment in review of Shelley Taylor, *Positive Illusions*, Amazon.com, posted November 11, 2011.
34. Helplessness: Maier and Seligman 1976. Psychological junk food: Peterson 2000: 49.
35. Counteractive optimism: Zhang and Fishbach 2010: 16. Tyler and Rosier 2009 give evidence of strategic pessimism to avoid reputational loss.
36. Achen and Bartels 2016: 283. On Iraq, see Hochschild and Einstein 2015.
37. Within-county partisanship is in Tyson 2020. ZIP codes: Newport 2020. 2020 survey results: author's analysis of *Economist*/YouGov polls, March through December, 2020. Using different evidence, Leonhardt 2021 finds roughly symmetrical over-optimism and over-pessimism about Covid-19 among Republicans and Democrats respectively.
38. Pew Research Center 2015b.
39. Quoted in Vogel 2012: 19.
40. Convention on Biological Diversity 2019.
41. Transfer genes: National Academy of Sciences 1987. For varying assessments of GMO risks, see Juma 2016; Lynas 2018; Krimsky 2019a; Carey 2020.

    The FDA also epitomizes the effort to balance risks and benefits of new technologies that directly affect bodily well-being. As the BiDil case shows, it is subject to intense political and economic pressures in the context of seeking to evaluate mixed and sometimes uncertain scientific evidence. Political scientist Daniel Carpenter (2010) examines the FDA's struggles to avoid both Scylla and Charybdis.
42. Kennedy and Thigpen 2020.
43. Kemp is quoted on WSB-TV Atlanta 2, May 19, 2020. Cuomo in quoted in the *New York Post*, May 8, in Hogan and Musumeci 2020.
44. Juma 2016: 5.
45. Tocqueville (1848) [1966]: 420.
46. Almond and Verba 1963: 214.
47. The question is: "Do you think . . . biotechnology and genetic engineering . . . will have a positive, a negative, or no effect on our way of life in the next 20 years?"
48. Jasanoff and Kim 2009; Vogel 2012; Nuffield Council on Bioethics 2018: ch 4.
49. Mary-Claire King: King 2006. Editors: Couzin and Kaiser 2007. Midst of a revolution: Kleiner 2008. Broad Institute Covid-19 testing: email from Stacey Gabriel to Broad Institute members, July 3, 2020. Complex information: Zwart 2007.
50. Lander is quoted in Fallows 2014. Along with co-authors NIH director Francis Collins, Nobel laureate Jennifer Doudna, and NIH principal investigator Charles Rotimi,

Lander uses terms such as "breathtaking," "paradigm-shifting," "transform[ing]," and "profound" in a 2021 summary of genomics' history (Collins et al. 2021).

51. Venter is quoted in Nave 2016.
52. Familial searches: Granja and Machado 2019. Genome editing: Nuffield Council on Bioethics 2018. Livestock traits: Department of Agriculture n.d. Plants and microbes: Department of Energy n.d.
53. BGI describes its services at https//www.bgi.com, accessed June 28, 2018. The Saudi Biobank 2019 is a good example of a national biobank; Gottweis and Petersen 2008 and Tsai and Lee 2020 provide analyses.
54. Quotation from *Renmin Net* and other information: Dirks and Leibold 2020: 1. See also Wee 2020.
55. Screwworm: Esvelt 2019. Doudna and Sternberg 2018, Carey 2019, Lovett and Ronai 2020, and Kahn 2020 all provide useful nontechnical overviews of gene drive. They tend to support it under certain conditions, while the United Nations Convention on Biological Diversity has considered a moratorium (Callaway 2018).
56. Elliptical eugenic uses: Duster 2003: 129, 130–131. Flynn 2019 and Church 2019 explain Church's proposed dating app; criticisms appear in Shanks 2019. "~Super~ racist": Regalado 2019.
57. "Frankenbugs" is the title of one segment in McMaster 2019. "Frankenjournalism" is in Kloor 2015.
58. Baltimore's comment: McMaster 2019. Feldman and Lowe (2008) present a useful history of Cambridge's engagement with recombinant DNA, with the hindsight of several decades; Krimsky 1982 offers an earlier, rawer history. The most official description of recominbinant DNA and its controversy is in National Research Council 1987.
59. That summary is by law lecturer Patrick Foong (2019). The movie *Rampage* (2018) portrays a primatologist, a cuddly silverback gorilla, and CRISPR gone terribly wrong. Geneticist Arun Sharma and colleagues (2015) offer a brief Enthusiastic counterargument to Skepticism about chimeric research.
60. GeneWatch UK's principles: http//www.genewatch.org/sub-396416. Philosopher of science Sheldon Krimsky provides both early (1977) and recent (2019b) discussions of the need for lay oversight of scientists.
61. Copeland 2020; Anderson 2019. Krasner (2019) gives a useful compendium of views on privacy in relation to direct-to-consumer genetic testing.
62. Without publicity: DNA Forensics 2020. Draw up guidelines: Rainey 2018.
63. Vogel 2012: 2, emphases in original.
64. Quoted in Pollack 2010: 21.
65. The possible explanations for mental illness offered to respondents were: bad character, bad luck, stress, the way he/she was raised, genetic or inherited problem, God's will, brain disease or disorder, and the normal ups-and-downs of life. "A genetic problem" correlates in respondents' answers with "brain disease" at $r^2 = 0.30$ but is not linked to other explanations. In a stronger cluster, "bad character" correlates at $r^2 = .38$ with "the way he/she was raised" and at $r^2 = .29$ with "the normal ups-and-downs of life." In a third cluster, "God's will" correlates with "bad luck" ($r^2 = .29$), and each is moderately related to bad character ($r^2 = .24$, and .22, respectively). These

results suggest that some Americans have a strain of depressive fatalism—linking the deity, bad luck, and poor choices—to which I return in the discussion of Rejection.

The GSS asked a similar question with similar lists in 1996, 2006, and 2018, though without "luck." The patterns of association are comparable in each case.

66. Explanatory power: Yengo et al. 2018; Locke et al. 2015. Heine 2017: 65 quotes *The Onion*.
67. Biologically embody: Krieger 2014: 645, 653. Human action: Krieger 2005: 2159. See also Sankar et al. 2004.
68. National Institutes of Health 2002: 1.
69. National Institutes of Health 2002: 1, 4, italics in original. NIH's overarching emphasis arguably has shifted toward Enthusiasm; Krieger observes that the NIH 2016–2020 *Strategic Plan* uses variants of "gene" fifty-eight times and "disparities" six times, the latter mostly in one paragraph (Krieger 2016).
70. Heine 2017: 258 (citations omitted). GeneWatch UK (n.d.) is more pugnacious: "An over-emphasis on genetic explanations and solutions to these problems . . . ["as diverse as hunger, crime, climate change and cancer"] can mean that underlying social, economic and environmental issues are ignored."
71. Spelke: Edge.org 2005.
72. Rosa et al. 2015: 2, 3, 9, emphasis in original.
73. Joyner et al. 2016.
74. The research is in Perry et al. 2016. Both quotations are in Newman 2016.
75. Scully et al. 2020.
76. All quotations in this paragraph appear in Boyd et al. 2020. Whether or not because of criticism, Azar and her co-authors later backed off from their suggestion of a genetic or biological element in racial disparities in Covid-19 severity. An editor's note in Boyd et al.'s blog post reports that "*the final published version of the [Azar et al.] paper has been revised to clarify the authors' conclusion that the disparities are most likely explained by societal factors*" (emphasis in original). The published article is Azar et al. 2020.
77. Quotes are from, respectively, Slow Food 2015 and Slow Food 2016. See also Boony 2003.
78. Sandel 2007: 9, 26.

Chapter 3

1. War on science: Mooney 2005. Index entries: Otto 2016. GSS analyses: Gauchat 2012: 177. Gauchat's analysis ends in 2010, but the evidence in later years offers no reason to question his conclusion. See also Blank and Shaw 2015.
2. Mitchell et al. 2020.
3. On fracking: CBS/*New York Times*, March 2011; United Technologies/*National Journal*, May 17–20, 2012; CBS News Poll, December 2012. On nuclear power: Gallup, March 2011; CNN, March 2011; CBS News, March 2011; Heimlich

2011. On genetically modified foods: CBS News/*New York Times*, June 2012; CBS News, January 2013; Funk and Kennedy 2016.

Useful analyses include Taverne 2007, Berezow and Campbell 2014, Nisbet et al. 2015, and Rainie and Funk 2015.

4. CNN/ORC International 2012. In the same survey, more liberals (or Democrats) than conservatives (or Republicans) agree that landing "a robotic explorer on Mars that is sending back pictures and data" is a major achievement. Why the difference? I do not see a partisan valence to that issue.
5. Very serious diseases: Harvard School of Public Health, June 2001. Terri Schiavo: Harvard School of Public Health, April 1–5, 2005. Astrology: in addition to the GSS, see Lundgren 2014.
6. Thigpen and Funk 2020. The proportions had changed little by December 2020, even as access to a Covid-19 vaccine moved toward reality (YouGov December 19–22, 2020).
7. 2018 partisan gap: Blank and Shaw 2015; Newport and Dugan 2017; Brenan and Saad 2018.
8. Identity-protective cognition is in Kahan 2016: 26.
9. Lanham 2017 offers one analysis of Inhofe's stunt. On Sheriff Woods, see Miller 2020. Woods's own explanation: "In light of the current events when it comes to the sentiment and/or hatred toward law enforcement in our country today, this is being done to ensure there is clear communication and for identification purposes of any individual walking into a lobby." In the same article Miller reports, without comment, that "Marion County set a single-day record on Tuesday for the most deaths related to Covid-19."
10. *Nova* article is quoted in Eschner 2017. IVF might be unethical: Princeton Survey Research Associates/*Newsweek* Poll, August 24–25, 2006. 2013 survey: Pew Research Center, March 21–April 8, 2013. Pew's most recent survey (Funk et al. 2020b) shows a quarter of Americans disapproving of "new technologies that will help women get pregnant." That is slightly above the median of 21 percent across twenty countries.
11. On smoking: AARP Views on Smoking Survey, Gallup Poll, July 2011; September 2013. On alternative medicine: 60 Minutes/*Vanity Fair*, October 2011. On space: Pew Research Center/American Association for the Advancement of Science, April 28–May 12, 2009; 60 Minutes/*Vanity Fair*, March 31–April 3, 2011; CBS News, June 2011; CNN, July 18–20, 2011. On genetically modified food: Funk and Kennedy 2016.
12. IVF controversy: Jasanoff and Metzler 2018: 1012.
13. On partisanship in Congress, see Voteview.com n.d. On the American public, see Pew Research Center 2014.
14. Holder: Office of Public Affairs 2010. Harris is quoted in Dolan 2011. Cuomo is quoted in Campbell 2011: 299.
15. Approve BiDil: Food and Drug Administration 2005: 203, 206–7, 209. Precision medicine: Office of the Press Secretary 2016.
16. Subcommittee on Oversight and Investigations of the House Energy and Commerce Committee 2010.
17. Targeted News Service 2017.

18. In 2013, the FDA halted 23andMe's sale of direct-to-consumer health information derived from genetic testing because of "concern about the public health consequences of inaccurate results from the P.G.S. [Personal Genome Service] device" (Pollack 2013). After negotiations, data analysis, user studies, and new "special controls," in 2017 the FDA permitted 23andMe to market tests for ten diseases or conditions (Food and Drug Administration 2017). As a small but telling indicator of genomics nonpartisanship, note that although the Obama administration's FDA ruling against 23andMe held through the 2016 presidential election, its founder, Susan Wojcicki, endorsed Democratic candidate Hillary Clinton.
19. Trump is quoted in Allen 2018.
20. A reader of an earlier draft points out that this is not the full story. It took "years of bipartisan disagreement and negotiating that led to the bill that finally passed. And . . . there are gaping holes in the bill that were left there because of bipartisan disagreement on what should be covered by GINA." Both are important points, but they do not undermine my claim that federal legislation about genomic science, however it is accomplished, is nonpartisan. It does not stand out for that reason; as this reader also points out, "most federal law, including landmark enactments, continue to garner substantial bipartisan support" (see Curry and Lee 2019). Nor does GINA stand out because of its "gaping holes;" very little legislation achieves all of its sponsors' aims. Those observations place GINA in context, but it remains the case that the sole federal law about genomic science passed consensually.
21. On Florida, see Downey 2020. For state laws, see National Human Genome Research Institute 2020b.
22. National Conference of State Legislatures 2014.
23. National Conference of State Legislatures 2018.
24. Compensation for exoneration: Innocence Project 2020b.
25. See Simoncelli and Steinhardt 2006 on "function creep."
26. State laws: Rainey 2018a. Don't look for partial matches: Field and Debus-Sherrill 2017; see also Debus-Sherrill and Field 2019. "Last resort method": Murray et al. 2017: 2. The FBI takes pains to point out on its website that states use "specially-designed software (not CODIS software)" to conduct familial searches.
27. Roberts 2011: 267.
28. National Council of State Legislatures 2014.
29. National Academies of Sciences, Engineering, and Medicine (2017a) describes governance principles, U.S. laws and regulations for gene editing research and therapy, and some other countries' governance regimes.
30. For example, see Joh 2006; Kaye 2012; Paciocco 2015.
31. Paciocco 2015.
32. Martin and Quinn 2002.
33. They were all civil liberties cases, in which the four dissenters argued for more individual autonomy or privacy and less state power than was contained in the majority decision. My thanks to Nicholas Short for this analysis.
34. 569 U.S. 435 (2013).

35. Liberals see more genetic influence: Suhay and Jayaratne 2012; Garretson and Suhay 2015; Schneider et al. 2018. Conservatives see more genetic influence: Suhay and Jayaratne 2012; Morin-Chassé et al. 2017. No clear impact: Shostak et al. 2009; Suhay and Jayaratne 2012; Suhay et al. 2017.

Chapter 4

1. Typologies are in, respectively, Hirschman 1970; Mayhew 1974; Berlin 2013; Kahneman 2013.
2. Dulbecco is quoted in Smith 2004: 52. Hockfield is quoted in McMaster 2019.
3. Anderson and Domenici are quoted in Cook-Deegan 1994: 152, 174.
4. Altman 2008.
5. The FDA also approved a genetically modified pig in December 2020. The animal is a potential source of organs for transplantation into humans, and of food or medications for people with alpha-gal allergies, since alpha-gal has been edited out of the pig's genome.
6. The FDA's approval is at Food and Drug Administration 2020. The database is at ClinicalTrials.gov 2020. Degenerative brain disease: Kolata 2017. Replacement skin: Grady 2017. Davies 2020, Part II, offers a history and current analysis of gene therapy.
7. Sudden reversals: Lowe 2020. Breakthrough: Food and Drug Administration 2019. Blessed: personal communication with the author, October 23, 2019.
8. Ledford 2020. Davies (2020) and Isaacson (2021) off useful explorations of the development and use of CRISPR gene editing.
9. Among dozens of analyses, Broad Institute scientists David Liu and Feng Zhang (2020) offer a useful and accessible overview. "Incredibly profound" is in Corn 2015. Comment about "beautiful technology" was made at a conference on New Transformations of the Human, Center for Advanced Studies in the Behavioral Sciences, Palo Alto, CA, May 31, 2018.
10. Mayor Vellucci is quoted in Feldman and Lowe 2008: 405.
11. Moderna: Divine 2020. Quotes about Roche are in Miller 2019. Davies (2020) reviews genomics-related startup companies, and the billions of dollars changing hands.
12. The quote is from Grehen, in Stein 2020. Grehen's excitement is readily understood. His severe hemophilia A left him at risk of a fatal brain hemorrhage or severe disability from bleeding joints. "It was quite difficult, because I was always getting these bleeds in the ankles. Walking around was getting more and more troublesome. Internally, there could have been a lot worse situations" (Stein 2020).

    The FDA unexpectedly denied approval to BioMarin's hemophilia gene therapy in August 2020. It will be available commercially in 2022 at the earliest, after further trials and analysis. The phenomenon of an astronomically expensive, one-time-only, gene therapy nonetheless persists; other firms are pursuing it for use in other disorders.

13. Both quoted in Stein 2020. Pian et al. 2020 discuss paying for life-changing therapies.
14. Genomics economic impact: Battelle Institute 2013. Sample preparation: Maliwal 2020. Gene editing products and services: Bergin 2019a. DNA sequencing: Bergin 2019b. Genome editing: Fan 2019. Proteomics: Joshi 2018.

    Analogous to genomics, proteomics is the study of all of the proteins in a cell, tissue, or organism. Proteins are created, loosely speaking, at the direction of genes; their nature and function may vary across cells and over time. Again loosely speaking, proteins create and regulate all of an organism's structures and functions—so they are potentially the source of huge medical innovation, and comparable profits.
15. Patent applications: Duffin 2020. Bioscience patents: Mikulic 2020. For detailed data on patents, see World Intellectual Property Organization 2020. Biotech R & D: OECD 2019. See also Pian et al. 2020.
16. Pinker 2018: 386.
17. Asplen is quoted in Lazer 2004: 8.
18. I say "in most cases" because identical twins almost always have the same genotype, and it is statistically possible for non-twins to have the same genetic markers on the loci in forensic DNA databases.
19. National Research Council 2009: 100.
20. Lazer 2004.
21. Wang and Wein 2018: 1114. The authors note, without comment, that as of 2018, at least 200,000 sexual assault kits with biological evidence remained unanalyzed in American storage facilities.
22. New York's low percentage is still higher than the number of hits from ballistic imaging that resulted in convictions during the period of this study. See Wang and Wein 2018.
23. Doleac 2017: 166. She notes that this analysis does not account for privacy concerns about forensic DNA databases.
24. Anker et al. 2018: 5.
25. Bieber is quoted in Gerber 2016.
26. Mnookin is quoted in Gerber 2016. Liberty 2015 and Epstein 2009 support familial searching; see next section, on Skepticism, for opposition.
27. Innocence Project 2020a.
28. Criticism of Innocence Project: Smith 2010. Challenge convictions: Walsh et al. 2017; see also Garrett 2011.
29. Policing Project: Slater 2020. Grants for proposals: Bureau of Justice Assistance n.d. (c. 2020). Federal government spent: USASpending 2020.
30. Testimonials are from Pastor Kenneth Edward Copeland and Kimberly Elise, respectively, in African Ancestry 2019.
31. Hochschild and Sen 2015b.
32. Amazon Prime Day: Steelman 2018. DNA market doubled: Genetics Digest 2019.
33. *The Onion* 2018.
34. Affiliative self-fashioning: Nelson 2016. Choosing selectively: Roth and Ivemark 2018.
35. Information on AncestryDNA comes from Ancestry Corporate 2020. Allen is quoted in Copeland 2020: 63, emphasis in original. Blackstone acquisition: Blackstone 2020.

36. Asymptotic rise: Regalado 2018. Prediction for 2022: Buchholz 2019. Estimate for 2018: Biospace 2020. Illumina's view: Farr 2019.
37. However, respondents in Maya Sen's and my survey experiment anticipated slightly less gratification from DNA ancestry tests showing racial mixture than from those showing a single biogeographical line of descent (Hochschild and Sen 2015b).
38. Gates 2006.
39. Nelson 2018: 535.
40. Quoted in Bliss 2012: 107. See also Conley and Fletcher 2017: ch. 5; Hammonds and Herzig 2008: part 9.
41. Burchard 2003; Mak et al. 2018; Demenais et al. 2018. Among many disagreements, see Fullwiley 2007; Williams 2011; and much of Koenig et al. 2008.
42. Warfarin: Limdi et al. 2015. Lung cancer: AACR Communications Office 2020. Post-traumatic stress disorder: Stein et al. 2021. See also Ortega and Meyers 2014; Collins et al. 2021.

    These and similar findings imply an important institutional, scientific, financial, and statistical paradigm shift. Rather than only seeking to develop drugs suitable for a wide swath of the population—statins for everyone with high cholesterol—pharmacogenomicists and pharmaceutical firms are moving to investigate drugs appropriate for genetically distinctive, perhaps small, tranches of the population.
43. See also Satel 2002.
44. A Harvard anthropologist's 1939 textbook included a wonderful map of the eighteen "races of Europe." In an elaborately overlapping swarm of lines, dots, dashes, and hashmarks, Carleton Coon showed how the "Partially Mongoloid," "Lappish," "Brünn strain, Tronder etc., unreduced, only partly brachycephalized," "Pleistocene Mediterranean Survivor," "Neo-Danubian," "Nordic" and (separately) "Noric," along with eleven other races, were arrayed among those whom we now designate as White (Coon 1939).
45. Lewontin 1972. As one demonstration of Lewontin's claim, genomicist Stephen Schuster and his co-authors show that the genetic structures of indigenous hunter-gatherer peoples of southern Africa tend to differ more from each other than Europeans differ from Asians (Schuster et al. 2010).
46. Quotations are from Cavalli-Sforza et al. 1994: 19, emphases added.
47. Cavalli-Sforza et al. 1994: 136. Examples include article titles such as "Whole Genome Distribution and Ethnic Differentiation of Copy Number Variation in Caucasian and Asian Populations" (2009) or "Population Histories of the United States Revealed Through Fine-scale Migration and Haplotype Analysis" (2020).
48. Forum on Neuroscience and Nervous System Disorders 2020.
49. Reich 2018: 259–60.
50. Reich 2018: 258–59.
51. Conley and Fletcher 2017: 110.
52. Christmas: Zimmer 2015. Rural health care: Trivedi 2017. Infectious diseases: Gardy and Loman 2018. Passenger pigeons: Knapton 2017. Sewage into electricity: Adee 2016.
53. Pinker 2018: 77.
54. Silver 2012: 7; Clapper 2016: 9.

55. Achievement: Wade 2000. Disillusionment: Wade 2009. World Socialists: Gaglioti 2003.
56. Evans et al. 2011.
57. O'Neal 2011.
58. "Gifted": Sandel 2007. NHGRI warning: National Human Genome Research Institute 2020a.
59. Future threat: Roberts 2010: 441. GINA is weak: Clayton et al. 2019.
60. Cook 2018.
61. Blocher 2000.
62. Gene editing: Nuffield Council on Bioethics 2018. Green light: Knoepfler 2018.
63. Lander et al. 2019. Isaacson (2021) and Davies (2020) discuss this article and, more generally, arguments about a moratorium on germline gene editing.
64. Quoted in Stein 2020.
65. Fundamentally entwined, and right to own: Reardon 2017: 9. Business plan: Rajan 2006. Pharmocracy: Rajan 2017.
66. Specter 2012: 43.
67. Quotations in this paragraph are from Specter 2012.
68. "Lack control mechanisms" through "caution": Esvelt and Gemmell 2017. Bioerror: Garthwaite 2016. See also Greely 2019, and citations in Chapter 2, note 55.

    In August 2020, Florida authorities permitted release of 750 million genetically modified *Aedes aegypti* mosquitoes (LaMotte 2020).
69. Line not to be crossed: Collins 2015. Foundational unit: Foht 2015. Parental choice: Caplan et al. 1999: 1285. Common good: Darnovsky 2018.
70. Shaer 2016. See also, among others, Thompson 2013; Hsu 2015; Starr 2016.
71. Collateral consequences: Jannetta and Okeke 2017: 15. Expungement: Joh 2015: 51, emphasis in original.
72. Persons of interest: a spokesperson for ANDE, in Harkins 2019. "Unsolved crimes" and not authorized: FBI n.d. (c. 2019), emphasis in original.
73. Edelman and Stanley 2019.
74. Demographic trends: Grimm 2007: 1164. Overrepresented in databases: Rohlfs et al. 2013.
75. Familial searching: Murphy 2010: 291, 319. Too high a price: May 2018. On privacy, see also Copeland 2020; Kolenc 2019; Putnam 2020.
76. Holes et al. 2018. See also Molteni 2018, 2019b; Scutti 2018. GEDmatch data have subsequently been used in roughly 150 cold case arrests.
77. Erlich et al. 2018.
78. Shelby is quoted in Selk 2018; the headline is from the same source. Survey data: Auxier et al. 2019. Three-fifths of respondents over age sixty-five endorsed the strategy, compared with two-fifths of those under age thirty.
79. Hidden data: United Data Connect 2020. Genetic privacy: Bala 2019.
80. Roberts 2011: 264–65 (emphasis added), 269, 277. On disproportionate harm to men of color, see Ossorio and Duster 2005; Murphy and Tong 2019. Although Skeptics pay less attention to the gender, age, and immigration status characteristics of forensic DNA databases, these disproportionalities may be even greater than those of race.

81. No scientific value: Duster 2017; Biell and Hunter 2019. Astrology: Robert Green and Adam Rutherford, both quoted in Brown 2018. Botched testing: Brown 2018. (For more of both kinds of stories, see Copeland 2020.) Financial incentives: Bolnick et al. 2007.
82. Backdoor to eugenics: Duster 2003. Morning is quoted in Zhang 2016; see also Morning 2014. Innate abilities: Roth et al. 2020.
83. White supremacy: Panofsky and Donovan 2019. Geneticized racial identities: Roth and Ivemark 2018: 176.
84. Genocide and imperialism: Garrison n.d. Collective impact: Blanchard et al. 2019: 637. Power dynamics: Leroux 2018: 80. See also Walajahi et al. 2019; Tallbear 2013.

Chapter 5

1. National Institutes of Health 2009: 15, 16.
2. Paradigm-shifting areas: National Human Genome Research Institute 2019b. Human disparities: Green et al. 2011: 210.
3. National Human Genome Research Institute 2016.
4. Liberal implications: Rothstein et al. 2009; see also Loi et al. 2013; Rozek et al. 2014. Conservative responsibility: Chiapperino 2018: 54.
5. Robison 2015.
6. "Charles Socarides Dies" 2006.
7. Chan School: Obesity Prevention Source 2020. Cohen is quoted in Green 2015
8. Cultural tropes: Maurer and Sobal 1999.
9. United Health Foundation 2019.
10. Beaver and Schwartz 2016.
11. Terrie Moffitt is quoted in Cohen 2011. Burt and Simons 2014 call for "an end to heritability studies" in criminology, but offer instead "a more useful biosocial research agenda" that accords with "recent advances in our understanding of gene function and developmental plasticity" (quote on p. 223).
12. Evidence-based consensus: Wasserman 1996: 107. Pushing genocide: Roush 1995: 1808.
13. Wasserman 1996: 107.
14. 2015 survey: Schneider et al. 2018. Free will: Public Policy Research Institute 1996, in iPoll.
15. Ancestry testing: Balding and Innocent 2013. Interpretive arc: Nelson 2008: 259.
16. Sociopolitical standpoint: Roth and Ivemark 2018: 173. Claiming heritage: Duncan 2004.
17. Rotimi 2004: 543. He is writing here about "large-scale genotyping projects," not individual direct-to-consumer ancestry testing, but the aspiration is the same.
18. Foeman is quoted in "DNA Tests" 2017. African ancestry: Lawton et al. 2018. See the DNA Discussion Project, from which this evidence comes. www.wcupa.edu/DNADiscussion/.

19. NPR et al. 2015.
20. The quotations are examples, not comments by indentifiable individuals. On victims of environment, see Gould 1996; Jayaratne et al. 2006; Heine 2017. On weakness of will, see Suhay and Jayaratne 2012; Joslyn and Haider-Markel 2016; Schneider et al. 2018.
21. Quoted in Friend 2003. In a 2020 Pew Research Center survey of residents of twenty countries around the world, support for animal cloning ranges from a high of 48 percent in India to a low of 14 percent in France. A third of American respondents approve, somewhat above the median across all countries of 27 percent. Men, younger adults, those with higher levels of education, and the less religious tend to approve of animal cloning at higher rates, in the United States as well as elsewhere (Funk et al. 2020b).
22. Benefit the many: Joyner et al. 2016. Disappointment to clinicians: Limandri 2019: 4.
23. Kvaale et al. 2013: 782, statistical tests are omitted.
24. Duster 2014: 5–6, internal citations not included.
25. Montoya 2011: 102, 103, emphasis in original.
26. Berryessa et al. 2013.
27. Biological culture: Berryessa et al. 2013. Biological dispositions: Rose 2000: 5, 7.
28. Racial oppression: Roberts 2011: 255. At odds with race: Ossorio and Duster 2005: 121. Black Power activist: Harmon 2005.
29. Essentialist identity: quoted in Schwartz 2019.
30. Blog post on Reich: Kahn 2018. Testing Uighurs: Wee 2019; Byler 2019.
31. Scientific fatalism: Hughes 2013: 18. Remain on track: Talbott 2017, emphasis in original.
32. Hirschman 1991. My reflections on this book are in Hochschild 2018.

Chapter 6

1. Death of expertise: Nichols 2017. See also Lederman and Lederman 2014; Rynes et al. 2018; Jacoby 2018. Look at the mess: Gass 2016. Expertise: Rocco 2017.
2. Anti-intellectualism: Hofstadter 1962. University types: Cramer 2016: 131.
3. Pew Research Center, June 27–July 9, 2017; see also Berkman and Plutzer 2009. Americans are sometimes right to mistrust experts; "expert political judgment" is mistaken a disconcerting amount of the time (Tetlock 2005, 2015).
4. Levin 2020.
5. Supreme Court justices: Fox News/Opinion Dynamics Poll, July 12–13, 2005. College majors: Trachta 2019
6. Funk et al. 2020a.
7. A robust scholarly literature argues that analyses even in the hard sciences and the social science disciplines nearest to them are inevitably based on and reveal the author's values (see Latour and Woolgar 1986; Jasanoff 2004; Douglas 2009). Nonetheless, I instructed research assistants to code articles as having a valence only if the author

was fairly explicit about it, in order not to push either this claim or the coders' own lack of expertise in genomics too far.

8. The ELSI Research Program is institutionalized mainly through a set of federally funded, mostly university-based research centers, often focused around bioethics. The ELSI Research Program "fosters basic and applied research on the ethical, legal and social implications of genetic and genomic research for individuals, families and communities" (National Human Genome Research Institute 2020c).
9. No disciplines fall clearly into the Hope quadrant. Our selection criteria explain that absence, since the coding project was set up to capture the valences of only articles that engage explicitly with genetic influence or with societal uses of genomics. Similarly, the coding project does not reflect the huge number of articles that implicitly rebuff genomics by never mentioning it even when it could plausibly be relevant.
10. 2006 survey of academics: Gross 2013; see also Ladd and Lipset 1975; Rosenkranz 2014; Shields 2018. Identified as liberal, and proportions of Republicans: Gross and Simmons 2014. Radicals or far left: Gross 2013: 46–47.
11. To be symmetrical, the fourth category here should correspond to Hope in the basic framework, but it does not match that quadrant's substantive meaning—and is, in fact, mostly incoherent. I include it here for the sake of completeness; except for law, few articles fall in that group.
12. Dates after each quotation refer to the year of the online survey.
13. Phenylketonuria (PKU) results from an inherited genetic mutation that leads to decreased metabolism of a particular amino acid, and therefore to its accumulation in the body. If not recognized and treated soon after birth, PKU can lead to severe cognitive loss, seizures, and behavioral or mental problems. It cannot be cured, but it can be effectively managed by a stringent, low-protein diet. Researchers and medical professionals describe discovery and treatment of PKU as one of the strongest and clearest demonstrations of the benefits of medical genetics.
14. Auxier 2020.
15. Church and Regis 2012: 240–241.

Chapter 7

1. Doudna and Sternberg 2018: xii.
2. This chapter builds on, but does not replicate, Hochschild and Sen 2015a. Special thanks to Meredith Dost for superb data analyses and displays.
3. *CSI* effect: Schweitzer et al. 2007: 357; Shelton et al. 2006-2007.
4. In 2011, a quarter of GKAP respondents had heard about DNA ancestry testing (the question was not repeated in 2017). Too few respondents had direct experiences with such testing, however, for me to be able to analyze its impact on them.
5. A robust example is the Personal Genetics Education Program, which offers classroom curricula, programs for community and faith-based engagement, and other

publicly oriented resources (pged.org/mission/). See genome.gov/dna-day for National DNA Day.

6. Strictly speaking, neither response option is correct. Mature red blood cells and cornified cells in skin, nails, and hair destroy their nuclei, so they lack nuclear DNA. I expect, however, that very few nongeneticists know that fact, and even with that knowledge, one would probably not choose an option other than "in every cell in the human body."
7. Fifty percent of White, compared with 24 percent of African American, 28 percent of Latinx, and 43 perent of "other" (mostly Asian American) respondents in GKAP 1 correctly stated that Blacks and Whites share more than half of their DNA. Results in GKAP 2 were similar, with correct answers from 52 percent of Whites, 31 percent of Blacks, 38 percent of Latinx, and 53 percent of "others." In logit regression analyses, age made little (in 2011) or no (in 2017) substantive difference in the likelihood of a correct answer; in both years, people with more education were more likely to give the right answer than people with less. Even with controls for age and education, however, Blacks were less likely than Whites to agree that Blacks and Whites share more than half of their DNA. Hispanics and Asians resembled Blacks in this regard in GKAP 1 but not in GKAP 2.
8. Showing a slight increase in genetics knowledge in 2017, on GKAP 2, 18 percent answered none correctly, 37 percent answered one correctly, 27 answered two correctly, and 17 percent answered all three correctly.

   Correlations among these three knowledge items ranged from a low of 0.172 (DNA in which cells × mouse-human DNA sharing, 2017) to a high of 0.475 (mouse-human DNA sharing × Black-White DNA sharing, 2017).
9. Nonattitudes: Bishop 1986; Neijens 2004; Zizzo 2010. Meaningful response: Zaller 1992; Schuman and Presser 1996: 159; Sturgis and Smith 2010; Blumenthal and Swanson 2013; Dost et al. 2021.
10. Hochschild and Sen 2015a; Dost et al. 2021.
11. Freese and Shostak 2009 review these attitudes through 2008.
12. Geneticists now generally agree that even susceptibility to the flu has a genetic component, and that even eye color is not strictly Mendelian. I ignore these refinements on the grounds that they are almost certainly unknown to the public.
13. Schneider et al. 2018 and Parrott et al. 2003 also analyze laypeople's judgments about the impact of genetics on diseases, traits, or behaviors. See also Condit and Bates 2005; Jayaratne et al. 2006; Singer et al. 2007; Sturgis et al. 2010; Freese and Shostak 2009; Shostak et al. 2009; Garretson and Suhay 2015; Morin-Chassé et al. 2017; and Komisarchik and Hochschild 2020—all of which address public opinion about genetic influence.

    These studies mostly show different levels of attribution to genetic causes. Results may vary depending on the race of the survey respondent or of the subject of the question, or on different measurements of education levels, ideology, phenotypes, and other variables. Sample selection and analytic strategies differ across studies. Gernetic attributions are most likely not deeply fixed in the same way as, for example, judgments about abortion or Donald Trump. Thus comparing genetic explanations for various phenotypes within one survey is more trustworthy than seeking to determine the public's "real" genetic attribution for any given phenotype across surveys.

14. One could probe GKAP respondents' evaluations of sexual orientation more fully by comparing genetically knowledgeable liberals' views of its causes with genetically knowledgeable conservatives' views. That is tangential to my focus in *Genomic Politics*, however, so I have not done so.
15. Statements were presented separately for somatic and germline gene therapy (with other questions between the two sets). I focus here on views about risks and benefits, so I merge relevant items from both batteries. Chapter 9 considers each policy arena separately.
16. The rather odd formulation of "racially inflected genetic disease" seeks to capture the possible view that racial categories include a genetic component rather than being only socially constructed.
17. The subsample size of each quadrant is unweighted; percentages within each quadrant are weighted to be representative of the American adult population.
18. See Appendix to Chapter 7 for discussion of the regression analyses.
19. The corresponding figure for 2017 (not shown) demonstrates that the positive association between genetics knowledge and Enthusiasm persists in 2017, though it is slightly weaker than in 2011.
20. At every level of knowledge, Democrats are slightly more likely than Republicans are to be Enthusiasts; however, we see even less partisan difference in the other three quadrants.
21. One anomaly is worth pondering: in 2017, though not in 2011, the older the respondents, the more likely they are to be Enthusiasts regardless of levels of genetics knowledge. In parallel, knowledgeable young adults were more likely to be Skeptics in 2017 than their counterparts were in 2011.

    I have no explanation for why in this arena, unlike in most, optimism about a new technology is greater (and growing) among older people than among their juniors. Whatever the explanation for senior citizens' Enthusiasm, it is not indifference about privacy—concern about which is the hallmark of Skepticism. Older adults are more likely to worry about the security of their personal information and to be more attentive to privacy-related news; they are less likely to perceive benefits from having data collected about them (Auxier et al. 2019).
22. Causal direction: Miller 2004; Allum et al. 2008; Snow and Dibner 2016. It is, of course, possible that the more one knows about a scientific topic, the less enthusiasm one feels, whichever way the causal arrow runs. That is, roughly speaking, the pattern in the Skeptical quadrant. Skeptics are a small minority among GKAP respondents, however, so this section of the text focuses on the dominant pattern of association between knowledge and optimism.
23. More abstractly, how and how much do other independent variables (a.k.a. respondent characteristics) such as religiosity, values, or family status mediate between science knowledge and Enthusiasm? This question emerges out of scholars' criticism about the focus on scientific literacy as too cognitive, rationalist, or elitist. (Its critics

dub the focus on scientific literacy the "deficit model.") For reviews or compendia, see Snow and Dibner 2016; Committee on the Science of Science Communication 2017; Jamieson et al. 2017.

24. Cultural commitments: Kahan 2015, 2017.
25. Views on climate change dramatically demonstrate the power of partisan cultural commitment, but as GKAP shows, not all science issues follow the same trajectory. Pew Research Center analyses show that, controlling for other variables, ideology or partisanship is strongly associated with attitudes toward six of twenty-one issues under consideration, weakly associated with another nine, and not statistically associated with the remaining six issues. Parallel analyses focusing on science knowledge or education show similarly mixed impact: strong associations with views on three science issues, moderate associations with ten more, and no association with nine. Ideology matters most on issues other than those showing the strongest links to knowledge. Age has as much impact as partisanship, being strongly associated with views on ten issues; race or ethnicity is strongly related to five, religion or religiosity to two, and gender to one. (All of these results are from logistic regression analyses that control for the other variables.)

    Partially contradicting GKAP results, in the Pew survey, neither science knowledge nor political identity is as predictive as age or race/ethnicity for the six questions about biomedical science and technology (Pew Research Center 2015a). In light of these disparate results, I fall back on scholarly analyses' almost inevitable final sentence: "It's complicated; more research is needed."
26. Almost all of the coding was done before I developed the basic framework, and coders knew nothing of it. The coding buckets emerged after multiple readings of the comments as a means to systematize them and identify prominent themes. Coding categories underwent a lot of verbal mitosis and meiosis as we sought a framework that was both parsimonious and capacious. In short, *Genomic Politics*'s theory did not influence the coding. Coders were not otherwise involved in this project.
27. Zaller 1992.
28. ~4,300 GKAP 1 respondents × 2 questions + ~1800 GKAP 2 respondents × 2 questions = ~12,000 blank text boxes.

Chapter 8

1. After each survey asks what actors are best (and, separately, worst) suited to making policy in the respondent's arena of genomics expertise, and why, it offers a list of examples. They include elected officials, scientists, judges, corporations, nonprofits, regulatory bodies, deliberative bodies, and the general public (the 2013 survey did not include "deliberative bodies"). Although I did not intend it, it is possible that the experts' answers were influenced by this list. However, each survey also explicitly invites respondents to identify actors who are not policymakers with regard to genomics but should be (without offering examples). Many did so.

2. As in earlier chapters, I edit responses for obvious typographical or grammatical mistakes; I do not always include all of a given response.
3. Nuclear power: Conant 2006; Bird and Sherwin 2006. Catch up to last ten years: Prewitt et al. 2012 develop that possibility. Isaacson 2021 and Greely 2021 offer useful discussions of how leading genomicists are or might be working to develop governance structures.
4. See Juma 2016 for a fascinating history.
5. Jamieson et al. 2017; American Academy of Arts and Sciences 2019.
6. Contrary to Maya Sen's and my expectation when designing GKAP 1, Americans do not see public funding and regulation as opposite—that is, proactive versus precautionary—strategies for technology management. In 2011, correlations between support for government funding and for regulation reached 0.48 for group-inflected disease, 0.58 for individual inherited disease, and fully 0.59 for group-inflected traits. Support for funding and for regulation with regard to medical and forensic biobanks correlated in each case at 0.26, lower but still respectable.
7. Both proponents of the scientific literacy theory and those suspicious of the deficit model can find supporting evidence here. On one hand, reinforcing the idea that knowledge is associated with science appreciation, only 26 percent of Rejecters in 2011 have heard or read at least "some" about forensic DNA databases, compared with half or more in the other three quadrants. Twice as many Rejecters (53 percent in 2011) as in the other three quadrants have never heard of them. Results are similar in 2017. At the other end of the scale, a higher proportion of Enthusiasts than of the others have heard at least something about forensic biobanks, and a smaller proportion know nothing. So the people most hostile to this technology know the least about it, and the people most welcoming of this technology know the most about it.

   On the other hand, also in 2011, the same shares of Hopeful and Skeptics have heard of forensic biobanks—but the two quadrants hold diametrically opposite views about both genetic influence and technology optimism. So for people in these quadrants, presumably values, not knowledge, are dispositive. These patterns too are almost identical in 2017.
8. GKAP 2 asked for views on funding and regulation with regard only to medical biobanks, not the other two medical uses of genomics included in GKAP 1.
9. Note that, on this and other figures, where possible I set the *y*-axis to run from 0 to less than 100 percent, in order to show the differences across bars (e.g., quadrants) as clearly as possible. So the height of the bars may not be comparable across the figures.
10. Silber Mohamed 2018: 474.
11. Unlike with forensic biobanks, subject-specific knowledge about medical biobanks is not associated with attitudes. That is because there is no such knowledge: no more than a tenth in any quadrant in either year have heard or read even "some" about medical biobanks, and at least 70 percent in all quadrants know "nothing at all" about them. By default, therefore, residence in different quadrants is associated with different values, not with different levels of information; scientific literacy plays no role here.
12. National Institutes of Health 2020b. See also Davies 2020.

13. International statements: Brokowski 2018. Ethicist and pastor Larry Locke examines objections to CRISPR's use, especially "the perils of 'playing God,'" in Locke 2020. As I write, Greely 2021 provides the most recent extensive consideration of the ethics, rules, and practice of germline gene editing.
14. Ethics dumping: *The Economist* 2019.
15. Doudna and Sternberg 2018.
16. Among many other articles and books, attorney and bioethicist Josephine Johnston (2019) summarizes American law on germline gene editing; Henry Greely (2021) exhaustively examines He Jiankui's actions and their aftermath. See also Davies 2020 and Isaacson 2021.
17. I generally treat "gene editing" and "gene therapy" as synonyms in the surveys and their surrounding discussion. That is not quite right, since editing is a variant of the broader category of therapy. But my goal in GKAP is to elicit whatever public awareness is available, and I judged that using two similar terms might cast a wider net among respondents than using only one. Items are randomized within each battery.
18. Funk et al. 2016; STAT and Harvard T. H. Chan School of Public Health 2016 (but note small sample size for relevant questions); Neergaard 2018; Funk et al. 2020b. Science writer Cary Funk and her colleagues at the Pew Research Center find an interesting twist on the GKAP 2 evidence of general support for gene editing along with anxiety about particular features of it: across twenty countries, including the United States, survey respondents were twice as opposed to general "scientific research on gene editing" than to specific uses of prenatal gene editing (i.e., to treat a disease with which a baby would otherwise be born or to reduce risk of a serious disease later in life). GKAP and Pew results are not directly contradictory since they address different types of general and particular conditions, but they point in different directions. As always, more research is needed.
19. Among the many works on framing, see Chong and Druckman 2007; Druckman and Bolsen 2011; Lull and Scheufele 2017.
20. Dost et al. 2021.
21. Funk 2016.
22. Funk 2016.
23. Kaiser Health tracking poll, November 2016; Bloomberg poll, July 2017; Politico/ Harvard Public Health poll, February 2019; Kaiser Health tracking poll, August 2018.
24. I am drawing an analogy here to the argument that many people who claimed that Barack Obama was born outside the United States, and therefore was not a constitutionally legitimate president, were really taking whatever opportunity availed itself to express their general hostility to his presidency rather than reporting developed views about where he was born (Langer 2010; Berinsky 2012).
25. Results are similar if we compare conservatives to liberals, rather than Republicans to Democrats.

    The 2016 STAT/ Harvard Chan School survey found similar results regarding public sector actors; only 16 percent of Democrats and 6 percent of Republicans agree that government officials should determine whether to permit germline gene therapy. This sample, however, was much more favorable toward "scientists, physicians, and other technological experts" as decision-makers. No partisan data are reported (STAT/Harvard T. H. Chan School of Public Health 2016).

26. Book jackets: quoted in Shiller 2019: 60, who also reports on dial telephones; other examples are in Juma 2016.
27. For examples of excitement about new genomics technologies, see Green et al. 2009; Church and Regis 2014; Silber Mohamed 2018; Marsh and Ronner 2019, and Collins et al. 2021.
28. These examples are selected on the dependent variable, so their generalizability is not certain. That is, technologies for which fears *are* realized, such as nuclear bombs and early attempts at human gene therapy, may disappear or are kept under strict control—at least, so one hopes. Thus the fact that people come to accept many technologies of which they are initially wary may simply indicate that technologies that people come to accept are the ones that have been able to persist. In brief, successful innovations such as IVF or smartphones cannot be taken as proof that innovations usually succeed if they survive long enough.

Chapter 9

1. Collins et al. 2021.
2. Promulgators of these metaphors include Cohen et al. 1972; Jordan 1981; Gais et al. 1984; Sabatier 1991; Campbell 2012.
3. The relative political invisibility of genomics in the United States might change. President Joe Biden has appointed Eric Lander, formerly director of the Broad Institute, to the new Cabinet-level position of director of the White House Office of Science and Technology Policy. Lander, a leader of the Human Genome Project at the turn of the twenty-first century, has long experience in the public arena. The social science expert in genomic science, Alondra Nelson, is the new deputy director. Thus genomics might be promoted in the Biden administration; political controversy will not be far behind.
4. National Academies of Sciences, Engineering, and Medicine 2017a. Other reports include Nuffield Council on Bioethics 2012, 2018; LaBarbera 2015; National Academies of Sciences, Engineering, and Medicine 2017b; and, most recently, National Academy of Medicine et al. 2020.
5. A few examples: Morning 2011 offers a caution about how not to teach genomics. Geneticist Ting Wu's Personal Genetics Education Project (pged.org/mission) provides material for high school and college educators, libraries, museums, and community forums. Science and technology scholar Lundy Braun and anthropologist Barry Saunders (Braun and Saunders 2017; Saunders and Braun 2017) address race-related genomics curricula in medical school.
6. American Academy of Arts and Sciences 2019; American Anthropological Association n.d. But see Sturgis 2014.

7. Berg et al. 1975; Berg and Singer 1995. See also Wright 1986.
8. On policy entrepreneurs, see Kingdon 1984; Doig and Hargrove 1990; Carpenter 2010.
9. Shiller 2019.
10. Caplan et al. 2006 provides a useful compendium of reports.
11. For Facebook posts, see https://www.facebook.com/joaocarllo/posts/3837332892958172 and https://www.facebook.com/steve.markevich/posts/10224137049434689 (spelling and punctuation are unchanged.) The CDC and other agencies vigorously refuted this argument. Nonetheless, Republicans were twice as likely as Democrats (36% to 18%) to agree that the vaccine changes DNA; Hispanics were slightly more likely than Blacks or Whites to agree. Survey results are in *The Economist*/YouGov poll of May 8–11, 2021.
12. Couronne 2019. See also Marsh and Ronner 2019 for an analysis of the "Wild West of IVF."
13. Sinsheimer 1969: 8. Although an Enthusiast, Sinsheimer did express caution: "The horizons of the new eugenics are in principle boundless—for we should have the potential to create new genes and new qualities yet undreamed. But of course the ethical dilemma remains. What are the best qualities, and who shall choose?" (13).
14. James Collins, quoted in Shaw 2020: 347, emphasis in original.
15. Unbounded good: Chargoff 1987. Reaches of the human mind: Jasanoff 2019: 178–179. Permissionless innovation: Lehrer and Mills 2019: 63–64. Steven Pinker operationalizes permissionless innovation thus: "Given this potential bonanza [resulting from biomedical research], the primary moral goal for today's bioethics can be summarized in a single sentence. Get out of the way" (Pinker 2015).
16. Cohen et al. 2008.
17. LaBarbera 2015.
18. Inspired largely by Hirschman 1991, I develop this argument in Hochschild 2017, 2018.
19. Berkowitz 2020.
20. As biologist Nessa Carey writes, a bit tendentiously, opposition to "golden rice" that might prevent blindness and starvation is predicated on the "philosophical logic" that "if you oppose GM on principle, then you must oppose all GM. Whether these opponents have ever sat down and explained that principle to a bereaved parent or a child who has avoidably and irreversibly lost their sight is something you might be interested to know about" (Carey 2020: 45). Robert Sinsheimer anticipated this point, as he did so many others (Sinsheimer 1969: 13).
21. Benjamin 2019.
22. Muhammad 2011.
23. Gross 2019.
24. Astronomically expensive: tg 2019; Stein 2020; Hayden 2020. Prismatic effect: Conley and Fletcher 2017: 179, 180.

Appendix

1. Wood 2013.
2. We also coded 465 articles in the bench sciences for comparison, and as a rough check on the coding scheme and on the students' coding choices. I do not report those results.
3. Special thanks for heroic work to Chris Chaky and Ryan Zhang.
4. My thanks to Alex Crabill for the first survey, and Ryan Zhang for the second.
5. Ansolabehere et al. 2008.

References

23andMe (2018). "World Cup Fans, You Can Root for Your Roots.," May 3. https://blog.23andme.com/ancestry-reports/world-cup-fans-can-root-for-your-roots/.

AACR Communications Office (2020). "Native American Ancestry Associated with Lung Cancer Mutations in Latin American Patients." Broad Institute, December 4.

Achen, C., and L. Bartels (2016). *Democracy for Realists: Why Elections Do Not Produce Responsive Government*. Princeton University Press.

Adee, S. (2016). "Bacteria Made to Turn Sewage into Clean Water—and Electricity." *New Scientist*, July 27.

African Ancestry. (2019). "Testimonials." africanancestry.com/testimonials/.

Allen, J. (2018). "Trump Challenges 'Pocahontas' Warren to DNA Test to Prove She's Native American." NBC News, July 5.

Allum, N., et al. (2008). "Science Knowledge and Attitudes Across Cultures: A Meta-Analysis." *Public Understanding of Science* 17(1): 35–54.

Almond, G., and S. Verba (1963). *The Civic Culture: Political Attitudes and Democracy in Five Nations*. Princeton University Press.

Altman, R. (2008). "Genotyping Cost Is Asymptoting to Free." *Building Confidence*, October 10. http://rbaltman.wordpress.com/2008/10/08/genotyping-cost-is-asymptoting-to-free/.

American Academy of Arts and Sciences (2019). *Encountering Science in America*. American Academy of Arts and Sciences.

American Anthropological Association (1998). "Statement on Race." *American Anthropologist* 100: 712–713.

American Anthropological Association (n.d.). "Race: Are We So Different." www.americananthro.org/LearnAndTeach/Content.aspx?ItemNumber=2062.

Ancestry Corporate (2020). "The Many Ways Our Family Has Grown." www.ancestry.com/corporate/about-ancestry/company-facts.

Anderson, S. (2019). *A Broken Tree: How DNA Exposed a Family's Secrets*. Rowman & Littlefield.

Anker, A., et al. (2018). "The Effects of DNA Databases on the Deterrence and Detection of Offenders." Rockwool Foundation, Copenhagen. www.rockwoolfonden.dk/app/uploads/2018/04/RFF-Study-Paper-128_The-effects-of-DNA-databases-on-the-deterrence-and-detection-of-offenders-1.pdf

Ansolabehere, S., et al. (2008). "The Strength of Issues: Using Muliple Measures to Gauge Preference Stability, Ideological Constraint, and Issue Voting." *American Political Science Review* 102(2): 215–232.

Arnold, F. (2012). "What Is Life?" *Bulletin of the American Academy of Arts & Sciences*, Winter, 9–10.

Asch, A. (2000). "Why I Haven't Changed My Mind About Prenatal Diagnosis: Reflections and Refinements." In *Prenatal Testing and Disability Rights*, ed. E. Parens and A. Asch, 234–258. Georgetown University Press.

Asch, A., and D. Barlevy (2012). "Disability and Genetics: A Disability Critique of Pre-natal Testing and Pre-implantation Genetic Diagnosis (PGD)." *eLS*, May 15. https://doi.org/10.1002/9780470015902.a0005212.pub2.

Athanasiou, T., and M. Darnovsky (2018 [2002]). "The Genome as Commons." In *Beyond Bioethics: Toward a New Biopolitics*, ed. O. Obasogie and M. Darnovsky, 157–162. University of California Press.

Atzmon, G., et al. (2010). "Abraham's Children in the Genome Era: Major Jewish Diaspora Populations Comprise Distinct Genetic Clusters with Shared Middle Eastern Ancestry." *American Journal of Human Genetics* 86(6): 850–859.

Auxier, B. (2020). *What We've Learned About Americans' Views of Technology during the Time of COVID-19*. Pew Research Center.

Auxier, B., et al. (2019). *Americans and Privacy: Concerned, Confused, and Feeling Lack of Control over Their Personal Information*. Pew Research Center.

Azar, K., et al. (2020). "Disparities in Outcomes Among Covid-19 Patients in a Large Health Care System in California." *Health Affairs* 39(7): 1253–1262.

Baggini, J. (2015). "Do Your Genes Determine Your Entire Life?" *The Guardian, March* 19.

Bala, N. (2019) "We're Entering a New Phase in Law Enforcement's Use of Consumer Genetic Data." *R Street*, December 19.

Balding, D., and T. Innocent (2013). "Sense About Genetic Ancestry Testing." Sense About Science. http://www.senseaboutscience.org/data/files/resources/119/Sense-About-Genetic-Ancestry-Testing.pdf.

Battelle Institute (2013). "Updated Battelle Study Shows that the Genetics and Genomics Industry in the U.S. Has a Trillion Dollar Impact." http://www.battelle.org/media/press-releases/updated-battelle-study-genetics-and-genomics-industry.

Bayefsky, M. (2016). "Comparative Preimplantation Genetic Diagnosis Policy in Europe and the USA and Its Implications for Reproductive Tourism." *Reproductive Biomedicine and Society Online* 3: 41–47.

Bayefsky, M. (2018). "Who Should Regulate Preimplantation Genetic Diagnosis in the United States?" *AMA Journal of Ethics* 20(12): E1160–1167.

Bearman, Peter, ed. (2008). "Exploring Genetics and Social Structure." *American Journal of Sociology*. 114 (S1).

Beaver, K., and J. Schwartz (2016). "The Utility of Findings from Biosocial Research for Public Policy." In *Advancing Criminology and Criminal Justice Policy*, ed. T. Blomberg et al., 452–460. Routledge.

Benjamin, R. (2019). *Race After Technology: Abolitionist Tools for the New Jim Code*. Polity.

Berezow, A., and H. Campbell (2014). *Science Left Behind: Feel-Good Fallacies and the Rise of the Anti-Scientific Left*. Public Affairs.

Berg, P. (2008). "Meetings That Changed the World: Asilomar 1975: DNA Modification Secured." *Nature* 455: 290–291.

Berg, P., et al. (1975). "Summary Statement of the Asilomar Conference on Recombinant DNA Molecules." *Proceedings of the National Academy of Sciences* 72(6): 1981–1984.

Berg, P., and M. Singer (1995), "The Recombinant DNA Controversy: Twenty Years Later." *Proceedings of the National Academy of Science* 92: 9011–9013.

Bergin, J. (2019a). "Global DNA Read, Write and Edit Market." BCC Research LLC, Report Code BIO193A.

Bergin, J. (2019b). "Global DNA Sequencing: Research, Applied and Clinical Markets." BCC Research LLC., Report Code BIO045G.

Berinsky, A. (2012). "Rumors, Truths, and Reality: A Study of Political Misinformation." Political Science Department, MIT. http:// web.mit.edu/ berinsky/ www/ les/ rumor.pdf.

Berkman, M., and E. Plutzer (2009). "Scientific Expertise and the Culture War: Public Opinion and the Teaching of Evolution in the American States." *Perspectives on Politics* 7(3): 485–499.

Berkowitz, A. (2020). "Why Sequencing the Human Genome Failed to Produce Big Breakthroughs in Disease." *The Conversation*, February 11. theconversation.com/why-sequencing-the-human-genome-failed-to-produce-big-breakthroughs-in-disease-130568.

Berlin, I. (2013). *The Hedgehog and the Fox: An Essay on Tolstoy's View of History*. 2nd ed. Princeton University Press.

Berryessa, C., et al. (2013). "Ethical, Legal, and Social Issues Surrounding Research on Genetic Contributions to Anti-Social Behavior." *Aggression and Violent Behavior* 18(6): 605–610.

Biell, M., and M. Hunter (2019). "Direct-to-Consumer Genetic Testing's Red Herring: 'Genetic Ancestry' and Personalized Medicine." *Frontiers in Medicine* 6: 48.

Biospace (2020). "DNA Test Kits Market: Increase in Demand for Ancestry Testing to Drive Market." April 27. www.biospace.com/article/dna-test-kits-market-increase-in-demand-for-ancestry-testing-to-drive-market/.

Bird, K., and M. Sherwin (2006). *American Prometheus: The Triumph and Tragedy of J. Robert Oppenheimer*. Vintage.

Bishop, G., et al. (1986). "Opinions on Fictitious Issues: The Pressure to Answer Survey Questions." *Public Opinion Quarterly* 50(2): 240–250.

Blackstone. (2020). "Blackstone to Acquire Ancestry®, Leading Online Family History Business, for $4.7 Billion." Press release, August 5. www.blackstone.com/press-releases/article/blackstone-to-acquire-ancestry-leading-online-family-history-business-for-4-7-billion/.

Blanchard, J., et al. (2019). "'We Don't Need a Swab in Our Mouth to Prove Who We Are': Identity, Resistance, and Adaptation of Genetic Ancestry Testing Among Native American Communities." *Current Anthropology* 60(5): 637–655.

Blank, J., and D. Shaw (2015). "Does Partisanship Shape Attitudes Toward Science and Public Policy? The Case for Ideology and Religion." *Annals of the American Academy of Political and Social Science* 658 (March): 18–35.

Bliss, C. (2012). *Race Decoded: The Genomic Fight for Social Justice*. Stanford University Press.

Blocher, M. (2000). "What the New Genetic Technologies Mean for the Pro-Life Movement." At the Center: A Ministry of Life Matters Worldwide, October. http://www.atcmag.com/Issues/ID/27/What-the-New-Genetic-Technologies-Mean-for-the-Pro-Life-Movement.

Blumenthal, M., and E. Swanson (2013). "Beware: Survey Questions About Fictional Issues Still Get Answers." *Huffington Post Politics*, April 11.

Bolnick, D., et al. (2007). "The Science and Business of Genetic Ancestry Testing." *Science* 318: 399–400.

Boony, S. (2003). "Why Are Most Europeans Opposed to GMOs? Factors Explaining Rejection in France and Europe." *Electronic Journal of Biotechnology* 6(1).

Bowers, B. (2012). "Rick Santorum Says, 'Amniocentesis Does, in Fact, Result More Often than Not . . . in Abortion.'" PolitiFact, February 27.

Boyd, R., et al. (2020). "On Racism: A New Standard for Publishing on Racial Health Inequities." *Health Affairs*, July 2. www.healthaffairs.org/do/10.1377/hblog20200630.939347/full/.

Braun, L., and B. Saunders (2017). "Avoiding Racial Essentialism in Medical Science Curricula." *AMA Journal of Ethics* 19(6): 518–527.

Braveman, P., et al. (2011). "The Social Determinants of Health: Coming of Age." *Annual Review of Public Health* 32: 381–398.

Brenan, M., and L. Saad (2018). "Global Warming Concern Steady Despite Some Partisan Shifts." Gallup News: Politics, March 28. news.gallup.com/poll/231530/global-warming-concern-steady-despite-partisan-shifts.aspx.

Brokowski, C. (2018). "Do CRISPR Germline Ethics Statements Cut It?" *The CRISPR Journal* 1(2).

Brown, G. (2016). "Mom and Baby with Down Syndrome Mail Letter to Doctor Who Suggested Abortion." ABC News, June 7.

Brown, K. (2018). "How DNA Testing Botched My Family's Heritage, and Probably Yours, Too." Gizmodo, January 16.

Bucerius, S., and M. Tonry, eds. (2014). *The Oxford Handbook of Ethnicity, Crime, and Immigration*. Oxford University Press.

Buchholz, K. (2019). "Consumer Genetic Testing Grows in Popularity." Statista, November 18. www.statista.com/chart/19996/size-of-global-direct-to-consumer-gentic-testing-market/.

Bull, J., and H. Malik (2017). "The Gene Drive Bubble: New Realities." *PLOS Genetics* 13(7): e1006850.

Burchard, E., et al. (2003). "The Importance of Race and Ethnic Background in Biomedical Research and Practice." *New England Jounal of Medicine* 348(12): 1170–1175.

Bureau of Justice Assistance (n.d.). "Awards." bja.ojp.gov/funding/awards/list?awardee=&city=&combine_awards=DNA&field_award_status_value=All&field_fiscal_year_value=&field_funding_type_value=All&state=All&topic=All&page=1.

Burnett, F. H. (2010 [1911]). *The Secret Garden*. HarperCollins.

Burt, C., and R. Simons (2014). "Pulling Back the Curtain on Heritability Studies: Biosocial Criminology in the Postgenomic Era." *Criminology* 52(2): 223–262.

Byler, D. (2019). "China's Hi-Tech War on Its Muslim Minority." *The Guardian*, April 11.

Callaway, E. (2018). "UN Treaty Agrees to Limit Gene Drives but Rejects a Moratorium." *Nature*, November 29.

Campbell, A. (2012). "Policy Makes Mass Politics." *Annual Review of Political Science* 15: 333–351.

Campbell, L. (2011). "Non-conviction DNA Databases in the United States and England: Historical Differences, Current Convergences?" *International Journal of Evidence and Proof* 15(4): 281–310.

Caplan, A., et al. (1999). "What Is Immoral About Eugenics?" *British Medical Journal* 319: 1284–1285.

Caplan, A., et al., eds. (2006). *The Case of Terri Schiavo: Ethics at the End of Life*. Prometheus.

Carey, N. (2012). *The Epigenetics Revolution*. Icon Books.

Carey, N. (2019). *Hacking the Code of Life: How Gene Editing Will Rewrite Our Futures*. Icon Books.

Carpenter, D. (2010). *Reputation and Power: Organizational Image and Pharmaceutical Regulation at the FDA*. Princeton University Press.

Carpenter, D., and R. Nevin (2010). "Environmental Causes of Violence." *Physiology & Behavior* 99(2): 260–268.

Cavalli-Sforza, L., et al. (1994). *The History and Geography of Human Genes*. Princeton University Press.

Center for Drug Evaluation and Research (2005). *Cardiovascular and Renal Drugs Advisory Committee, Volume II*. Food and Drug Administration.

Chargaff, E. (1987). "Engineering a Molecular Nightmare." *Nature* 327(6119): 199–200.

"Charles Socarides Dies; Said He 'Cured' Gays" (2006). *Washington Post*, January 2.

Charlotte Lozier Institute (2015). "New Study: Abortion after Prenatal Diagnosis of Down Syndrome Reduces Down Syndrome Community by Thirty Percent." Charlotte Lozier Institute. lozierinstitute.org/new-study-abortion-after-prenatal-diagnosis-of-down-syndrome-reduces-down-syndrome-community-by-thirty-percent/.

Chastain, D., et al. (2020). "Racial Disproportionality in Covid-19 Clinical Trials." *New England Journal of Medicine* 383:e59.

Chiapperino, L. (2018). "Epigenetics: Ethics, Politics, Biosociality." *British Medical Bulletin* 128: 49–60.

Chong, D., and J. Druckman (2007). "Framing Theory." *Annual Review of Political Science* 10: 103–126.

Church, G. (2019). "Regarding the Potential Use of Genetic Information in Dating Apps as Mentioned in the Interview on *60 Minutes*." December 14. http://arep.med.harvard.edu/gmc/gen_faq.html.

Church, G., and E. Regis (2014). *Regenesis: How Synthetic Biology Will Reinvent Nature and Ourselves*. Basic Books.

Clapper, J. (2016). "Worldwide Threat Assessment of the US Intelligence Community." Office of the Director of National Intelligence. www.dni.gov/files/documents/SASC_Unclassified_2016_ATA_SFR_FINAL.pdf.

Clayton, E., et al. (2019). "The Law of Genetic Privacy: Applications, Implications, and Limitations." *Journal of Law and the Bioscience* 6(1): 1–36.

ClinicalTrials.gov (2020). [Database on gene therapy trials.] clinicaltrials.gov/ct2/results?term=%22gene+therapy%22&Search=Apply&recrs=a&recrs=f&recrs=d&age_v=&gndr=&type=&rslt=.

Cohen, M., et al. (1972). "A Garbage Can Model of Organizational Choice." *Administrative Science Quarterly* 17(1): 1–25.

Cohen, M., et al. (2008). *The Party Decides: Presidential Nominations Before and After Reform*. University of Chicago Press.

Cohen, P. (2011). "Genetic Basis for Crime: A New Look." *New York Times*, June 19.

Cohn, E., et al. (2017). "Self-Reported Race and Ethnicity of US Biobank Participants Compared to the US Census." *Journal of Community Genetics* 8: 229–238.

Collins, F. (2015). "Statement on NIH Funding for Research Using Gene-Editing Technologies in Human Embryos." National Institutes of Health, April 28. www.nih.gov/about-nih/who-we-are/nih-director/statements/statement-nih-funding-research-using-gene-editing-technologies-human-embryos.

Collins, F. et al. (2021). "Human Molecular Genetics and Genomics—Important Advances and Exciting Possibilities." *New England Journal of Medicine* 384 (1): 1–4.

Committee on the Science of Science Communication (2017). *Communicating Science Effectively: A Research Agenda*. National Academies Press.

Conant, J. (2006). *109 East Palace: Robert Oppenheimer and the Secret City of Los Alamos*. Simon & Schuster.

Condit, C., et al. (2009). "Believing in Both Genetic Determinism and Behavioral Action: A Materialist Framework and Implications." *Public Understanding of Science* 18(6): 730–746.

Condit, C., and B. Bates. (2005). "How Lay People Respond to Messages About Genetics, Health, and Race." *Clinical Genetics* 68(2): 97–105.

Conley, D., and J. Fletcher (2017). *The Genome Factor: What the Social Genomics Revolution Reveals About Ourselves, Our History, and the Future*. Princeton University Press.

Convention on Biological Diversity (2019). "The Cartagena Protocol on Biosafety." bch.cbd.int/protocol/.

Cook, M. (2018). "Two Cheers, at Least, for Utilitarianism." *BioEdge*, July 29. www.bioedge.org/pointedremarks/view/two-cheers-at-least-for-utilitarianism/12758/.

Cook-Deegan, R. (1994). *The Gene Wars: Science, Politics, and the Human Genome*. Norton.

Coon, C. (1939). *The Races of Europe*. Macmillan.

Copeland, L. (2020). *The Lost Family: How DNA Testing Is Upending Who We Are*. Abrams Press.

Corn, J. (2015). "The Future Utlility of DNA Science." Paper presented at the Symposium on the Past, Present, and Future of DNA, October 2, Radcliffe Institute for Advanced Study, Cambridge, MA.

Couronne, I. (2019). "In US, Relaxed IVF Laws Help Would-Be Parents Realize Dreams." MedicalXpress, June 27. medicalxpress.com/news/2019-06-ivf-laws-would-be-parents.html.

Couzin, J., and J. Kaiser (2007). "Closing the Net on Common Disease Genes." *Science* 316(5826): 820–822.

Cramer, K. (2016). *The Politics of Resentment: Rural Consciousness in Wisconsin and the Rise of Scott Walker*. University of Chicago Press.

Cullen, F., and P. Wilcox, eds. (2013). *The Oxford Handbook of Criminological Theory*. Oxford University Press.

Curry, J., and F. Lee (2019). "Non-Party Government: Bipartisan Lawmaking and Party Power in Congress." *Perspectives on Politics* 17(1): 47–65.

Darnovsky, M. (2018). "Genetically Modifying Future Children Isn't Just Wrong. It Would Harm All of Us." *The Guardian*, July 17.

Davies, K. (2020). *Editing Humanity: The CRISPR Revolution and the New Era of Genome Editing*. Pegasus Books.

De Graff, G., F. Buckley, and B. Skotko (2020). "Estimation of the Number of People with Down Syndrome in Europe." *European Journal of Human Genetics*. Published online October 31.

Debus-Sherrill, S., and M. Field (2019). "Familial DNA Searching—An Emerging Forensic Investigative Tool." *Science & Justice* 59(1): 20–28.

Demenais, F., et al. (2018). "Multiancestry Association Study Identifies New Asthma Risk Loci that Colocalize with Immune-Cell Enhancer Marks." *Nature Genetics* 50(1): 42–53.

Department of Agriculture (n.d.). "Roles of USDA Agencies in Biotechnology." www.usda.gov/topics/biotechnology/roles-usda-agencies-biotechnology.

Department of Energy (n.d.). "Genomic Science Program." genomicscience.energy.gov.

DeSalle, R., and M. (2005). *Welcome to the Genome: A User's Guide to the Genetic Past, Present, and Future*. Wiley-Liss.

Devaney, S. (2020). "All of Us." *Scientific American* 222(1): S15–S17.

DiEuliis, D., and J. Giordano (2017). "Why Gene Editors Like CRISPR/Cas May Be a Game-Changer for Neuroweapons." *Health Security* 15(3): 296–302.

Dirks, E., and J. Leibold (2020). "Genomic Surveillance: Inside China's DNA Dragnet." Australian Strategic Policy Institute. Policy Brief Report No. 34/2020. s3-ap-southeast-2.amazonaws.com/ad-aspi/2020-06/Genomic%20surveillance_1.pdf?QhPFyrNVaSjv blmFT24HRXSuHyRfhpml=

Divine, J. (2020). "Moderna Keeps Surging, MGM Stumbles." *U.S. News & World Report*, July 17.

DNA11 (2020). "From Life Comes Art." http://www.dna11.com/about-us.

DNA Forensics (2020). "States Using Familial Searches." http://www.dnaforensics.com/statesandfamilialsearches.aspx.

"DNA Tests, and Sometimes Surprising Results" (2017). *New York Times*, April 23.

Doig, J., and E. Hargrove, eds. (1990). *Leadership and Innovation: Entrepreneurs in Government*. Johns Hopkins University Press.

Dolan, M. (2011). "State to Double Crime Searches Using Family DNA." *Los Angeles Times*, May 9.

Doleac, J. (2017). "The Effect of DNA Databases on Crime." *American Economic Journal: Applied Economics* 9(1): 165–201.

Dost, M., et al. (2021). "Democratic Competence amid Scientific Advance: Public Views on Genetic Biobanks." Department of Government, Harvard University.

Doudna, J., and S. Sternberg (2018). *A Crack in Creation: Gene Editing and the Unthinkable Power to Control Evolution*. Mariner Books.

Douglas, H. (2009). *Science, Policy, and the Value-Free Ideal*. University of Pittsburgh Press.

Douglas, H. (2015). "Politics and Science: Untangling Values, Ideologies, and Reasons." *Annals of the American Academy of Political and Social Science* 658(1): 296–306.

Downey, R. (2020). "Florida Becomes First State to Protect DNA from Life, Disability Insurers." *Florida Politics*, July 1.

Druckman, J., and T. Bolsen (2011). "Framing, Motivated Reasoning, and Opinions About Emergent Technologies." *Journal of Communication* 61(4): 659–688.

Duffin, E. (2020). "Percentage of Patent Applications in the United States in 2018, by Top Fields of Technology." Statista, February 20. https://www.statista.com/statistics/256734/percentage-of-patent-applications-in-the-us-by-fields-of-technology/

Duncan, P. (2004). [Letter to the editor.] *New York Times*, July 5.

Duster, T. (2003). *Backdoor to Eugenics*. 2nd ed. Routledge.

Duster, T. (2014). "A Post-Genomic Surprise: The Molecular Reinscription of Race in Science, Law and Medicine." *British Journal of Sociology* 66(1): 1–27.

Duster, T. (2017). "Ancestry Testing and DNA: Uses, Limits—and Caveat Emptor." In *Genetics as Social Practice: Transdisciplinary Views on Science and Culture*, ed. B. Prainsack et al., 59–72. Routledge.

The Economist (2007). "Biology's Big Bang." June 16.

The Economist (2019). "Recent Events Highlight an Unpleasant Scientific Practice: Ethics Dumping." February 2.

The Economist (2021). "The Roaring 20s?" January 16.

Edelman, V., and J. Stanley (2019). "Rapid DNA Machines in Police Departments Need Regulation." ACLU, October 2. www.aclu.org/blog/privacy-technology/medical-and-genetic-privacy/rapid-dna-machines-police-departments-need.

Edge.org (2005). "The Science of Gender and Science: Pinker vs. Spelke, a Debate." Edge.org, May 16. www.edge.org/event/the-science-of-gender-and-science-pinker-vs-spelke-a-debate.

Eilenberg, Jon (2015). Chartgeist. *Wired*, August. http://contentviewer.adobe.com/s/Wired/5857345fd35d4d1f9a1f00273013f68a/WI0815_10_Folio/2080_2308AP_chartgeist.html.

Epstein, J. (2009). "'Genetic Surveillance'—The *Bogeyman* Response to Familial DNA Investigations." *Journal of Law, Technology, and Policy* 2009(1): 141–173.

Erlich, Y., et al. (2018). "Identity Inference of Genomic Data Using Long-Range Familial Searches." *Science* 362(6415): 690–694.

Eschner, K. (2017). "In Vitro Fertilization Was Once as Controversial as Gene Editing Is Today." Smithsonian.com, September 27. www.smithsonianmag.com/smart-news/vitro-fertilization-was-once-controversial-cloning-today-180964989/

Esvelt, K. (2019). "When Are We Obligated to Edit Wild Creatures?" *Leapsmag*, August 30. leapsmag.com/when-are-we-obligated-to-edit-wild-creatures/.

Esvelt, K., and N. Gemmell (2017). "Conservation Demands Safe Gene Drive." *PLOS: Biology*, November 16.

Eurobarometer (2010). Eurobarometer 73.1: "The European Parliament, Biotechnology, and Science and Technology." GESIS, Leibniz Institute for the Social Sciences. zacat.gesis.org/webview/.

Evans, J., et al. (2011). "Deflating the Genomic Bubble." *Science* 331(6019): 861–862.

Fallows, J. (2014). "When Will Genomics Cure Cancer?" *The Atlantic*. January/February.

Fan, M. (2019). "Genome Editing: Technologies and Global Markets." BCC Research LLC. Report Code BIO146B.

Farr, C. (2019). "Consumer DNA Testing Has Hit a Lull—Here's How It Could Capture the Next Wave of Users." CNBC, August 25.

Federal Bureau of Investigation (n.d.). "Rapid DNA: General Information." www.fbi.gov/services/laboratory/biometric-analysis/codis/rapid-dna.

Feldman, M., and N. Lowe (2008). "Consensus from Controversy: Cambridge's Biosafety Ordinance and the Anchoring of the Biotech Industry." *European Planning Studies* 16(3): 395–410.

Field, M., and S. Debus-Sherrill (2017). "Study of Familial DNA Searching Policies and Practices: National Survey of CODIS Laboratories Brief." National Criminal Justice Reference Service. www.ncjrs.gov/pdffiles1/nij/grants/251049.pdf.

Fine, C., et al. (2017). "Sex-Linked Behavior: Evolution, Stability, and Variability." *Trends in Cognitive Sciences* 21(9): 666–673.

Fischhoff, B., et al. (1978). "How Safe Is Safe Enough? A Psychometric Study of Attitudes Towards Technological Risks and Benefits." *Policy Sciences* 9(2): 127–152.

Flynn, M. (2019). "A Harvard Scientist Is Developing a DNA-Based Dating App to Reduce Genetic Disease. Critics Called It Eugenics." *Washington Post*, December 13.

Foht, B. (2015). "The Case Against Human Gene Editing." *National Review*, December 4.

Food and Drug Administration (2005). "FDA Approves BiDil Heart Failure Drug for Black Patients." wayback.archive-it.org/7993/20161024051521/http://www.fda.gov/NewsEvents/Newsroom/PressAnnouncements/2005/ucm108445.htm.

Food and Drug Administration (2017). "FDA Allows Marketing of First Direct-to-Consumer Tests That Provide Genetic Risk Information for Certain Conditions." www.fda.gov/news-events/press-announcements/fda-allows-marketing-first-direct-consumer-tests-provide-genetic-risk-information-certain-conditions.

Food and Drug Administration (2019). "FDA Approves New Breakthrough Therapy for Cystic Fibrosis." www.fda.gov/news-events/press-announcements/fda-approves-new-breakthrough-therapy-cystic-fibrosis.

Food and Drug Administration (2020). "FDA Continues Strong Support of Innovation in Development of Gene Therapy Products." News release. FDA. January 28. https://www.fda.gov/news-events/press-announcements/fda-continues-strong-support-innovation-development-gene-therapy-products.

Foong, P. (2019). "How Should Chimeric Embryo Research Be Regulated?" Bioedge, August 25. www.bioedge.org/indepth/view/how-should-chimeric-embryo-research-be-regulated/13188.

Forum on Neuroscience and Nervous System Disorders (2020). "Sex Differences in Brain Disorders: Emerging Transcriptomic Evidence and Implications for Therapeutic Development—A Workshop." National Academies of Science. https://www.nationalacademies.org/our-work/sex-differences-in-brain-disorders-emerging-transcriptomic-evidence-and-implications-for-therapeutic-development-a-workshop.

Freese, J. (2008). "Genetics and the Social Science Explanation of Individual Outcomes." *American Journal of Sociology* 114 supp.: S1–S35.

Freese, J., and S. Shostak (2009). "Genetics and Social Inquiry." *Annual Review of Sociology* 35: 107–128.

Friend, T. (2003). "The Real Face of Cloning." *USA Today*, January 17.

Fullwiley, D. (2007). "The Molecularization of Race: Institutionalizing Human Difference in Pharmacogenetics Practice." *Science as Culture* 16(1): 1–30.

Funk, C., et al. (2016). "U.S. Public Wary of Biomedical Technologies to 'Enhance' Human Abilities." July 26. Pew Research Center.

Funk, C., et al. (2020a). "Trust in Medical Scientists Has Grown in U.S., but Mainly Among Democrats." May 21. Pew Research Center.

Funk, C., et al. (2020b). "Biotechnology Research Viewed with Caution Globally, but Most Support Gene Editing for Babies to Treat Disease." December 10. Pew Research Center.

Funk, C., and B. Kennedy (2016). "The New Food Fights: U.S. Public Divides over Food Science." December 1. Pew Research Center.

Gaglioti, F. (2003). "Human Genome Project Completed: An Extraordinary Scientific Achievement." World Socialist Web Site, May 7. www.wsws.org/en/articles/2003/05/gene-m07.html.

Gais, T., et al. (1984). "Interest Groups, Iron Triangles and Representative Institutions in American National Government." *British Journal of Political Science* 14(2): 161–185.

Garde, D. (2017). "Fighting the Monster Inside." *Boston Globe*, November 26.

Gardy, J., and N. Loman (2018). "Towards a Genomics-Informed, Real-Time, Global Pathogen Surveillance System." *Nature Reviews: Genetics* 19: 9–20.

Garretson, J., and E. Suhay (2015). "Scientific Communication About Biological Influences on Homosexuality and the Politics of Gay Rights." *Political Research Quarterly* 69(1): 17–29.

Garrett, B. (2011). *Convicting the Innocent: Where Criminal Prosecutions Go Wrong.* Harvard University Press.

Garrison, N. (n.d.). "Cases of How Tribes Are Relating to Genetics Research." American Indian & Alaska Native Genetics Resource Center. http://genetics.ncai.org/what-do-tribes-think-about-genetics-research.cfm.

Garrison, V. (2012). "Disability, Prenatal Testing and the Case for a Moral, Compassionate Abortion." Rewire News Group, August 16. rewire.news/article/2012/08/16/disability-prenatal-testing-and-case-moral-compassionate-abortion/.
Garthwaite, J. (2016). "U.S. Military Preps for Gene Drives Run Amok." *Scientific American*, November 18.
Gass, N. (2016). "Trump: 'The Experts Are Terrible.'" Politico, April 4.
Gates, H. L., Jr. (2006). "My Yiddishe Mama." *Wall Street Journal*, February 1.
Gates, H. L., Jr. (2008). *In Search of Our Roots: How 19 Extraordinary African Americans Reclaimed Their Past*. Crown Publishers.
Gauchat, G. (2012). "Politicization of Science in the Public Sphere: A Study of Public Trust in the United States, 1974 to 2010." *American Sociological Review* 77(2): 167–187.
Genetics Digest (2019). "3 Mistakes to Avoid When Shopping for a DNA Test." https://www.geneticsdigest.com/best_ancestry_genealogy_dna_test/indexbng.html.
GeneWatch UK (n.d.). "About GeneWatch." http//www. genewatch.org/sub-396416.
Gerber, M. (2016). "The Controversial DNA Search That Helped Nab the 'Grim Sleeper' Is Winning over Skeptics." *Los Angeles Times*, October 25.
Gerstein, J. (2010). "Obama Talks DNA on 'America's Most Wanted': Transcript." Politico, March 9.
Gibson, L. (2008). "Long Way Home." *University of Chicago Magazine*, 100(3).
Gill, J., et al. (2018)."Direct-to-Consumer Genetic Testing: The Implications of the US FDA's First Marketing Authorization for BRCA Mutation Testing." *Journal of the American Medical Association* 319(23): 2377–2378.
Gore, A. (2013). *The Future: Six Drivers of Global Change*. Random House.
Gottweis, H., and A. Petersen (2008). *Biobanks: Governance in Comparative Perspective*. Routledge.
Gould, S. (1996). *The Mismeasure of Man*. Norton.
Grady, D. (2017). "Gene Therapy Creates Replacement Skin to Save a Dying Boy." *New York Times*, November 8.
Granja, R., and H. Machado (2019). "Ethical Controversies of Familial Searching: The Views of Stakeholders in the United Kingdom and in Poland." *Science, Technology & Human Values* 44(6): 1068–1092.
Greely, H. (2015). "The Future of Health Care." *Science* 349(6255): 1456.
Greely, H. (2019). "Combating Malaria by Modifying Mosquitoes Could Save Thousands of Lives. It's Also Risky." *Washington Post*, February 19.
Greely, H. (2021) *CRISPR People: The Science and Ethics of Editing Humans*. MIT Press.
Green, C. (2015). "It's Our Culture, Not an Obesity Gene, That Makes People Fat, Expert Says." *Healthline*, August 27.
Green, E., et al. (2011). "Charting a Course for Genomic Medicine from Base Pairs to Bedside." *Nature* 470 (February 9): 204–213.
Green, R., et al. (2009). "Disclosure of APOE Genotype for Risk of Alzheimer's Disease." *New England Journal of Medicine* 361(3): 245–254.
Grimm, D. (2007). "The Demographics of Genetic Surveillance: Familial DNA Testing and the Hispanic Community." *Columbia Law Review* 107(5): 1164–1194.
Gross, N. (2013). *Why Are Professors Liberal and Why Do Conservatives Care?* Harvard University Press.
Gross, N., and S. Simmons (2014). "The Social and Political Views of American College and University Professors." In *Professors and Their Politics*, ed. N. Gross and S. Simmons, 19–52. Johns Hopkins University Press.

Gross, T. (2019). "Historian Henry Louis Gates Jr. on DNA Testing and Finding His Own Roots." *Fresh Air*, NPR. January 21.

Guterl, F. (2013). "Rising Risks." *Scientific American* 308(2): 82.

Haga, S., and G. Ginsburg (2006). "Prescribing BiDil: Is It Black and White?" *Journal of the American College of Cardiology* 48(1): 12–14.

Halpern, D. (2011). *Sex Differences in Cognitive Abilities*. 4th ed. Psychology Press.

Hammonds, E., and R. Herzig, eds. (2008). *The Nature of Difference: Sciences of Race in the United States from Jefferson to Genomics*. MIT Press.

Harkins, P. (2019). "Rapid DNA Machines Have the Potential to Change Policing, But Not Everyone in Utah Is on Board." *Salt Lake Tribune*, July 21.

Harmon, A. (2005). "Blacks Pin Hope on DNA to Fill Slavery's Gaps in Family Trees." *New York Times*, July 25.

Hart Research Associates (2009). "Nanotechnology, Synthetic Biology, and Public Opinion." Hart Research Associates, Washington, DC.

Hayden, E. (2020). "If DNA Is Like Software, Can We Just Fix the Code?" *MIT Technology Review*, February 26.

Hazlett, A., et al. (2011). "Hoping for the Best or Preparing for the Worst: Regulatory Focus and Preferences for Optimism and Pessimism in Predicting Personal Outcomes." *Social Cognition* 29(1): 74–96.

Heimlich, R. (2011). "Partisan Divide over Alternative Energy Widens." Pew Research Center, November 11.

Heine, S. (2017). *DNA Is Not Destiny: The Remarkable, Completely Misunderstood Relationship Between You and Your Genes*. Norton.

Heisman, J. (2014). [Letter to the editor.] *New York Times Magazine*, June 29.

Hill, M., et al. (2017). "Has Noninvasive Prenatal Testing Impacted Termination of Pregnancy and Live Birth Rates of Infants with Down Syndrome?" *Prenatal Diagnosis* 37(13): 1281–1290.

Hirschman, A. (1970). *Exit, Voice, and Loyalty*. Harvard University Press.

Hirschman, A. (1991). *The Rhetoric of Reaction: Perversity, Futility, Jeopardy*. Harvard University Press.

Hjörleifsson, S., et al. (2008). "Decoding the Genetics Debate: Hype and Hope in Icelandic News Media in 2000 and 2004." *New Genetics and Society* 27(4): 377–394.

Hochschild, J. (2005). "Looking Ahead: Racial Trends in the U.S." *Daedalus* 134(1): 70–81.

Hochschild, J. (2017). "Left Pessimism and Political Science." *Perspectives on Politics*. 15(1): 6–19.

Hochschild, J. (2018). "How History Has Proven Albert Hirschman's Insight to Be Essential but Also Wrong: *The Rhetoric of Reaction: Perversity, Futility, Jeopardy*, by Albert Hirschman." *Social Research* 85(3): 597–611.

Hochschild, J., and K. Einstein (2015). *Do Facts Matter? Information and Misinformation in American Politics*. University of Oklahoma Press.

Hochschild, J., and M. Sen (2015a). "Genetic Determinism, Technology Optimism, and Race: Views of the American Public." *Annals of the American Academy of Political and Social Science* 661(1): 160–180.

Hochschild, J., and M. Sen (2015b). "To Test or Not? Singular or Multiple Heritage? Genomic Ancestry Testing and Americans' Racial Identity." *Du Bois Review* 12(2): 321–347.

Hochschild, J. Weaver, and T. Burch (2012). *Creating a New Racial Order: How Immigration, Multiracialism, Genomics, and the Young Can Remake Race in America*. Princeton University Press.

Hofstadter, R. (1962). *Anti-Intellectualism in American Life*. Vintage Books.

Hogan, B., and N. Musumeci (2020). "Gov. Cuomo's Coronavirus Optimism: 'We Have the Beast on the Run.'" *New York Post*, May 8.

Holes, P., et al. (2018). *Evil Has a Name*. Audible Originals.

Hsu, S. (2015). "FBI Notifies Crime Labs of Errors Used in DNA Match Calculations Since 1999." *Washington Post*, May 29.

Huckins, G. (2020). "A Study Finds Sex Differences in the Brain. Does It Matter?" *Wired*, July 27.

Hughes, A. (2013). "Me, My Genome, and 23andMe." *The New Atlantis* 40 (fall): 3–18.

"Illuminating BiDil." (2005). [Editorial.] *Nature Biotechnology* 23 (August): 903.

Innocence Project (2020a). "Exonerate the Innocent." www.innocenceproject.org/exonerate/.

Innocence Project (2020b). "Compensating the Wrongly Convicted." https://innocenceproject.org/compensating-wrongly-convicted/.

International Commission on Missing Persons (2012). "Over 7,000 Srebrenica Victims Have Now Been Recovered." http://www.ic-mp.org/press-releases/over-7000-srebrenica-victims-recovered/.

Isaacson, W. (2021). *The Code Breaker: Jennifer Doudna, Gene Editing, and the Future of the Human Race*. Simon & Schuster.

Jacoby, S. (2018). *The Age of American Unreason in a Culture of Lies*. Vintage.

James, S., et al. (1983). "John Henryism and Blood Pressure Differences among Black Men." *Journal of Behavioral Medicine* 6(3): 259–278.

Jamieson, K. H., et al. (2017). *The Oxford Handbook of the Science of Science Communication*. Oxford University Press.

Jannetta, J., and C. Okeke (2017). *Strategies for Reducing Criminal and Juvenile Justice Involvement*. Urban Institute. www.urban.org/sites/default/files/publication/94516/strategies-for-reducing-criminal-and-juvenile-justice-involvement_2.pdf.

Jaroff, L. (1989). "Science: The Gene Hunt." *Time*, March 20.

Jasanoff, S., ed. (2004). *States of Knowledge: The Co-production of Science and the Social Order*. Routledge.

Jasanoff, S. (2019). *Can Science Make Sense of Life?* Polity Press.

Jasanoff, S., and S.-H. Kim (2009). "Containing the Atom: Sociotechnical Imaginaries and Nuclear Power in the United States and South Korea." *Minerva* 47: 119–146.

Jasanoff, S., and I. Metzler (2018). "Borderlands of Life: IVF Embryos and the Law in the United States, United Kingdom, and Germany." *Science, Technology, and Human Values* 45(6): 1001–1037.

Jayaratne, T., et al. (2006). "White Americans' Genetic Lay Theories of Race Differences and Sexual Orientation." *Group Processes and Intergroup Relations* 9(1): 77–94.

Jayaratne, T., et al. (2009). "The Perennial Debate: Nature, Nurture, or Choice? Black and White Americans' Explanations for Individual Differences." *Review of General Psychology* 13(1): 24–33.

Joh, E. (2006). "Reclaiming 'Abandoned' DNA: The Fourth Amendment and Genetic Privacy." *Northwestern University Law Review* 100(2): 857–884.

Joh, E. (2015). "The Myth of Arrestee DNA Expungement." *University of Pennsylvania Law Review Online* 164: 51–60.

Johnston, J. (2019). "He Jiankui Is Going to Jail. Would the U.S. Criminally Prosecute a Rogue Gene-Editing Researcher?" STAT, December 31. www.statnews.com/2019/12/31/he-jiankui-jail-prosecute-rogue-gene-editing-researcher/.

Jordan, A. G. (1981). "Iron Triangles, Woolly Corporatism and Elastic Nets: Images of the Policy Process." *Journal of Public Policy* 1(1): 95–123.
Joseph, R. (2000). "The Evolution of Sex Differences in Language, Sexuality, and Visual-Spatial Skills." *Archives of Sexual Behavior* 29: 35–66.
Joshi, B. (2018). "Proteomics: Technologies and Global Markets." BCC Research LLC. Report Code BIO034E.
Joslyn, M., and D. Haider-Markel (2016). "Genetic Attributions, Immutability, and Stereotypical Judgments." *Social Science Quarterly* 97(2): 376–390.
Joyner, M., et al. (2016). "What Happens When Underperforming Big Ideas in Research Become Entrenched?" *Journal of the American Medical Association* 316(13): 1355–1356.
Juma, C. (2016). *Innovation and Its Enemies: Why People Resist New Technologies*. Oxford University Press.
Kahan, D. (2015). "Climate-Science Communication and the *Measurement Problem*." *Advances in Political Psychology* 36(S1): 1–43.
Kahan, D. (2016). "The Expressive Rationality of Inaccurate Perceptions." *Behavioral and Brain Sciences* 40: 26–28.
Kahan, D. (2017). "On the Sources of Ordinary Science Knowledge and Extraordinary Science Ignorance." In *Oxford Handbook of the Science of Science Communication*, ed. K. H. Jamieson et al., 35–50, Oxford University Press.
Kahn, J. (2012) "The Troubling Persistence of Race in Pharmacogenomics." *Journal of Law, Medicine and Ethics* 40(4): 873–885.
Kahn, J. (2013). *Race in a Bottle: The Story of BiDil and Racialized Medicine in a Post-Genomic Age*. Columbia University Press.
Kahn, J., et al. (2018). "How Not to Talk About Race and Genetics." Buzzfeed, March 30.
Kahn, J. (2020). "Unnatural Selection." *New York Times Magazine*, January 12.
Kahneman, D. (2013). *Thinking, Fast and Slow*. Farrar, Straus and Giroux.
Kaufman, J., et al. (2010). "Race, Medicine, and the Science Behind BiDil: How ACE-Inhibition Took the Fall for the First Ethnic Drug." *Review of Black Political Economy* 37(2): 115–130.
Kaye, D. (2012). "A Fourth Amendment Theory for Arrestee DNA and Other Biometric Databases." *University of Pennsylvania Journal of Constitutional Law* 15(4): 1095–1160.
Kelly, S., and H. Farrimond (2012). "Non-Invasive Prenatal Genetic Testing: A Study of Public Attitudes " *Public Health Genomics* 15(2): 73–81.
Kennedy, B., and C Thigpen (2020). "Many Publics Around World Doubt Safety of Genetically Modified Foods. " November 11. Pew Research Center.
King, M.-C. (2006). "Genomics, Race, and Medicine." Tanner Lectures on Human Values. Harvard University.
Kingdon, J. (1984). *Agendas, Alternatives, and Public Policies*. Scott, Foresman, and Co.
Kleiner, K. (2008). "Whole Genome Sequencing to Cost Only $1,000 by End of 2009." Singularity Hub, December 30. http://singularityhub.com/2008/12/30/whole-genome-sequencing-to-cost-only-1000-by-end-of-2009/.
Kloor, K. (2015). "Frankenjournalism at MSNBC." *Discover*, March 21.
Knapton, S. (2017). "Woolly Mammoth Will Be Back from Extinction Within Two Years, Say Harvard Scientists." *The Telegraph*, February 17.
Knoepfler, P. (2018). "Mixed Nuffield Council Report Too Aspirational on Human Genetic Modification." The Niche, July 19. ipscell.com/2018/07/mixed-nuffield-council-report-too-aspirational-on-human-genetic-modification/.
Koenig, B., et al., eds. (2008). *Revisiting Race in a Genomic Age*. Rutgers University Press.

Kolata, G. (2017). "In a First, Gene Therapy Halts a Fatal Brain Disease." *New York Times*, October 5.

Kolenc, A. (2019). "'23 And Plea': Limiting Police Use of Genealogy Sites After *Carpenter v. United States*." *West Virginia Law Review* 122(1): 53–106.

Komisarchik, M., and J. Hochschild (2020). "Genetics, Violence, Race and the Partisan Processing of Responsibility." Department of Political Science, University of Rochester.

Krasner, B. (2019). *DNA Testing and Privacy*. Greenhaven Publishing.

Krieger, N. (2001). "Theories for Social Epidemiology in the 21st Century: An Ecosocial Approach." *International Journal of Epidemiology* 30: 668–677.

Krieger, N. (2005). "Stormy Weather: Race, Gene Expression, and the Science of Health Disparities." *American Journal of Public Health* 95(12): 2155–2160.

Krieger, N. (2014). "Discrimination and Health Inequities." *International Journal of Health Services* 44(4): 643–710.

Krieger, N. (2016). "Epidemiologic Theories of Disease Distribution, Population Health and Health Inequities: Conceptual Frameworks and Empirical Applications—An Introductory Lecture." National Institutes of Minority Health and Health Disparities, Health Disparities Research Institute, Washington, DC, August 15.

Krimsky, N. (1977). "Public Must Regulate Recombinant Research." *Chemical and Engineering News*, May 30. sites.tufts.edu/sheldonkrimsky/files/2018/05/pub1977TheRecombinantDNAControversy.pdf.

Krimsky, S. (1982). *Genetic Alchemy: The Social History of the Recombinant DNA Controversy*. MIT Press.

Krimsky, S. (2019a). *GMOs Decoded: A Skeptic's View of Genetically Modified Foods*. MIT Press.

Krimsky, S. (2019b). *Conflicts of Interest in Science*. Hot Books.

Kvaale, E., et al. (2013). "The 'Side Effects' of Medicalization: A Meta-Analytic Review of How Biogenetic Explanations Affect Stigma." *Clinical Psychology Review* 33(6): 782–794.

LaBarbera, A. (2015). "Proceedings of the International Summit on Human Gene Editing: A Global Discussion—Washington D.C., December 1–3, 2015." *Journal of Assisted Reproduction and Genetics* 33:1123–1127.

Ladd, E., Jr., and S. M. Lipset (1975). *The Divided Academy: Professors and Politics*. W.W. Norton.

Lander, E., et al. (2019). "Adopt a Moratorium on Heritable Genome Editing." *Nature* 567 (March 14): 165–168.

Langer, G. (2010). "This I Believe." ABC News Online, August 30. http://blogs.abcnews.com/thenumbers/2010/08/this-i-believe.html.

Lanham, R. (2017). "Inhofe's Greatest Climate Change Denial Hits." HuffPost, December 6.

Lamotte, S. (2020). "750 Million Genetically Engineered Mosquitoes Approved for Release in Florida Keys." CNN Health, August 20.

Latour, B., and S. Woolgar (1986). *Laboratory Life: The Construction of Scientific Facts*. Princeton University Press.

Lawton, B. (2018). "Bridging Discussions of Human History: Ancestry DNA and New Roles for Africana Studies." *Genealogy* 2(1): 5.

Lazarus, R. (1983). "The Costs and Benefits of Denial." In *The Denial of Stress*, ed. S. Breznitz, 1–30. International Universities Press.

Lazer, D., ed. (2004). *DNA and the Criminal Justice System: The Technology of Justice.* MIT Press.

Lederman, N., and J. Lederman (2014). "The Death of Expertise." *Journal of Science Teacher Education* 25(6): 645–649.

Ledford, H. (2020). "CRISPR Treatment Inserted Directly into the Body for First Time." *Nature* 579(7798): 185.

Lehrer, E., and A. Mills (2019). "Fixing Science Policy." *National Affairs* 44 (fall).

Lei, H.-H. (2007). "DNA-Supported Ancestral Tourism." *Eye on DNA*, June 11. http://www.eyeondna.com/2007/06/11/dna-supported-ancestral-tourism/.

Leonhardt, D. (2021). "Covid's Partisan Errors." *New York Times*, March 18.

Leroux, D. (2018). "'We've Been Here for 2,000 Years': White Settlers, Native American DNA and the Phenomenon of Indigenization." *Social Studies of Science* 48(1): 80–100.

Levin, Y. (2020). "Tribalism Comes for Pandemic Science." *The New Atlantis*, June 5.

Levine, H., et al. (2008). "Drug Arrests and DNA: Building Jim Crow's Database." Council for Responsible Genetics. http://www.councilforresponsiblegenetics.org/pagedocuments/0rrxbggaei.pdf.

Levitt, M. (2007). "Forensic Databases: Benefits and Ethical and Social Costs." *British Medical Bulletin* 83(1): 235–248.

Lewontin, R. (1972). "The Apportionment of Human Diversity." In *Evolutionary Biology*, ed. T. Dobzhansky et al., 6: 381–398. Appleton-Century-Crofts.

Liberty, A. (2015). "Defending the Black Sheep of the Forensic DNA Family: The Case for Implementing Familial DNA Searches in Minnesota." *Hamline Law Review* 38(3): 466–517.

Limandri, B. (2019). "Pharmacogenetic Testing: Why Is It So Disappointing?" *Journal of Psychosocial Nursing and Mental Health Services* 57(4): 9–12.

Limdi, N., et al. (2015). "Race Influences Warfarin Dose Changes Associated with Genetic Factors." *Blood* 126(4): 539–545.

Liu, D., and F. Zhang (2020). "The Extraordinary Evolution of Genome Editing." Broad Institute. www.youtube.com/watch?v=oZ48YlKRUuI&feature=youtu.be.

Liu, S., et al. (2020). "Integrative Structural, Functional, and Transcriptomic Analyses of Sex-Biased Brain Organization in Humans." *PNAS* 117(31): 18788–18798.

Livingstone, K., et al. (2016). "FTO Genotype and Weight Loss: Systematic Review and Meta-Analysis of 9563 Individual Participant Data from Eight Randomised Controlled Trials." *British Medical Journal* 354: i4707.

Locke, A., et al. (2015). "Genetic Studies of Body Mass Index Yield New Insights for Obesity Biology." *Nature* 518(7538): 197–206.

Locke, L. (2020). "The Promise of CRISPR for Human Germline Editing and the Perils of 'Playing God.'" *The CRISPR Journal* 3(1).

Loi, M., et al. (2013). "Social Epigenetics and Equality of Opportunity." *Public Health Ethics* 6(2): 142–153.

Lopez, G. (2015). "*Lopez Tonight*—Snoop Dogg's DNA Test." January 15. www.youtube.com/watch?v=Exz0yNdvksg.

Lovett, B., and I. Ronai (2020). "Viewpoint: Battling Deadly Disease with Gene Drives Worth the Limited Risk." Genetic Literacy Project, August 7. geneticliteracyproject.org/2020/08/07/viewpoint-battling-deadly-disease-with-gene-drives-is-worth-the-limited-risk/?mc_cid=ecd769a9ba&mc_eid=c0d2d9ca48.

Lowe, D. (2020). "Coronavirus Vaccine Update, June 11." *In the Pipeline*. blogs.sciencemag.org/pipeline/archives/2020/06/11/coronavirus-vaccine-update-june-11.

Lull, R., and D. Scheufele (2017). "Understanding and Overcoming Fear of the Unnatural in Discussion of GMOs." In *Oxford Handbook of the Science of Science Communication*, ed. K. Jamieson et al., 409–419. Oxford University Press.

Lundgren, J. (2014). "Who Believes that Astrology Is Scientific?" Public Law Research Paper, No. 14–10.

Lynas, M. (2018). *Seeds of Science: Why We Got It So Wrong on GMOs*. Bloomsbury Sigma.

Ma, H., et al. (2017). "Correction of a Pathogenic Gene Mutation in Human Embryos." *Nature* 548 (August 24): 413–415.

Maier, S., and M. Seligman (1976). "Learned Helplessness: Theory and Evidence." *Journal of Experimental Psychology* 105(1): 3–46.

Mak, A., et al. (2018). "Whole-Genome Sequencing of Pharmacogenetic Drug Response in Racially Diverse Children with Asthma." *American Journal of Respiratory and Critical Care Medicine* 197(12): 1552–1564.

Maliwal, N. (2020). "Sample Preparation in Genomics, Proteomics, and Epigenomics: Global Markets." BCC Research LLC.

Marsh, M. and W. Ronner (2019). *The Pursuit of Parenthood: Reproductive Technology from Test-Tube Babies to Uterus Transplants*. Johns Hopkins University Press.

Martin, A., et al. (2019). "Clinical Use of Current Polygenic Risk Scores May Exacerbate Health Disparities." *Nature Genetics* 51: 584–591.

Martin, A., and K. Quinn (2002). "Dynamic Ideal Point Estimation via Markov Chain Monte Carlo for the U.S. Supreme Court, 1953–1999." *Political Analysis* 10(2): 134–153.

Mattei, J.-F. (2012). "Humanity and Human DNA." *European Journal of Medical Genetics* 55(10): 503–509.

Maurer, D., and J. Sobal, eds. (1999). *Interpreting Weight: The Social Management of Fatness and Thinness*. Walter de Gruyter.

Mayhew, D. (1974). *Congress: The Electoral Connection*. Yale University Press.

McMaster, J. (2019). *From Controversy to Cure: Inside the Cambridge Biotech Boom*. Joe McMaster, director. MIT Video Productions.

Mikulic, M. (2020). "Number of U.S. Bioscience-Related Patents from 2009 to 2019." Statista, July 14.

Miller, J. (2004). "Public Understanding of, and Attitudes toward, Scientific Research: What We Know and What We Need to Know." *Public Understanding of Science* 13(3): 273–294.

Miller, J. (2019). "Roche 'Steps Up' for Gene Therapy with $4.3 Billion Spark Bet." *Reuters*, February 25.

Miller, A. (2020). "Florida Sheriff Forbids Employees, Visitors to Wear Masks: If They Do, 'They Will Be Asked to Leave.'" *USA Today*, August 12.

Mills, M., and F. Tropf (2020). "Sociology, Genetics, and the Coming of Age of Sociogenomics." *Annual Review of Sociology* 46: 553–581.

Mitchell, A., et al. (2020). "Three Months In, Many Americans See Exaggeration, Conspiracy Theories and Partisanship in Covid-19 News." Pew Research Center.

Molteni, M. (2018). "The Creepy Genetics Behind the Golden State Killer Case." *Wired*, April 27.

Molteni, M. (2019a). "Crispr Gene Editing Is Coming for the Womb." *Wired*, April 17.

Molteni, M. (2019b). "What the Golden State Killer Tells Us About Forensic Genetics." *Wired*, April 24.

Montoya, M. (2011). *Making the Mexican Diabetic: Race, Science, and the Genetics of Inequality*. University of California Press.

Mooney, C. (2005). *The Republican War on Science*. Basic Books.
Moore, David. (2107). *The Developing Genome: An Introduction to Behavioral Epigenetics*. Oxford University Press.
Morin-Chassé, A., et al. (2017). "Discord over DNA: Ideological Responses to Scientific Communication about Genes and Race." *Journal of Race, Ethnicity, and Politics* 2(2): 260–299.
Morning, A. (2011). *The Nature of Race: How Scientists Think and Teach about Human Difference*. University of California Press.
Morning, A. (2014). "Does Genomics Challenge the Social Construction of Race?" *Sociological Theory* 32(3): 189–207.
Morning, A., et al. (2019). "Socially Desirable Reporting and the Expression of Biological Concepts of Race." *Du Bois Review* 16(2): 439–455.
Morton, J. (2015). "Women Have the Right to Prenatal Genetic Testing—and to Choose Abortion." *The Guardian*, March 26. www.theguardian.com/commentisfree/2015/mar/26/women-right-to-prenatal-genetic-testing-abortion.
Muhammad, K. (2011). *The Condemnation of Blackness: Race, Crime, and the Making of Modern America*. Harvard University Press.
Murphy, E. (2010). "Relative Doubt: Familial Searches of DNA Databases." *Michigan Law Review* 109(3): 291–348.
Murphy, E., and Tong, J. (2020). "The Racial Composition of Forensic DNA Databases." *California Law Review* 108(6).
Murray, A., et al. (2017). "Familial DNA Testing: Current Practices and Recommendations for Implementation." *Investigative Sciences Journal* 9(4).
National Academies of Sciences, Engineering, and Medicine (2017a). *Human Genome Editing: Science, Ethics, and Governance*. National Academies Press.
National Academies of Sciences, Engineering, and Medicine (2017b). *An Evidence Framework for Genetic Testing*. National Academies Press.
National Academy of Medicine et al. (2020). *Heritable Human Genome Editing*. National Academies Press.
National Conference of State Legislatures (2014). "Forensic Science Laws Database." http://www.ncsl.org/research/civil-and-criminal-justice/dna-laws-database.aspx.
National Conference of State Legislatures (2018). "DNA Arrestee Laws." www.ncsl.org/Documents/cj/Arrestee_DNA_Laws.pdf.
National Human Genome Research Institute (2016). "Epigenomics Fact Sheet." www.genome.gov/about-genomics/fact-sheets/Epigenomics-Fact-Sheet.
National Human Genome Research Institute (2019a). "DNA Sequencing Costs: Data." www.genome.gov/about-genomics/fact-sheets/DNA-Sequencing-Costs-Data.
National Human Genome Research Institute (2019b). "Establishing a 2020 Vision for Genomics." www.genome.gov/about-nhgri/strategic-plan. www.genome.gov/about-nhgri/strategic-plan.
National Human Genome Research Institute (2020a). "Genetic Discrimination." www.genome.gov/about-genomics/policy-issues/Genetic-Discrimination.
National Human Genome Research Institute (2020b). "Genome Statute and Legislation Database." www.genome.gov/about-genomics/policy-issues/Genome-Statute-Legislation-Database.
National Human Genome Research Institute (2020c). "Ethical, Legal and Social Implications Research Program." www.genome.gov/Funded-Programs-Projects/ELSI-Research-Program-ethical-legal-social-implications.

National Institutes of Health (2002). *Strategic Research Plan and Budget to Reduce and Ultimately Eliminate Health Disparities, Volume I: Fiscal Years 2002–2006*. www.nimhd.nih.gov/docs/2002_2006__vol1_031003ed_rev.pdf.

National Institutes of Health (2009). *NIH Health Disparities Strategic Plan and Budget: Fiscal Years 2009–2013*. www.nimhd.nih.gov/docs/2009-2013nih_health_disparities_strategic_plan_and_budget.pdf.

National Institutes of Health (2020a). "Appropriations History by Institute/Center (1938 to Present)." officeofbudget.od.nih.gov/approp_hist.html.

National Institutes of Health (2020b). "Genetics Home Reference: Gene Therapy." ghr.nlm.nih.gov/primer/therapy/availability.

National Research Council (1987). *Introduction of Recombinant DNA-Engineered Organisms into the Environment: Key Issues*. National Academies Press.

National Research Council (2009). *Strengthening Forensic Science in the United States: A Path Forward*. National Academies Press.

Nave, K. (2016). "How Craig Venter Is Fighting Ageing with Genome Sequencing." *Wired*, May 3.

Neergaard, L. (2018). "AP-NORC Poll: Edit Baby Genes for Health, Not Smarts." Associated Press, December 29.

Neijens, P. (2004). "Coping with the Nonattitudes Phenomenon: A Survey Research Approach." In *Studies in Public Opinion: Attitudes, Nonattitudes, Measurement Error, and Change*, ed. W. Saris and P. Sniderman, 295–313. Princeton University Press.

Nelson, A. (2008). "The Factness of Diaspora: The Social Sciences of Genetic Genealogy." In *Revisiting Race in a Genomic Age*, ed. B. Koenig et al., 253–270. Rutgers University Press.

Nelson, A. (2016). *The Social Life of DNA: Race, Reparations, and Reconciliation After the Genome*. Beacon Press.

Nelson, A. (2018). "The Social Life of DNA: Racial Reconciliation and Institutional Morality After the Genome." *British Journal of Sociology* 69(3): 522–537.

Nelson, A. (2019). "The Return of Eugenics." *Nature* 576 (December 19–26): 375–376.

Newman, C. (2016). "Largest-Ever Study Reveals Environmental Impact of Genetically Modified Crops." PhysOrg, September 16. phys.org/news/2016-09-largest-ever-reveals-environmental-impact-genetically.html.

Newport, F. (2020). "The Partisan Gap in Views of the Coronavirus." Gallup Organization, May 15.

Newport, F., and A. Dugan (2017). "Partisan Differences Growing on a Number of Issues." Gallup Organization, August 3.

Nichols, T. (2017). *The Death of Expertise: The Campaign Against Established Knowledge and Why It Matters*. Oxford University Press.

Nisbet, E., et al. (2015). "The Partisan Brain: How Dissonant Science Messages Lead Conservatives and Liberals to (Dis)Trust Science." *Annals of the American Academy of Political and Social Science* 658 (March): 36–66.

NPR et al. (2015). *What Shapes Health*. Harvard T. H. Chan School of Public Health. media.npr.org/documents/2015/feb/What_Shapes_Health_Report.pdf.

Nuffield Council on Bioethics (2012). *Novel Techniques for the Prevention of Mitochondrial DNA Disorders: An Ethical Review*. Nuffield Council on Bioethics.

Nuffield Council on Bioethics (2018). *Genome Editing and Human Reproduction: Social and Ethical Issues*. Nuffield Council on Bioethics.

Obesity Prevention Source (2020). "Genes Are Not Destiny." Harvard T. H. Chan School of Public Health. www.hsph.harvard.edu/obesity-prevention-source/obesity-causes/genes-and-obesity/.

OECD (2019). "Key Biotechnology Indicators." October. www.oecd.org/innovation/inno/keybiotechnologyindicators.htm.

Office of Public Affairs (2010). "Attorney General Issues Memoranda to Improve Use of DNA Evidence." *Justice News*, November 18. www.justice.gov/opa/pr/attorney-general-issues-memoranda-improve-use-dna-evidence.

Office of the Press Secretary (2016). "Fact Sheet: Obama Administration Announces Key Actions to Accelerate Precision Medicine Initiative." February 25. obamawhitehouse.archives.gov/the-press-office/2016/02/25/fact-sheet-obama-administration-announces-key-actions-accelerate.

O'Neal, S. (2011). "NASA Names *2012* as the Silliest Sci-Fi Film." AV Club. January 6. news.avclub.com/nasa-names-2012-as-the-silliest-sci-fi-film-1798223478.

The Onion (2018). "Pros and Cons of Genetic Testing for Ancestry." October 26. www.theonion.com/pros-and-cons-of-genetic-testing-for-ancestry-1830027181.

Ortega, V., and D. Meyers (2014). "Pharmacogenetics: Implications of Race and Ethnicity on Defining Genetic Profiles for Personalized Medicine." *Journal of Allergy and Clinical Immunology* 133(1): 16–26.

Ossorio, P., and T. Duster (2005). "Race and Genetics: Controversies in Biomedical, Behavioral, and Forensic Sciences." *American Psychologist* 60(1): 115–128. [See also erratum in *American Psychologist* 60(4).]

Otto, S. (2016). *The War on Science: Who's Waging It, Why It Matters, What We Can Do About It*. Milkweed Editions.

Paciocco, P. (2015). "Abandoning Abandoned DNA: Reconsidering How the Fourth Amendment Abandonment Doctrine Is Applied to DNA Samples." *Criminal Law Bulletin* 51(6): 1386–1428.

Panofsky, A., and J. Donovan (2019). "Genetic Ancestry Testing among White Nationalists: From Identity Repair to Citizen Science." *Social Studies of Science* 49(5): 653–681.

Parrott, R., et al. (2003). "Diversity in Lay Perceptions of the Sources of Human Traits: Genes, Environments, and Personal Behaviors." *Social Science & Medicine* 56(5): 1099–1109.

Perry, E., et al. (2016). "Genetically Engineered Crops and Pesticide Use in U.S. Maize and Soybeans." *Science Advances* 2(8): e1600850.

Pew Research Center (2014). "Political Polarization in the American Public." Pew Research Center, June 12.

Pew Research Center (2015a). "Public and Scientists' Views on Science and Society." Pew Research Center, January 29.

Pew Research Center (2015b). "Major Gaps Between the Public, Scientists on Key Issues." Pew Research Center, July 1.

Pian, J., et al. (2020). "The Past, Present, and (Near) Future of Gene Therapy and Gene Editing." *NEJM Catalyst* 1(5).

Pinker, S. (2015). "The Moral Imperative for Bioethics." *Boston Globe*, August 1.

Pinker, S. (2018). *Enlightenment Now: The Case for Reason, Science, Humanism, and Progress*. Penguin Random House.

Plomin, R. (2019). *Blueprint: How DNA Makes Us Who We Are*. MIT Press.

Pollack, A. (2013). "F.D.A. Orders Genetic Testing Firm to Stop Selling DNA Analysis Service." *New York Times,* November 25.

Pollack, H. (2010). "Put to the Test." *American Prospect,* November, 21–24.

Prewitt, K., et al., eds. (2012). *Using Science as Evidence in Public Policy.* National Academies Press.

Putnam, A. (2020). "A Genetic Panopticon of Our Own Making: How the Fourth Amendment Applies to Commercial Genealogy DNA Testing." *Criminal Law Bulletin* 56(2).

Quillian, L., and D. Pager (2010). "Estimating Risk: Stereotype Amplification and the Perceived Risk of Criminal Victimization." *Social Psychology Quarterly* 73(1): 79–104.

Rajan, K. (2006). *Biocapital: The Constitution of Postgenomic Life.* Duke University Press.

Rajan, K. (2017). *Pharmocracy: Value, Politics, and Knowledge in Global Biomedine.* Duke University Press.

Rainey, J. (2018). "Familial DNA Puts Elusive Killers behind Bars. But Only 12 States Use It." NBC News, April 28.

Rainie, L., and C. Funk (2015). "Americans, Politics and Science Issues." Pew Research Center, July 1.

Ray, T. (2019). "23andMe Health Report Concerns Linger amid Incremental Acceptance of DTC Testing Model." Genome Web, March 7. www.genomeweb.com/cancer/23andme-health-report-concerns-linger-amid-incremental-acceptance-dtc-testing-model#.XaTfZehKg2z%20.

Reardon, J. (2017). *The Postgenomic Condition: Ethics, Justice, and Knowledge after the Genome.* University of Chicago Press.

Regalado, A. (2018). "2017 Was the Year Consumer DNA Testing Blew Up." *MIT Technology Review,* February 12.

Regalado, A. (2019). "Here Are Some Actual Facts about George Church's DNA Dating Company." *MIT Technology Review,* December 11.

Reich, D. (2018). *Who We Are and How We Got Here: Ancient DNA and the New Science of the Human Past.* Pantheon.

Roberts, D. (2011). *Fatal Invention: How Science, Politics, and Big Business Re-create Race in the Twenty-first Century.* New Press.

Roberts, D., and O. Rollins (2020). "Why Sociology Matters to Race and Biosocial Science." *Annual Review of Sociology* 46: 195–214.

Roberts, J. (2010). "Preempting Discrimination: Lessons from the Genetic Information Nondiscrimination Act." *Vanderbilt Law Review* 63(2): 439–490.

Robison, S. (2015). "The Trans-Ideological Potential of Epigenetics." The Nexus of Epigenetics, June 1. nexusofepigenetics.com/2015/06/01/the-trans-ideological-potential-of-epigenetics/.

Rocco, P. (2017). "Most Presidents Rely on Expertise. Trump Treats Experts Like the Enemy." *The Monkey Cage,* July 24.

Rohlfs, R., et al. (2013). "The Influence of Relatives on the Efficiency and Error Rate of Familial Searching." *PLOS One* 8(8): e70495.

Roose, K. (2014). "Marc Andreessen in Conversation." *New York,* October 20–November 2.

Rosa, E., et al. (2015). *The Risk Society Revisited: Social Theory and Risk Governance.* Temple University Press.

Rose, N. (2000). "The Biology of Culpability: Pathological Identity and Crime Control in a Biological Culture." *Theoretical Criminology* 4(1): 5–34.

Rosen, C. (2003). "Liberty, Privacy, and DNA Databases." *The New Atlantis*, spring, 37–52.
Rosenkranz, N. (2014). "Intellectual Diversity in the Legal Academy." *Harvard Journal of Law and Public Policy* 37: 137–143.
Roth, W., and B. Ivemark (2018). "Genetic Options: The Impact of Genetic Ancestry Testing on Consumers' Racial and Ethnic Identities." *American Journal of Sociology* 124(1): 150–184.
Roth, W., et al. (2020). "Do Genetic Ancestry Tests Increase Racial Essentialism? Findings from a Randomized Controlled Trial." *PLOS One* 15(1): e0227399.
Rothstein, M., et al. (2009). "Ethical Implications of Epigenetics Research." *Nature Review: Genetics* 10: 224.
Rotimi, C. (2004) "Are Medical and Nonmedical Uses of Large-Scale Genomic Markers Conflating Genetics and 'Race'?" *Nature Genetics* 36(11): 543–547.
Roush, W. (1995). "Conflict Marks Crime Conference." *Science* 269(5232): 1808–1809.
Rozek, L., et al. (2014). "Epigenetics: Relevance and Implications for Public Health." *Annual Review of Public Health* 35: 105–122.
Rynes, S., et al. (2018). "When the 'Best Available Evidence' Doesn't Win: How Doubts About Science and Scientists Threaten the Future of Evidence-Based Management." *Journal of Management* 44(8): 2995–3010.
Sabatier, P. (1991). "Toward Better Theories of the Policy Process." *PS: Political Science & Politics* 24(2): 147–156.
Salkin, A. (2008). "When in Doubt, Spit It Out." *New York Times*, September 12.
Sandel, M. (2007). *The Case Against Perfection: Ethics in the Age of Genetic Engineering*. Harvard University Press.
Sankar, P., et al. (2004). "Genetic Research and Health Disparities." *Journal of the American Medical Association* 291(24): 2985–2989.
Satel, S. (2002). "I Am a Racially Profiling Doctor." *New York Times Magazine*, May 5.
Saudi Biobank (2019). "Saudi Biobank Overview." kaimrc.med.sa/?page_id=1454. kaimrc.med.sa/?page_id=1454.
Saunders, B., and L. Braun (2017). "Reforming the Use of Race in Medical Pedagogy." *American Journal of Bioethics* 17(9): 50–52.
Schickler, E. (2016). *Racial Realignment: The Transformation of American Liberalism, 1932–1965*. Princeton University Press.
Schneider, S., et al. (2018). "Genetic Attributions: Sign of Intolerance or Acceptance?" *Journal of Politics* 80(3): 1023–1027.
Schuman, H., and S. Presser (1996). *Questions and Answers in Attitude Surveys*. Sage Publications.
Schuster, J., and M. Finkelstein (2006). *The American Faculty: The Restructuring of Academic Work and Careers*. Johns Hopkins University Press.
Schuster, S., et al. (2010). "Complete Khoisan and Bantu Genomes from Southern Africa." *Nature* 463: 943–947.
Schwartz, O. (2019). "What Does It Mean to Be Genetically Jewish?" *The Guardian*, June 13.
Schweitzer, N., and M. Saks (2007). "The CSI Effect: Popular Fiction About Forensic Science Affects the Public's Expectations About Real Forensic Science." *Jurimetrics* 47(3): 357–364.
Scully, E., et al. (2020). "Considering How Biological Sex Impacts Immune Responses and Covid-19 Outcomes." *Nature Reviews Immunology* 20: 442–447.
Scutti, S. (2018). "What the Golden State Killer Case Means for Your Genetic Privacy." CNN Health, April 27.

Selk, A. (2018). "The Ingenious and 'Dystopian' DNA Technique Police Used to Hunt the 'Golden State Killer' Suspect." *Washington Post*, April 28.

Shaer, M. (2016). "The False Promise of DNA Testing." *The Atlantic*, June.

Shanks, P. (2019). "Scientist on the Loose: George Church Strays into Eugenics—Again." Center for Genetics and Society, December 11. www.geneticsandsociety.org/biopolitical-times/scientist-loose-george-church-strays-eugenics-again.

Sharkey, P. (2013). *Stuck in Place: Urban Neighborhoods and the End of Progress toward Racial Equality*. University of Chicago Press.

Sharma, A., et al. (2015). "Lift NIH Restrictions on Chimera Research." *Science* 350(6261): 640.

Shaw, J. (2020). "Engineering Life: Synthetic Biology and the Frontiers of Technology." *Harvard Magazine*, February.

Shelton, D., et al. (2006–2007) "A Study of Juror Expectations and Demands Concerning Scientific Evidence: Does the CSI Effect Exist?" *Vanderbilt Journal of Entertainment and Technology Law* 9(2): 331–368.

Sher, L., and N. Karlinsky (2010). "New Technique of Using Family's DNA Led Police to 'Grim Sleeper' Suspect." ABC News, July 8.

Sheridan, K. (2017a). "Patchwork Salvation." *Newsweek*, December 8.

Sheridan, K. (2017b). "Sight for Sore Eyes." *Newsweek*, November 3.

Shields, J. (2018). "The Disappearing Conservative Professor." *National Affairs*, fall, 138–150.

Shiller, R. (2019). *Narrative Economics: How Stories Go Viral and Drive Major Economic Events*. Princeton University Press.

Shostak, S., et al. (2009). "The Politics of the Gene: Social Status and Beliefs about Genetics for Individual Outcomes." *Social Psychology Quarterly* 72(1): 77–93.

Silber Mohamed, H. (2018). "Embryonic Politics: Attitudes about Abortion, Stem Cell Research, and IVF." *Politics and Religion* 11(3): 459–497.

Silver, N. (2012). *The Signal and the Noise*. Penguin Press.

Simoncelli, T., and B. Steinhardt (2006). "California's Proposition 69: A Dangerous Precedent for Criminal DNA Databases." *Journal of Law, Medicine and Ethics* 33(2): 199–213.

Singer, E., et al. (2007). "Beliefs about Genes and Environment as Determinants of Behavioral Characteristics." *International Journal of Public Opinion Research* 19(3): 331–353.

Sinsheimer, R. (1969). "The Prospect of Designed Genetic Change." *Engineering and Science Magazine* 32(7): 8–13.

Slater, A. (2020). "Rapid DNA." Policing Project Five-Minute Primers. January 24. www.policingproject.org/news-main/2020/1/23/policing-project-five-minute-primers-rapid-dna.

Slovic, P. (1987). "Perception of Risk." *Science* 236(4749): 280–285.

Slow Food (2015). "Why We Are against GMOs." www.slowfood.com/what-we-do/themes/gmos/why-we-are-against-gmos/.

Slow Food (2016). "Slow Food Position Paper on Genetically Modified Organisms." December. n4v5s9s7.stackpathcdn.com/sloweurope/wp-content/uploads/ING_position_paper_OGM-2.pdf.

Smith, A. (2010). "In Praise of the Guilty Project: A Criminal Defense Lawyer's Growing Anxiety about Innocence Projects." *University of Pennsylvania Journal of Law and Social Change* 13: 315–329.

Smith, G. (2004). *The Genomics Age: How DNA Technology Is Transforming the Way We Live and Who We Are*. HarperCollins.

Snow, C., and K. Dibner (2016). *Science Literacy: Concepts, Context, and Consequences*. National Academies Press.

Specter, M. (2012). "The Mosquito Solution." *New Yorker*, July 9–16.

Specter, M. (2014). "Seeds of Doubt." *New Yorker*, August 25.

Starr, D. (2016). "Forensics Gone Wrong: When DNA Snares the Innocent." *Science*, March 7.

STAT and Harvard T. H. Chan School of Public Health (2016). "The Public and Genetic Editing, Testing, and Therapy." January. cdn1.sph.harvard.edu/wp-content/uploads/sites/94/2016/01/STAT-Harvard-Poll-Jan-2016-Genetic-Technology.pdf.

Steelman, L. (2018). "This Was the Best-Selling Product on Amazon Prime Day 2017." *Real Simple*, July 16.

Stein, M., et al. (2021). "Genome-wide Association Analyses of Post-Traumatic Stress Disorder and Its Symptom Subdomains in the Million Veteran Program." *Nature Genetics* 53: 174–184.

Stein, R. (2017). "Scientists Precisely Edit DNA in Human Embryos to Fix a Disease Gene." *All Things Considered*, NPR, August 2.

Stein, R. (2020). "Gene Therapy Shows Promise for Hemophilia, But Could Be Most Expensive U.S. Drug Ever." NPR, July 20.

Steinhauer, J. (2010). "'Grim Sleeper' Arrest Fans Debate on a DNA Use." *New York Times*, July 9.

Stevens, J. (2008). "The Feasibility of Government Oversight on NIH-Funded Population Genetics Research." In *Revisiting Race in a Genomic Age*, ed. B. Koenig et al., 320–341. Rutgers University Press.

Sturgis, P. (2014). "On the Limits of Public Engagement for the Governance of Emerging Technologies." *Public Understanding of Science* 23(1): 38–42.

Sturgis, P., et al. (2010). "Public Attitudes to Genomic Science: An Experiment in Information Provision." *Public Understanding of Science* 19(2): 166–180.

Sturgis, P., and P. Smith (2010). "Fictitious Issues Revisited: Political Interest, Knowledge and the Generation of Nonattitudes." *Political Studies* 58(1): 66–84.

Subcommittee on Oversight and Investigations of the House Energy and Commerce Committee (2010). "Direct-to-Consumer Genetic Testing and the Consequences to the Public Health." U.S. House of Representatives, Washington, DC.

Suhay, E., and T. Jayaratne (2012). "Does Biology Justify Ideology? The Politics of Genetic Attribution." *Public Opinion Quarterly* 77(2): 497–521.

Suhay, E., et al. (2017). "Lay Belief in Biopolitics and Political Prejudice." *Social Psychological and Personality Science* 8(2): 173–182.

Summers, L. (2005). "Remarks at NBER Conference on Diversifying the Science & Engineering Workforce." Office of the President, Harvard University.

Talbott, S. (2017). "Evolution and the Purposes of Life." *The New Atlantis* (winter): 63–91.

Tallbear, K. (2013). *Native American DNA: Tribal Belonging and the False Promise of Genetic Science*. University of Minnesota Press.

Targeted News Service (2017). "Sen. Schumer Reveals: Popular at Home DNA Test Kits Are Putting Consumer Privacy at Great Risk." November 28.

Taverne, D. (2007). *The March of Unreason: Science, Democracy, and the New Fundamentalism*. Oxford University Press.

Taylor, A., et al. (2004). "Combination of Isosorbide Dinitrate and Hydralazine in Blacks with Heart Failure." *New England Jounal of Medicine* (351): 2049–2057.

Taylor, S. (1991). *Positive Illusions: Creative Self-Deception and the Healthy Mind*. Basic Books.

Taylor, S. and J. Brown. (1988). "Illusion and Well-Being: A Social Psychological Perspective on Mental Health." *Psychological Bulletin* 103(2): 193–210.

Tetlock, P. (2005). *Expert Political Judgment: How Good Is It? How Can We Know?* Princeton University Press.

Tetlock, P. (2016). *Superforecasting: The Art and Science of Prediction*. Broadway Books.

tg (2019). "Most Expensive Gene Therapy Globally Approved." *European Biotechnology*, May 27. european-biotechnology.com/up-to-date/latest-news/news/most-expensive-gene-therapy-globally-approved.html.

Thigpen, C., and C. Funk. (2020). "Most Americans Expect a Covid-19 Vaccine within a Year; 72% Say They Would Get Vaccinated." Pew Research Center, May 21.

Thompson, W. (2013). "Forensic DNA Evidence: The Myth of Infallibility." In *Genetic Explanations: Sense and Nonsense*, ed. S. Krimsky, 227–255. Harvard University Press.

Tiger, L. (1979). *Optimism: The Biology of Hope*. Simon & Schuster.

Tocqueville, A. de ([1848] 1966). *Democracy in America*. Harper & Row.

Tonry, M., ed. (2009). *The Oxford Handbook of Crime and Public Policy*. Oxford University Press.

Tonry, M., ed. (2013). *The Oxford Handbook of Crime and Criminal Justice*. Oxford University Press.

Toom, V. (2010). "Producing Suspects: The Politics of the National DNA Database of England and Wales." *Science as Culture* 19(3): 387–391.

Trachta, A. (2019). "The Most Popular College Majors." Niche, August 27. www.niche.com/blog/the-most-popular-college-majors/.

Trivedi, B. (2017). "Medicine's Future?" *Science* 358(6362): 436–440.

Tsai, Y-Y., and W.-J. Lee (2020). "An Imagined Future Community: Taiwan Biobank, Taiwanese Genome, and Nation-Building." *BioSocieties*, January 3.

Turkheimer, E. (2000). "Three Laws of Behavior Genetics and What They Mean." *Current Directions in Psychological Science* 9(5): 160–164.

Turkheimer, E., et al. (2003). "Socioeconomic Status Modifies Heritability of IQ in Young Children." *Psychological Science* 14(6): 623–628.

Tyler, J., and J. Rosier (2009). "Examining Self-Presentation as a Motivational Explanation for Comparative Optimism." *Journal of Personality and Social Psychology* 97(4): 716–727.

Tyson, A. (2020). "Republicans Remain Far Less Likely than Democrats to View Covid-19 as a Major Threat to Public Health." Pew Research Center, July 22.

United Data Connect (2020). "Forensic Genetic Genealogy Analysis." www.uniteddataconnect.com/forensic-genetic-genealogy-analysis.

United Health Foundation (2019). "Annual Report 2019." United Health Foundation. assets.americashealthrankings.org/app/uploads/ahr_2019annualreport.pdf.

U.S. Atomic Energy Commission (1954). *In the Matter of J. Robert Oppenheimer: Transcript of Hearing before Personnel Security Board*. Government Printing Office. ia700700.us.archive.org/5/items/unitedstatesatom007206mbp/unitedstatesatom007206mbp.pdf.

USASpending (2020). "Spending Explorer." Bureau of the Fiscal Service, Department of the Treasury. www.usaspending.gov/#/keyword_search.

Vogel, D. (2012). *The Politics of Precaution: Regulating Health, Safety, and Environmental Risks in Europe and the United States*. Princeton University Press.

Voteview.com (n.d.). "Parties Overview." voteview.com/parties/all.

Voyer, D., et al. (1995). "Magnitude of Sex Differences in Spatial Abilities: A Meta-Analysis and Consideration of Critical Variables." *Psychological Bulletin* 117(2): 250–270.

Wade, N. (2000). "Genetic Code of Human Life Is Cracked by Scientists." *New York Times*, June 26.

Wade, N. (2009). "Genes Show Limited Value in Predicting Diseases." *New York Times*, April 16.

Walajahi, H., et al. (2019). "Constructing Identities: The Implications of DTC Ancestry Testing for Tribal Communities." *Genetics in Medicine* 21: 1744–1750.

Walsh, K., et al. (2017). *Estimating the Prevalence of Wrongful Convictions*. Urban Institute. www.ncjrs.gov/pdffiles1/nij/grants/251115.pdf.

Wang, C., and L. Wein (2018). "Analyzing Approaches to the Backlog of Untested Sexual Assault Kits in the U.S.A." *Journal of Forensic Sciences* 63(4): 1110–1121.

Wasserman, D. (1996). "Research into Genetics and Crime: Consensus and Controversy." *Politics and the Life Sciences* 15(1): 107–109.

Wee, S.-L. (2019). "China Uses DNA to Track Its People, With the Help of American Expertise." *New York Times*, February 21.

Wee, S.-L. (2020). "China Is Collecting DNA from Tens of Millions of Men and Boys, Using U.S. Equipment." *New York Times*, June 19.

Weinstein, N. (1989). "Optimistic Biases about Personal Risks." *Science* 246(4935): 1232–1233.

Williams, J. (2011). "They Say It's in the Genes: Decoding Racial Ideology in Genomics." *Journal of Contemporary Ethnography* 40(5): 550–581.

Willing, R. (2006). "DNA Rewrites History for African-Americans." *USA Today*, February 1.

Wolf, Tom. 2015. "Governor Tom Wolf Announces a Moratorium on the Death Penalty in Pennsylvania," February 13. https://www.governor.pa.gov/newsroom/moratorium-on-the-death-penalty-in-pennsylvania/

Wood, B. (2013). "Four-Field Anthropology: A Perfect Union or a Failed State?" *Society* 50: 152–155.

Worcel, M., and J. Cohn (2012). [Letter to the editor.] *New York Times*, December 31.

World Intellectual Property Organization (2020). "WIPO IP Statistics Data Center—PCT." April. www3.wipo.int/ipstats/editSearchForm.htm?tab=pct.

Wright, S. (1986). "Molecular Biology or Molecular Politics: The Production of Scientific Consensus on the Hazards of Recombinant DNA Technology." *Social Studies of Science* 16(4): 593–620.

WSB-TV Atlanta 2 (2020). "Exclusive: Gov. Kemp Speaks Candidly about Coronavirus Numbers." May 19. www.wsbtv.com/news/georgia/exclusive-gov-kemp-speaks-candidly-about-coronavirus-numbers-reopening-georgia/TIT7LL66LBHHJMXYSOWQ2PSM5E/.

Yengo, L., et al. (2018). "Meta-analysis of Genome-Wide Association Studies for Height and Body Mass Index in ~700,000 Individuals of European Ancestry." *Human Molecular Genetics* 27(20): 3641–3649.

YouGov (2014). "Poll Results: Genetics." January 21. today.yougov.com/news/2014/01/21/poll-results-genetics/

YouGov (n.d.). "Explore What America Thinks: Surveys." https://today.yougov.com/topics/overview/survey-results.

Zakrzewski, S. (2020). "No, 'Racial Genetics' Aren't Affecting Covid-19 Deaths." *Sapiens*, June 11. www.sapiens.org/body/covid-race-genetics/.

Zaller, J. (1992). *The Nature and Origins of Mass Opinion*. Cambridge University Press.

Zhang, S. (2016). "Will the Alt-Right Promote a New Kind of Racist Genetics?" *The Atlantic*, December 29.

Zhang, Y., and A. Fishbach (2010). "Counteracting Obstacles with Optimistic Predictions." *Journal of Experimental Psychology: General* 139(1): 16–31.

Zimmer, C. (2015). "Is Most of Our DNA Garbage?" *New York Times Magazine*, March 5.

Zizzo, D. (2010). "Experimenter Demand Effects in Economic Experiments." *Experimental Economics* 13: 75–98.

Zwart, H. (2007). "Genomics and Self-Knowledge: Implications for Societal Research and Debate." *New Genetics and Society* 26(2): 181–202.

Index

For the benefit of digital users, indexed terms that span two pages (e.g., 52–53) may, on occasion, appear on only one of those pages.

Figures and tables are indicated by *f* and *t* after the page number.